Thermal Engineering: Concepts and Applications

Thermal Engineering: Concepts and Applications

Editor: Abigail Davis

MURPHY & MOORE
www.murphy-moorepublishing.com

Published by Murphy & Moore Publishing
1 Rockefeller Plaza,
New York City, NY 10020, USA
www.murphy-moorepublishing.com

Thermal Engineering: Concepts and Applications
Edited by Abigail Davis

International Standard Book Number: 978-1-63987-535-1 (Hardback)

Cataloging-in-Publication Data

Thermal engineering : concepts and applications / edited by Abigail Davis.
p. cm.
Includes bibliographical references and index.
ISBN 978-1-63987-535-1
1. Heat engineering. 2. Thermodynamics. 3. Power (Mechanics). I. Davis, Abigail.
TJ260 .T44 2022
621.402 2--dc23

Contents

Preface

I am honored to present to you this unique book which encompasses the most up-to-date data in the field. I was extremely pleased to get this opportunity of editing the work of experts from across the globe. I have also written papers in this field and researched the various aspects revolving around the progress of the discipline. I have tried to unify my knowledge along with that of stalwarts from every corner of the world, to produce a text which not only benefits the readers but also facilitates the growth of the field.

Thermal energy is produced by heating up molecules and atoms causing them to move fast and collide with each other. This energy can be transferred between two mediums or transformed into other forms of energy such as chemical, mechanical, or electrical energy. Thermal engineering is a sub-discipline of mechanical and chemical engineering. It uses concepts from various disciplines such as fluid mechanics, heat transfer, and thermodynamics. It is used for various industrial purposes such as heating, ventilation and air conditioning technology, cooling of computer chips, boiler design, process fired heaters, solar heating, combustion engines, thermal power plants, cooling systems, and heat exchangers. As this field is emerging at a fast pace, this book will help the readers to better understand the concepts of thermal engineering. It traces the progress of this field and highlights some of its key concepts and applications. This book will help new researchers by foregrounding their knowledge in this branch.

Finally, I would like to thank all the contributing authors for their valuable time and contributions. This book would not have been possible without their efforts. I would also like to thank my friends and family for their constant support.

Editor

1

ANALYTICAL DECOMPOSITION SOLUTIONS FOR HEAT TRANSFER ON STRAIGHT FINS WITH TEMPERATURE DEPENDENT THERMAL CONDUCTIVITY AND INTERNAL HEAT GENERATION

Akinbowale T. Akinshilo[1*]

ABSTRACT

This paper analyses heat transfer across straight convecting fins with temperature dependent thermal conductivity and internal heat generation using the Adomian decomposition method (ADM). The ADM is the preferred analytical scheme adopted to provide approximate solutions to nonlinear equations arising from the dependence of thermal conductivity and heat transfer coefficient on temperature distribution. The effect of parameters such as internal heat generation, thermo geometric and thermal conductivity on the temperature profile and heat flux is studied. Where results reveal that thermo geometric parameter and thermal conductivity causes a significant increase in heat transfer across fin base. This study provides useful insight to fins operational performance in applications such as radiators, boilers, refrigeration devices, oil pipelines amongst others. Comparison of solutions with existing works in literature forms good agreement.

Keywords: *Fins, Temperature Dependent Viscosities, Convection, Heat Generation, Adomian Decomposition Method*

INTRODUCTION

Overtime fins have found useful applications in heat transfer equipments due to its relatively large area which enhances heat transfer. Therefore serving as a means of effective heat exchange medium in applications such as automobiles, steam power plants,boilers, air-conditioning and refrigeration devices amongst others. Many research effort have been expended over the years in the study, analysis and experimental investigations of the fins which is of vast importance in the field of science and engineering.

In efforts to study fins Atay and Coskun [1] presented a comparative analysis of fin efficiency using analytical and numerical techniques to better ascertain the accuracy of the analytical solutions while Chowdhury et al. [2] compared analytical solutions for non-linear fin problems applying the homotopy analysis method (HAM) and homotopy pertubation method (HPM) where they showed that the HAM proves to be more accurate. Kraus [3] et al. studied heat transfer in extended surfaces for the purpose of optimising heat transfer in large media. Ghasemi et al. [4] analysed convective fins with temperature dependent thermal conductivity and heat generation applying the differential transform method(DTM) where results confirms the accuracy and convieniency of the DTM . The efficiency of fully wet semi-spherical porous fins was studied by Hatami et al. [5] where the effect of porosity is investigated on fins efficiency shortly after numerical and experimental investigation of heat exchangers was presented by Hatami et al. [6-8] where they optimised the operating condition and were also able to recover waste heat from the heat exchangers. Analytical solutions was developed by Fernandez et al. [9] to improve thermal performance on fins using nonlinear models. In the bid to efficiently analyse straight fins Sadris et al. [10] studied convective heat transfer in fins with thermal conductivity and variable heat transfer. Differential transform method was applied to the energy equation of a thermal conductive straight fins with three profiles by Moradi and Almadikia [11] while Joneidi et al. [12] studied fin efficiency with temperature dependent thermal conductivity using the same method analyzing the effect of physical paramters on heat transfer. Thermal behaviour of longitudinal fins under convective and radiative heat transfer was studied by Hatami and Ganji [13] considering different shapes and ceramic materials where they proved Silicon nitride is a very good material as a result of its high amount of heat transfer shortly after they [14] extended their study to fully wet porous fins considering

This paper was recommended for publication in revised form by Regional Editor Jaap Hoffman
[1] *Department of Mechanical Engineering, University of Lagos, Akoka- Yaba, Lagos, Nigeria*
**E-mail address: ta.akinshilo@gmail.com*

temperature and humidity differences as driving forces of heat and mass transfer. Approximate analytical solution was utilised by Arslanturk [15] to investigate efficiency of fins where fins temperature distribution was determined. Aziz and Enamul [16] presented analytical solutions developed by pertubations scheme for straight convecting fins to study its thermal performance.Shortly after Aziz [17] investigated convecting fins with internal heat generation still adopting the pertubation solutions.

In recent times, the study of fins with temperature dependent thermal conductivity has proven an exciting area to researchers [18-27] due to fins having large temperature difference between the base and tip.Otherwise thermal conductivity is assumed constant which is the case in ordinary fins. Temperature equation governing the heat transfer problem becomes nonlinear due to its dependence on coefficient of heat transfer and thermal conductivity which makes the exact solutions difficult to obtain. Therefore the use of numerical and analytical approximate solutions was applied by researchers [1-27]. Methods of solutions utilised include the pertubation method (PM),homotopy analysis method (HAM), homotopy pertubation method (HPM), variational iteration method (VIM) and garlerkin method of weighted residuals. Methods such as PM are limited owing to the problems of weak nonlinearities and artificial pertubation parameter. The need to find a initial condition to satisfy the boundary condition makes methods such as HPM,VIM,DTM,HAM requires computational tools in handling a solution of large parameters resulting to large computational cost and time [28-35]. The garlerkin's finite element of weighted residual, no doubt a powerful analytical method requires the weighted residuals to satisfy weighting functions which may be arbitrary [27] . The method of solutions of decomposing nonlinear coupled equations into linear and nonlinear terms makes the Adomian Decomposition Method (ADM) a powerful,yet relatively simplistic method which is not limited by any artificial parameter or initial guess term. These makes ADM an interesting scheme in providing analytical solutions to nonlinear problems in science and engineering as often employed by researchers .

Motivated by past works this paper aim to investigate heat transfer in convective straight fins with thermal dependent conductivity and internal heat generation. Adomian decomposition method is adopted in analyzing the coupled nonlinear equation which is used to investigate temperature distribution, heat flux at fins base and effiency.

MODEL DEVELOPMENT AND ANALYTICAL SOLUTIONS

A straight fin with constant heat transfer and temperature is considered as described in the model diagram figure 1. The fin thermal conductivity and internal heat generation are temperature dependent, both sides of the fins are exposed to ambient environmental condition. It is assumed that the fin base as a uniform temperature, surrounding media temperature is uniform,tip of the fin is insulated, there's no contact resistance between fin base and prime surface and heat transfer from fin edges and temperature gradient across fin thickness is negligible. Therefore going by the above assumptions the governing equation can be reduced to the following pair of ordinary differential equation which is introduced as :

$$\frac{d}{dx}\left[K(T)\frac{dT}{dx}\right]-\frac{h}{A_c}P(T-T_\infty)-q(T)=0 \tag{1}$$

with boundary considition considered as,

$$x=0 \quad \frac{dT}{dx}=0 \tag{2}$$

$$x=L \quad T=T_b$$

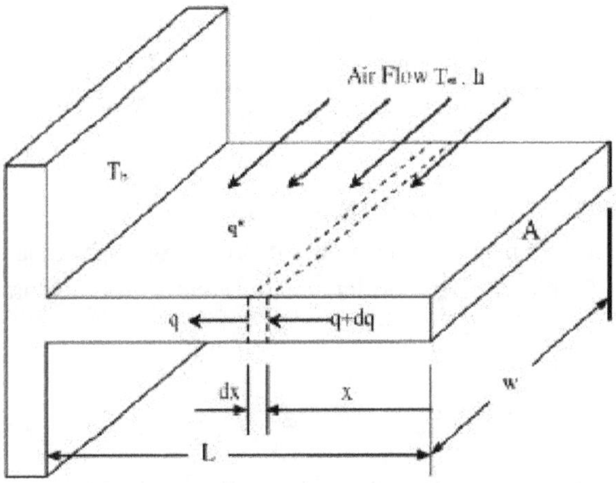

Figure 1. Physical model of problem [4]

Temperature dependent thermal properties is given by,

$$K(T) = K_a[1 + \lambda(T - T_\infty)] \tag{3}$$

Internal heat generation is given by

$$q(T) = q_a[1 + \psi(\mathrm{T} - \mathrm{T}_\infty)] \tag{4}$$

Substituting equations (3) and (4) into (1) yields

$$\frac{d}{dx}\left[K_a[1 + \lambda(T - T_\infty)]\frac{dT}{dx}\right] - \frac{h_b P(T - T_\infty)}{A_C} + q_a[1 + \psi(T - T_\infty)] = 0 \tag{5}$$

with the dimensionless parameters stated in the nomenclature equation (5) can be expressed as [4]:

$$\frac{d^2\theta}{dx^2} + \beta\theta\frac{d^2\theta}{dx^2} + \beta\left(\frac{d\theta}{dx}\right)^2 - M^2\theta + M^2Q + M^2Q\gamma\theta = 0 \tag{6}$$

The appropriate boundary condition is given as

$$x = 0 \quad \frac{d\theta}{dx} = 0 \tag{7}$$

$$x = 1 \quad \theta = 1$$

ADOMIAN DECOMPOSITION METHOD PRINCIPLES

The adomian decomposition method is used for providing approximate or semi approximate analytical solutions to a wide range of problems involving integral,differential and algebraic equations. Here the non-linear equation is described as,

$$Lu + Ru + Nu = g \tag{8}$$

where L is the highest order, R is the linear term, N is the nonlinear term and g is the source term. Applying an inverse operator L^{-1} yields,

$$u = f(x) - L^{-1}(Ru) - L^{-1}(Nu) \tag{9}$$

where $f(x)$ is the term obtained from integrating the source term. The non linear operator is expressed as an infinite series referred to as the Adomian polynomials, $Nu = F(u)$

$$F(u) = \sum_{n=0}^{\infty} A_n \tag{10}$$

By performing the series expansion of A_n. Higher order Polynomials are generated.The adomian method defines the solution u(x) by the series

$$u = \sum_{n=0}^{\infty} u_n \tag{11}$$

The series expansion of $F(u)$ is an infinite series, which is expressed using the Taylor series as,

$$F(u) = \mathrm{F(u_0)} + \frac{dF}{dX}\mathrm{(u_0)(u-u_0)} + \frac{d^2F}{dX^2}(u_0)\frac{(u-u_0)^2}{2!} + \frac{d^3F}{dX^3}(u_0)\frac{(u-u_0)^3}{3!} + \ldots\ldots \tag{12}$$

The above is rewritten as $u - u_0 = u_1 + u_2 + u_3 + \ldots\ldots$, substituting equation (12) then equating the expressions for $F(u)$ in equation (10) defines the formulae for the Adomian polynomial as [15]:

$$F(u) = A_1 + A_2 + \ldots. = F(u_0) + \frac{dF}{dX}(u_0)(\mathrm{u}_1 + u_2 + \ldots) + \frac{d^2F}{dX^2}(u_0)\frac{(u_1 + u_2 + \ldots..)^2}{2!} + \ldots.. \tag{13}$$

APPLICATION OF ADM

The adomian decomposition method (ADM) is adopted in generating solutions to the coupled ordinary nonlinear second order differential equation which may be expressed as:

$$L_{xx}(\theta)=-\beta\theta\frac{d^2\theta}{dx^2}-\beta\left(\frac{d\theta}{dx}\right)^2-M^2\theta+M^2Q-M^2Q\gamma\theta \tag{14}$$

In order to escape difficult integrations the highest order differential operator was taken as $L_{xx}=\frac{d^2}{dx^2}$ for the coupled equation.Thus inverting L_{xx} gives L_{xx}^{-1}. On applying L_{xx}^{-1} to equation (8) gives,

$$\theta=C_1+C_2x+L_{xx}^{-1}\left[-\beta\theta\frac{d^2\theta}{dx^2}-\beta\left(\frac{d\theta}{dx}\right)^2-M^2\theta-M^2Q\gamma\theta\right]+L_{xx}^{-1}M^2Q \tag{15}$$

with respect to ADM, θ may be expressed in the form,

$$\theta=\sum_{n=0}^{\infty}\theta_n \tag{16}$$

Utilising equations (16) in (15) yields,

$$\theta=C_1+C_2x+L_{xx}^{-1}(M^2Q)+L_{xx}^{-1}\left[\left(-\beta\sum_{n=0}^{\infty}\theta_n\frac{d^2}{dx^2}\sum_{n=0}^{\infty}\theta_n\right)-\left(\beta\frac{d}{dx}\sum_{n=0}^{\infty}(\theta_n)^2\right)+\left(M^2\sum_{n=0}^{\infty}\theta_n\right)-M^2Q\gamma\sum_{n=0}^{\infty}\theta_n\right] \tag{17}$$

The nonlinear terms will be explored in the form of Λ_n in the Adomian polynomials which yields,

$$\sum_{n=0}^{\infty}\Lambda_n=\left[\left(-\beta\sum_{n=0}^{\infty}\theta_n\frac{d^2}{dx^2}\sum_{n=0}^{\infty}\theta_n\right)-\left(\beta\frac{d}{dx}\sum_{n=0}^{\infty}(\theta_n)^2\right)+\left(M^2\sum_{n=0}^{\infty}\theta_n\right)-M^2Q\gamma\sum_{n=0}^{\infty}\theta_n\right] \tag{18}$$

Utilising equation (18), the equation (17) may be expressed as,

$$\theta = C_1 + C_2 x + L_{xx}^{-1}(M^2 Q) + L_{xx}^{-1}\left(\sum_{n=0}^{\infty}\Lambda_n\right) \tag{19}$$

where boundary conditions takes the form,

$$\text{x=0} \quad \sum_{n=0}^{\infty}\theta_n = 0 \tag{20}$$

$$\text{x=1} \quad \sum_{n=0}^{\infty}\theta_n = 1$$

The zeroth order can be obtained from the recursive relations equations (19) and (20),

$$\theta_0 = C_1 + C_2 x - L_{xx}^{-1} M^2 Q \tag{21}$$

with leading order boundary condition expressed as,

$$x = 0 \quad \frac{d\theta_0}{dx} = 0 \tag{22}$$

$$x = 1 \quad \theta_0 = 1$$

The remaining order of the solutions is given as,

$$u_{j+1} = L_{xx}^{-1}(\Lambda_j), \quad j \ge 0 \tag{23}$$

with boundary condition expressed as,

$$\sum_{n=0}^{\infty}\theta_n = 0 \text{ at x=0} \tag{24}$$

$$\sum_{n=0}^{\infty}\theta_n = 1 \text{ at x=1}$$

From equation (18) the Adomian polynomians Λ can be obtained as,

$$\Lambda_0 = M^2\theta_0 - \beta\theta_0\frac{d^2\theta_0}{dx^2} - \beta\left(\frac{d\theta_0}{dx}\right)^2 - M^2 Q\gamma\theta_0 \tag{25}$$

$$\Lambda_1 = M^2\theta_1 - \beta\theta_0 \frac{d^2\theta_1}{dx^2} - \beta\theta_1 \frac{d^2\theta_0}{dx^2} - 2\beta \frac{d\theta_0}{dx}\frac{d\theta_1}{dx} - M^2 Q\gamma\theta_1 \tag{26}$$

Zeroth order solution can be obtained by simplifying the recursive relation equation (21) using the leading order boundary condition equation (22) which yields.

$$\theta_0 = 1 - \frac{M^2 Q}{2}(1-x^2) \tag{27}$$

First order solution can be obtained from equation (23) and (24) which is expressed as,

$$\theta_1 = L_{xx}^{-1}(\Lambda_0) \tag{28}$$

with the first order boundary condition as follows,

$$x=0 \quad \frac{d\theta_1}{dx} = 0 \tag{29}$$

$$x=1 \quad \theta_1 = 1$$

Substituting equation (25) into (28) . Upon simplifying with the aid of first order boundary condition equation (29) can be easily shown as,

$$\theta_1 = -\frac{M^2}{2}(1-x^2) + \frac{M^4 Q}{24}\left(5-6x^2+x^4\right) + \frac{\beta M^2 Q}{2}(1-x^2) - \frac{\beta M^4 Q^2}{24}\left(5-6x^2+x^4\right) + \frac{\beta M^4 Q^2}{30}(1-x^6) + \frac{M^2 Q\gamma}{2}(1-x^2) - \frac{M^4 Q^2\gamma}{24}\left(5-6x^2+x^4\right) \tag{30}$$

Second order solution can be obtained from equation (23) and (24) which is expressed as,

$$\theta_2 = L_{xx}^{-1}(\Lambda_1) \tag{31}$$

$$x=0 \quad \frac{d\theta_2}{dx} = 0 \tag{32}$$

$$x=1 \quad \theta_2 = 1$$

Substituting equation (26) into (31) .Upon simplifying with the aid of second order boundary condition equation (32) can be easily shown as,

$$\theta_2=\left[M^2-M^2Q\gamma-\beta MQ\right][\frac{M^2}{24}(5-6x^2+x^4)-\frac{M^4Q}{720}(61-75x^2+15x^4-x^6)-$$
$$\frac{\beta MQ}{24}(5-6x^2+x^4)+\frac{\beta M^4Q^2}{720}(61-75x^2+15x^4-x^6)-\frac{\beta M^4Q^2}{1680}(27-28x^2+x^8)-$$
$$\frac{M^2Q\gamma}{24}(5-6x^2+x^4)+\frac{M^4Q^2\gamma}{720}(61-75x^2+15x^4-x^6)]+\frac{\beta M^2}{2}(1-x^6)+\frac{\beta M^4Q}{24}(5+6x^2-x^4)$$
$$-\frac{\beta^2M^2Q}{2}(1-x^2)-\frac{\beta^2M^4Q^2}{24}(5+6x^2-x^4)-\frac{\beta^2M^4Q^2}{30}(1-x^6)-\frac{\beta M^2Q\gamma}{2}(1-x^2)-$$
$$\frac{\beta M^4Q^2\gamma}{24}(5+6x^2-x^4)+\frac{\beta M^6Q^2}{120}(11+15x^2-5x^4-x^6)-\frac{\beta^2M^4Q^2}{24}(5+6x^2-x^4)-$$
$$\frac{\beta^2M^6Q^3}{120}(11-15x^2+5x^4-x^6)+\frac{\beta^2M^6Q^3}{1680}(13-28x^6+15x^8)+\frac{\beta M^4Q^2\gamma}{24}(5-6x^2+x^4)$$
$$-\frac{\beta M^6Q^3\gamma}{120}(11-15x^2+5x^4-x^6)+\frac{\beta M^4Q}{6}(1-x^4)-\frac{\beta M^6Q^2}{180}(13-15x^4+2x^6)-$$
$$\frac{\beta^2M^4Q^2}{6}(1-x^4)+\frac{\beta^2M^6Q^3}{360}(13-15x^4+2x^6)-\frac{\beta^2M^6Q^3}{140}(1-x^8)-\frac{\beta M^4Q^2\gamma}{2}(1-x^4)+$$
$$\frac{\beta M^6Q^3\gamma}{180}(13-15x^4+2x^6)$$

(33)

The summations of equations (27),(30) and (33) gives the ADM solutions for the temperature profile which is expressed as

$$\theta=1-\frac{M^2Q}{2}(1-x^2)-\frac{M^2}{2}(1-x^2)+\frac{M^4Q}{24}\left(5-6x^2+x^4\right)+\frac{\beta M^2Q}{2}(1-x^2)-$$
$$\frac{\beta M^4Q^2}{24}\left(5-6x^2+x^4\right)+\frac{\beta M^4Q^2}{30}(1-x^6)+\frac{M^2Q\gamma}{2}(1-x^2)-\frac{M^4Q^2\gamma}{24}\left(5-6x^2+x^4\right)+$$
$$\left[M^2-M^2Q\gamma-\beta MQ\right][\frac{M^2}{24}(5-6x^2+x^4)-\frac{M^4Q}{720}(61-75x^2+15x^4-x^6)-$$
$$\frac{\beta MQ}{24}(5-6x^2+x^4)+\frac{\beta M^4Q^2}{720}(61-75x^2+15x^4-x^6)-\frac{\beta M^4Q^2}{1680}(27-28x^2+x^8)-$$
$$\frac{M^2Q\gamma}{24}(5-6x^2+x^4)+\frac{M^4Q^2\gamma}{720}(61-75x^2+15x^4-x^6)]+\frac{\beta M^2}{2}(1-x^6)+\frac{\beta M^4Q}{24}(5+6x^2-x^4)$$
$$-\frac{\beta^2M^2Q}{2}(1-x^2)-\frac{\beta^2M^4Q^2}{24}(5+6x^2-x^4)-\frac{\beta^2M^4Q^2}{30}(1-x^6)-\frac{\beta M^2Q\gamma}{2}(1-x^2)-$$
$$\frac{\beta M^4Q^2\gamma}{24}(5+6x^2-x^4)+\frac{\beta M^6Q^2}{120}(11+15x^2-5x^4-x^6)-\frac{\beta^2M^4Q^2}{24}(5+6x^2-x^4)-$$
$$\frac{\beta^2M^6Q^3}{120}(11-15x^2+5x^4-x^6)+\frac{\beta^2M^6Q^3}{1680}(13-28x^6+15x^8)+\frac{\beta M^4Q^2\gamma}{24}(5-6x^2+x^4)$$
$$-\frac{\beta M^6Q^3\gamma}{120}(11-15x^2+5x^4-x^6)+\frac{\beta M^4Q}{6}(1-x^4)-\frac{\beta M^6Q^2}{180}(13-15x^4+2x^6)-$$
$$\frac{\beta^2M^4Q^2}{6}(1-x^4)+\frac{\beta^2M^6Q^3}{360}(13-15x^4+2x^6)-\frac{\beta^2M^6Q^3}{140}(1-x^8)-\frac{\beta M^4Q^2\gamma}{2}(1-x^4)+$$
$$\frac{\beta M^6Q^3\gamma}{180}(13-15x^4+2x^6)$$

(34)

Conductive heat transfer at the base of the fin is obtained from the generalised fourier conduction heat equation which is expressed as

$$q_b = A_c k(T)\frac{dT}{dx} \tag{35}$$

Non-dimensionalizing equation (35), heat transfer rate at fins base is defined as [27]

$$\overline{q_b} = (1+\beta\theta)\frac{d\theta}{dx} \tag{36}$$

Upon substituting parameters from the analysed temperature profile equation (34) in (36) . The heat transfer at the fin base may simply be shown as

The ratio of actual heat transfer from fins surface to heat transfer from the surface if the entire fin is at the same temperature as the base is regarded as the fins efficiency which is derived as [27]:

$$\eta = \frac{Q}{Q_{ideal}} = \frac{\int_0^b \mathrm{P(T-T_{\mathit{a}})}dx}{Pb(T_b - T_a)} = \int_0^1 \theta(x)dx \tag{38}$$

Therefore fins efficiency can be obtained upon simplifying the equation 38 which can be easily shown as,

$$\begin{aligned}\eta = {} & M^2Q + M^2Q - \frac{M^4Q}{3} - \beta M^2Q + \frac{\beta M^4Q^2}{3} - \frac{\beta M^4Q^2}{5} - M^2Q\gamma + \frac{M^4Q^2\gamma}{3} + \left[M^2 - M^2Q\gamma + \beta MQ\right] * \\ & \left[-\frac{M^2}{3} + \frac{96M^4Q}{720} + \frac{\beta MQ}{3} - \frac{96\beta M^4Q^2}{720} + \frac{49\beta M^4Q^2}{1680} + \frac{M^2Q\gamma}{3} - \frac{96M^4Q^2\gamma}{720}\right] - 3\beta M^2 + \frac{\beta M^4Q}{3} + \beta^2M^2Q \\ & - \frac{\beta^2M^4Q^2}{3} + \frac{\beta^2M^4Q^2}{5} + \beta M^2Q\gamma - \frac{\beta M^4Q^2\gamma}{3} + \frac{\beta M^6Q^2}{30} - \frac{\beta^2M^4Q^2}{3} + \frac{2\beta^2M^6Q^3}{15} - \frac{\beta^2M^6Q^3}{35} - \\ & \frac{\beta M^4Q^2\gamma}{3} + \frac{16\beta M^6Q^3\gamma}{120} - \frac{2\beta M^4Q}{3} + \frac{48\beta M^6Q^2}{120} + \frac{2\beta^2M^4Q^2}{3} - \frac{48\beta^2M^6Q^3}{360} + \frac{8\beta^2M^6Q^3}{140} + 2\beta M^4Q^2\gamma \\ & - \frac{48\beta M^6Q^3\gamma}{180}\end{aligned} \tag{39}$$

$$\overline{q_b}=\left[\begin{array}{l}
1+\beta+\frac{\beta M^2Q}{2}(x^2-1)-\frac{\beta M^2}{2}(1-x^2)+\frac{\beta M^4Q}{24}\left(5-6x^2+x^4\right)+\frac{\beta^2M^2Q}{2}(1-x^2)-\\
\frac{\beta^2M^4Q^2}{24}\left(5-6x^2+x^4\right)+\frac{\beta^2M^4Q^2}{30}(1-x^6)+\frac{\beta M^2Q\gamma}{2}(1-x^2)-\frac{\beta M^4Q^2\gamma}{24}\left(5-6x^2+x^4\right)+\\
\left[\beta M^2-\beta M^2Q\gamma-\beta^2MQ\right][\frac{M^2}{24}(5-6x^2+x^4)-\frac{M^4Q}{720}(61-75x^2+15x^4-x^6)-\\
\frac{\beta MQ}{24}(5-6x^2+x^4)+\frac{\beta M^4Q^2}{720}(61-75x^2+15x^4-x^6)-\frac{\beta M^4Q^2}{1680}(27-28x^2+x^8)-\\
\frac{M^2Q\gamma}{24}(5-6x^2+x^4)+\frac{M^4Q^2\gamma}{720}(61-75x^2+15x^4-x^6)]+\frac{\beta^2M^2}{2}(1-x^6)+\frac{\beta^2M^4Q}{24}(5+6x^2-x^4)\\
-\frac{\beta^3M^2Q}{2}(1-x^2)-\frac{\beta^3M^4Q^2}{24}(5+6x^2-x^4)-\frac{\beta^3M^4Q^2}{30}(1-x^6)-\frac{\beta^2M^2Q\gamma}{2}(1-x^2)-\\
\frac{\beta^2M^4Q^2\gamma}{24}(5+6x^2-x^4)+\frac{\beta^2M^6Q^2}{120}(11+15x^2-5x^4-x^6)-\frac{\beta^3M^4Q^2}{24}(5+6x^2-x^4)-\\
\frac{\beta^3M^6Q^3}{120}(11-15x^2+5x^4-x^6)+\frac{\beta^3M^6Q^3}{1680}(13-28x^6+15x^8)+\frac{\beta^2M^4Q^2\gamma}{24}(5-6x^2+x^4)\\
-\frac{\beta^2M^6Q^3\gamma}{120}(11-15x^2+5x^4-x^6)+\frac{\beta^2M^4Q}{6}(1-x^4)-\frac{\beta^2M^6Q^2}{180}(13-15x^4+2x^6)-\\
\frac{\beta^3M^4Q^2}{6}(1-x^4)+\frac{\beta^3M^6Q^3}{360}(13-15x^4+2x^6)-\frac{\beta^3M^6Q^3}{140}(1-x^8)-\frac{\beta^2M^4Q^2\gamma}{2}(1-x^4)+\\
\frac{\beta^2M^6Q^3\gamma}{180}(13-15x^4+2x^6)
\end{array}\right]$$

$$\left(\begin{array}{l}
M^2Qx+M^2x-\frac{M^4}{6}(3x-x^3)-\beta M^2Qx+\frac{\beta M^4Q^2}{6}(3x-x^3)-\frac{\beta M^4Q^2x^5}{6}-M^2Q\gamma x+\frac{M^4Q^2\gamma}{6}(3x-x^3)\\
+(M^2-M^2Q\gamma-\beta MQ)[\frac{M^2}{6}(3x-x^3)+\frac{M^4Q}{120}(25x-10x^3+x^5)-\frac{\beta MQ}{6}(3x-x^3)-\\
\frac{\beta M^4Q^2}{120}(25x-10x^3+x^5)+\frac{\beta M^4Q^2}{210}(7x-x^7)-\frac{M^2Q\gamma}{6}(3x-x^3)-\frac{M^4Q^2\gamma}{120}(25x-10x^3+x^5)]-\\
\beta M^2x+\frac{\beta M^4Q}{6}(3x-x^3)+\beta^2M^2Qx-\frac{\beta^2M^4Q^2}{6}(3x-x^3)+\frac{\beta^2M^4Q^2x^5}{5}+\beta M^2Q\gamma x-\\
\frac{\beta M^4Q^2\gamma}{6}(3x-x^3)+\frac{\beta M^4Q}{6}(3x-x^3)-\frac{\beta M^6Q^2}{60}(15x-10x^3+3x^5)-\frac{\beta^2M^4Q}{6}(3x-x^3)+\\
\frac{\beta^2M^6Q^3}{60}(15x-10x^3+3x^5)-\frac{\beta^2M^6Q^3}{70}(7x^5-5x^7)-\frac{\beta M^4Q^2\gamma}{6}(3x-x^3)+\frac{\beta M^6Q^3\gamma}{60}(15x-10x^3+3x^5)\\
-\frac{2\beta M^4Qx^3}{3}+\frac{\beta M^6Q^2}{15}(5x^3-x^5)+\frac{2\beta^2M^4Q^2x^3}{3}-\frac{\beta^2M^6Q^3}{30}(5x^3-x^5)+\frac{2\beta^2M^6Q^3x^7}{35}+\\
\frac{2\beta M^4Q^2\gamma x^3}{3}-\frac{\beta M^6Q^3\gamma}{15}(5x^3-x^5)
\end{array}\right)$$

)

(37

Table 1. Comparison of various values of X for temperature profiles. When M=0.5, γ=β=Q=0.

X	θ(X) Joneidi et al. [12]	Present work
0.00	0.8868188841	0.8880
0.05	0.8870960292	0.8883
0.10	0.8879276383	0.8891
0.15	0.8893142310	0.8905
0.20	0.8912566748	0.8924
0.25	0.8937561827	0.8949
0.30	0.8968143173	0.8979
0.35	0.9004329897	0.9015
0.40	0.9046144623	0.9056
0.45	0.9093613475	0.9103
0.50	0.9146766135	0.9155
0.55	0.9205635842	0.9213
0.60	0.9270259345	0.9277
0.65	0.9340677074	0.9347
0.70	0.9416933025	0.9422
0.75	0.9499074872	0.9504
0.80	0.9587153946	0.9591
0.85	0.9681225298	0.9684
0.90	0.9781347745	0.9783
0.95	0.9887583840	0.9889
1.00	1.0000000000	1.0000

RESULT AND DISCUSSIONS

This section shows the effect of various parameters on heat transfer on the straight fin exposed to convecting environmental condition where both thermal conductivity and internal heat generation are temperature dependent. The temperature distribution across the fin, heat flux at the fins base and efficiency is reported graphically as seen in figures 2-8.The figure 2 illustrates the effect of the thermo-geometric parameter (M) on the temperature distribution across the fin length which plays an important role as it determines the amount of heat dissipated from the fin to the environment . It can be depicted that as M increases the temperature distribution decreases across the fins length which can be physically explained by the rapid drop in fins temperature as convective heat transfer to the environment increases.

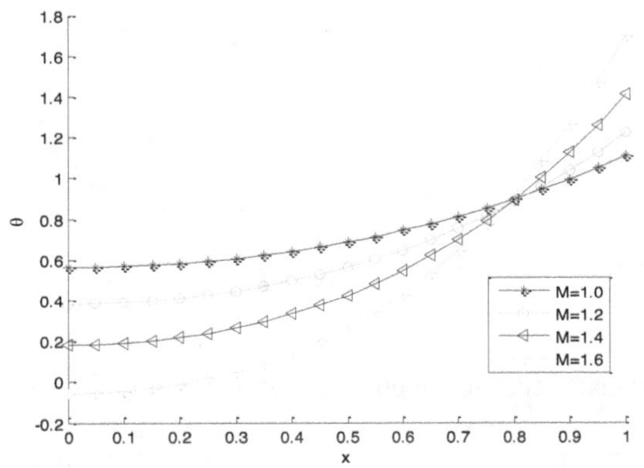

Figure 2. Effect of thermo geometric parameter on temperature distribution, where Q=0.8, γ=0.4, β=0.8

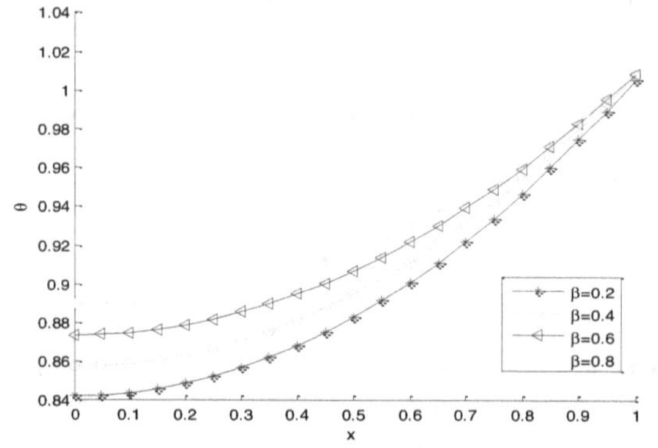

Figure 3. Effect of thermal conductivity parameter on temperature distribution, where Q=0.8, γ=0.4, M=0.5

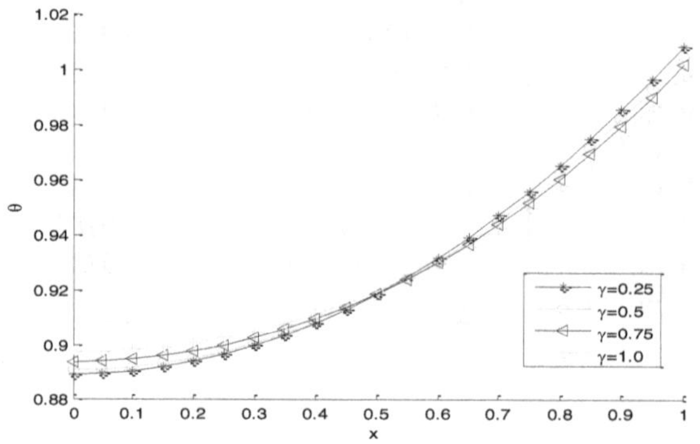

Figure 4. Effect of internal heat generation on temperature distribution, where Q=0.8, β=0.8, M=0.5

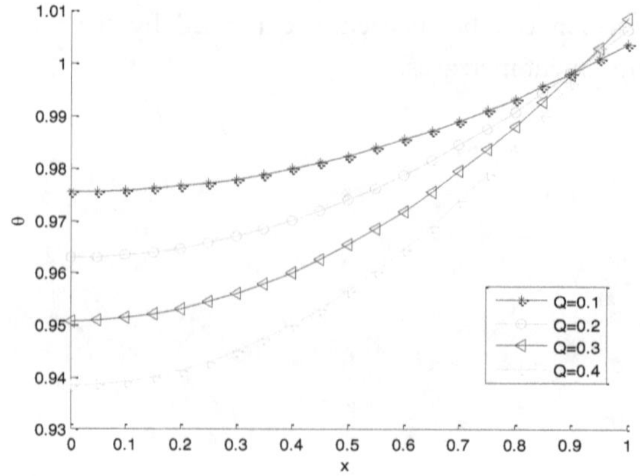

Figure 5. Effect of heat transfer coefficient on temperature distribution, where γ=0.4, β=0.8, M=0.5

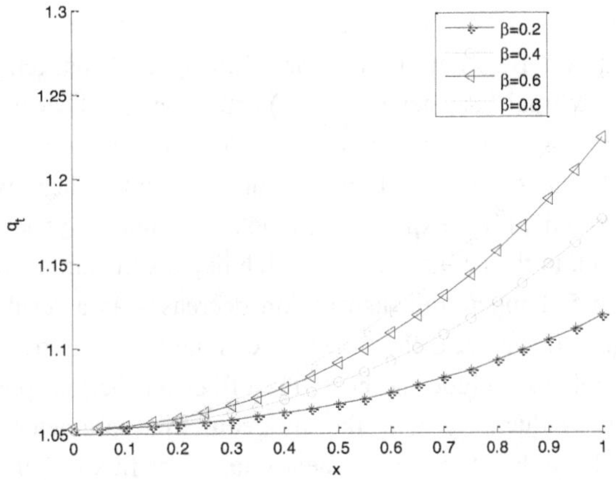

Figure 6. Effect of thermal conductivity parameter on heat flux, where Q=0.8, γ=0.4, M=0.5

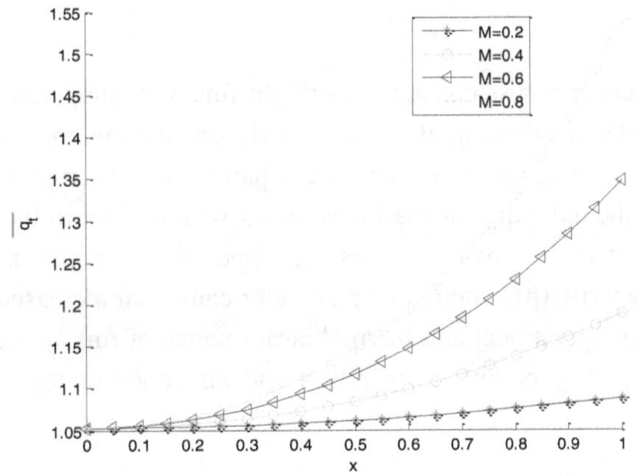

Figure 7. Effect of thermo-geometric parameter on heat flux, where Q=0.8, β=0.8, γ=0.5

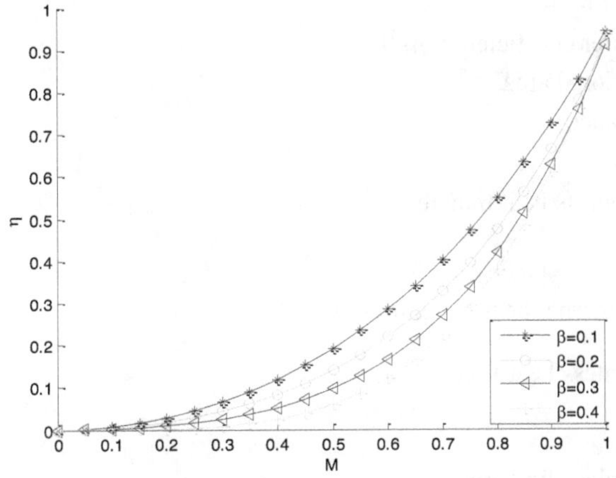

Figure 8. Effect of thermal conductivity on fins efficiency, where Q=γ=1.

Effect of thermal conductivity (β) on heat transfer across the fin can be seen in figure 3. Increase in fins thermal conductivity causes an increased heat transfer across fins surface as a result of increased heat been conducted from the prime surface to the fin where it is observed in figure 3 that temperature distribution increases with corresponding increase in β . Internal heat generation (γ) effect on fins heat transfer is illustrated in figure 4 where it is shown that temperature distribution increases near the fin base and as we move towards the fin tip temperature distribution decreases because gradient of temperature is such that heat extracted from the prime source alongside internal heat generated from the fin is expelled into the environment in such a manner heat is not stored in the fin so as to prevent heat flow back to the prime surface, which defeats the aim of fin. As heat transfer coefficient (Q) increases as observed in figure 5, temperature distribution decreases as a result of enhanced convective heat transfer due to flow caused by large temperature difference between the fin and surrounding environment.

Major analyses in fins problems include the heat flux effect on thermal performance where it is obvious that parameters such as thermal conductivity and thermo-geometric parameter have high influence on the functionality of the fins since they have direct impact on heat rate. Heat flux effect with respect to the fins base is observed in figure 6 where it is seen that with increasing thermal conductivity (β) heat flux increase significantly at the base of the fin. In figure 7 effect of thermo geometric parameter (M) is demonstrated, where an increase in (M) causes a corresponding increase in heat flux as depicted from the plot. The fins efficiency is observed in figure 8 where it is illustrated that at increasing thermal conductivity, efficiency improves due to increasing amount of heat transferred to the environment.

CONCLUSION

In this paper, the heat transfer across convective straight fins with temperature dependent conductivity and internal heat generation is considered adopting the adomian decomposition method. The obtained approximate solution is used to study the effect of temperature dependent parameters during heat transfer on the temperature profile. Also the heat flux across the axial distance is investigated with respect to the fins base. It is observed from results that heat loss is significant at fins base towards the upper tip at increasing effect of thermo geometric parameter (M) and thermal conductivity (β). Though M parameter causes an increased heat loss compared to β. The study provides useful insight to the operational and thermal performance of fins application in heat exchange media such as radiator, gas and steam plants, boiler, refrigeration and air conditioning equipment's and oil pipe lines amongst others.

NOMENCLATURE

Bi	Biot number
L	Length of fin, m
h	Heat transfer coefficient,$wm^{-2}k^{-1}$
A	Cross sectional area, m^2
x	Axial distance, m
$Q=\frac{q_a A_c}{h_b p(T_b-T_\infty)}$	Dimensionless heat transfer
q_i	Uniform internal heat generation, w/m^3
$\overline{q_b}$	Dimensionless heat flux
$X=\frac{x}{b}$	Dimensionless fin length

$H=\dfrac{h}{h_b}$	Dimensionless heat transfer coefficient at fin base
P	Perimeter of fin, m
$K=\dfrac{k}{k_a}$	Dimensionless thermal conductivity
$M^2=\dfrac{Ph_aL_2}{A_ck_a}$	Dimensionless geometric fin parameter
$\gamma=\psi(T_b-T_\infty)$	Dimensionless internal heat generation Parameter
δ	Fin thickness, m
$\beta=\lambda(T_b-T_\infty)$	Thermal conductivity parameter
$\theta=\dfrac{T-T_\infty}{T_b-T_\infty}$	Dimensionless temperature
b	Fin base
∞	Ambient temperature

REFERENCES

[1] Atay, M.T., Coskun, S.B. (2007). Comparative analysis of power law fin type problems using variational iteration method and finite element method. Mathemathical Problem in Engineering, article ID 42072.

[2] Chowdbury, M.S.H., Hashim, I., Abdulaziz, O. (2009).Comparison of homotopy analysis and homotopy perturbation method for purely nonlinear fin type problems. Communication in Nonlinear science and Numerical simulations, 14, 371-378.

[3] Kraus, A.D., Aziz, A., Welty, J.R. (2002). Extended surface heat transfer, John wiley and sons, New York.

[4] Ghasemi, S.E., Hatami, M., Ganji, D.D. (2014). Thermal analysis of convective fins with temperature dependent thermal conductivity and heat generation. Case studies in Thermal Engineering, 4, 1-8.

[5] Hatami, M., Mehdizadeh Ahangar, G.H.R, Ganji, D.D., Boubakar, K. (2014). Refrigeration analysis for fully wet semi-spherical porous fins. Energy Conversion and Management, 84,533-540.

[6] Hatami, M., Ganji, D.D., Gorji-Bandpy, M. (2014). Numerical study of finned type heat exchangers for ICEs exhaust waste heat recovery. Case Studies in Thermal Engineering, 4, 53-64.

[7] Hatami, M., Ganji, D.D., Gorji-Bandpy, M. (2014). Experimental and thermodynamical analysis of diesel exhausts vortex generator heat exchanger for optimizing its operating condition. Applied Thermal Engineering, 30, 1-12.

[8] Hatami, M., Jafaryar, M, Ganji, D.D., Gorji-Bandpy, M. (2014). Optimization of finned tube heat exchangers for diesel exhaust waste recovery using CFD and CCD techniques. International Communication in Heat and Mass Transfer, 57, 254-263.

[9] Fernandez, A. (2009). On some approximate methods for nonlinear models. Applied Mathematical Computation, 215, 168-174.

[10] Sadris, S., Raveshi, M.R., Amiri, S. (2012).Efficiency analysis of straight fin with variable heat transfer coefficient and thermal conductivity. Journal of Mechanical Science and Techonology, 26, 1283-1290.

[11] Moradi, A., Almadikia, H. (2011).Investigation of effect of thermal conductivity on straight fin performance with DTM. International Journal of Engineering and Applied science, 1, 42-54.

[12] Joneidi, A.A., Ganji, D.D., Babaelah, M. (2009). Differential transform method to determine fin efficiency of convective straight fins with temperature dependent thermal conductivity. International Communication in Heat and Mass Transfer, 39, 757-762.

[13] Hatami, M., Ganji, D.D. (2014). Thermal behavior of longitudinal convective-radiative porous fins with different section, shapes and ceramic materials (SiC and Si_3N_4). Ceramic International, 40, 6765-6775.
[14] Hatami, M., Ganji, D.D. (2014). Investigation of refrigeration efficiency for fully wet circular porous fins with variable sections by combined heat and mass transfer analysis. International Journal of refrigeration, 40, 140-151.
[15] Arslanturk, A. (2005). A decomposition method for fin efficiency of convective straight fin with temperature dependent thermal conductivity. International Communication in Heat and Mass Transfer, 32, 831-841.
[16] Aziz, A., Enamul-Huq, S.M. (1973).Perturbation solution for convecting fin with temperature dependent thermal conductivity. Journal of Heat Transfer, 97, 300-310.
[17] Aziz, A. (1977). Perturbation solution for convective fin with internal heat generation and temperature dependent thermal conductivity. International Journal of Heat and Mass Transfer, 20, 1253-1255.
[18] Mosayebidorcheh, S., Ganji, D.D., Masoud, F. (2014). Approximate solutions of nonlinear heat transfer equation of a fin with power law temperature dependent thermal conductivity and heat transfer coefficient. Propulsion and Power Research, 14, 41-47.
[19] Ganji, D.D., Dogonchi, A.S. (2014). Analytical investigation of convective heat transfer coefficient and heat generation. International Journal of Physical Science, 9, 466-474.
[20] Aziz, A., Bouaziz, M.N. (2011). A least squares method for a longitudinal fins with temperature dependent internal heat generation. Energy Conversion and Management, 52, 2876-2882.
[21] Hosseini, K., Daneshian, B., Amanifard, N., Ansari, R. (2012) .Homotopy analysis method for a fin with temperature dependent thermal conductivity and internal heat generation. International Journal of Nonlinear Science, 14, 210-210.
[22] Moitheki, R.J., Hayat, T., Malik, M.Y. (2010). Some exact solutions of the fin problem with power law temperature dependent thermal conductivity. Nonlinear Analysis for Real World Application, 11, 3287-3294.
[23] Khan, F., Ahmadzadeh, R.M., Hamedi, N.H. (2009). Analytical solutions and efficiency of nonlinear fin problem with temperature dependent thermal conductivity and heat transfer co-efficient. Communication in Nonlinear Science and Numerical Simulation, 14, 3327-3338.
[24] Domairry, G., Fazeli, M. (2009). Homotopy analysis method to determine the fin efficiency of convective straight fin with temperature dependent thermal conductivity. Communication in Nonlinear Science and Numerical Simulation, 14, 489-499.
[25] Cosun, S.B., Atay, M.T. (2008). Fin efficiency analysis of convective straight fin with temperature dependent thermal conductivity using variational iteration method. Applied Thermal Engineering, 28, 2345-2352.
[26] Languri, E.M., Ganji, D.D., Jamshidi, N. (2008). Variational iteration and homotopy perturbation methods for fin efficiency of convective straight fins with temperature dependent thermal conductivity. 5th WSEAS International Conference on Fluid Mechanics, Acapulco, Mexico.
[27] Oguntala, G., Sobamowo G. (2016). Garlerkin method of weighted residuals for convective straight fins with temperature dependent conductivity and internal heat generation. International Journal of Engineering and Technology, 6, 432-442.
[28] Filobello-Niño, U., Vazquez-Leal, H., Boubaker, K., Khan, Y., Perez-Sesma, A., Sarmiento Reyes, A., Jimenez-Fernandez, V.M., Diaz-Sanchez, A., Herrera-May, A., Sanchez-Orea J., Pereyra-Castro, K. (2013). Perturbation Method as a Powerful Tool to Solve Highly Nonlinear Problems: The Case of Gelfand's Equation. Asian Journal of Mathematics and Statistics, DOI: 10.3923 /ajms.2013.
[29] Lim, C.W., Wu, B.S. (2002). A modified Mickens procedure for certain non-linear oscillators. Journal of Sound and Vibration, 257, 202-206.
[30] Cheung, Y.K., Chen, S.H., Lau S.L. (1991). A modified Lindsteadt-Poincare method for certain strongly non-linear oscillators. International Journal of Non-Linear Mechanics, 26, 367-378.
[31] Lewis, R.W., Nithiarasu, P., Seatharamu, K.N. (2004). Fundamentals of the finite element method for heat and fluid flow, Antony Rowe Ltd, Wiltshire, Great Britain.
[32]Mercan, H., Atalik, K. (2018). Numerical investigation of blood flow features in intracranial saccular aneurysms. Journal of Thermal Engineering, 4 (7), 1867-1878.

[33] Tokgoz,N., Alic, E., Kaska, O., Aksoy, M.M. (2018). The numerical study of heat transfer enhancement using AL2O3-water nanofluid in corrugated duct application. Journal of Thermal Engineering, 4(3), 1984-1997.
[34] Belhadj, A., Boucherafa, R., Saim, R. (2018). A numerical study of forced convective flow in microchannel heat sinks with periodic-expansion in construction cross section. Journal of Thermal Engineering, 4(3), 1912-1925.
[35] Kilic, M. (2018). Numerical investigation of heat transfer from porous plate with transpiration cooling. Journal of Thermal Engineering, 4(1), 1632-1647.

2

BIOHEAT EQUATION WITH FOURIER AND NON-FOURIER HEAT TRANSPORT LAWS: APPLICABILITY TO HEAT TRANSFER IN HUMAN TISSUES

N. Kizilova[1,2,*], A. Korobov[3]

ABSTRACT

The paper is dedicated to mathematical problem formulations for the heat propagation in biological tissues based on the Fourier and non-Fourier laws at different boundary conditions. The heating of the tissues is provided by external heat sources like low intensity lasers or light-emitting diodes which are widely used in contemporary medical care. Numerical computations on the standard Pennes bioheat equation with Fourier heat conduction give the temperature curves for both heating and thermal relaxation processes that do not correspond to the in vivo measurement data on human skin tissue. It is shown the modified bioheat equation based on the Guyer-Krumhansl heat conduction with correct formulation of the boundary conditions produces realistic temperature curves when the distributed heat sources and sinks in the tissue are accounted for. The former corresponds to the metabolic heat and temperature dependent chemical reactions, while the latter is provided by the heat convection with blood microcirculation system. The proposed model gives realistic two time temperature curves. The perspective applications of the novel mathematical formulation are discussed.

Keywords: *Pennes Bioheat Equation, Non-Fourier Heat Transport Laws, Heat Transfer in Biological Tissues, Mathematical Modeling*

INTRODUCTION

During the last decades different methods of phototherapy and photodynamic therapy based on the coherent optical radiation of low-level lasers (LLL) and non-coherent radiation of light-emitting diodes (LED) had been introduced, tested and approved for routine treatment purposes [1-3]. Optical radiation improves local microcirculation in the tissues, produce stable dilatation of the blood and lymphatic vessels, and produce anti-inflammatory, analgesic and lymph-depleting effects [4]. Personal-specific application of the therapy needs detailed quantitative estimations of the time, intensity and dose of the irradiation. For the surface or internal (i.e. via optic fibers or during an open surgery) radiation the correspondent models of the tissues accounting for their individual geometry, dimensions and material parameters are needed. The multilayer models of human skin have been developed for the LLL treatment [5-7]. Similar approaches have been proposed in dentistry [8] and general surgery [9]. Such modeling is based on the heat generation and propagation in the biological tissues due to the photoinduced biochemical reactions and the correspondent variations in the local blood microcirculation.

The very first heat balance equation for biological tissues has been formulated by H. Pennes [10]. It is a heat balance equation accounted for the metabolic heat sources q_{met} and heat exchange with the blood microcirculation. The latter is modeled as continuously distributed sources with intensities proportional to the local temperature differences between the blood and tissue. The Pennes bioheat equation has the form

$$\rho c \frac{\partial T}{\partial t} = \nabla \cdot (k \nabla T) + q_{met} + q_{rad} - \rho_b c_b w_b (\mathrm{T} - T_b) \tag{1}$$

This paper was recommended for publication in revised form by Regional Editor N. Filiz Özdil
[1]Warsaw University of Technology, Warsaw, Poland
[2]V.N. Karazin Kharkov National University, Faculty of Mathematics and Informatics, Kharkov, Ukraine
[3]Kharkov National University, Faculty of Radiophysics, Biomedical Electronics and Computer Systems, Ukraine
**E-mail address: n.kizilova@gmail.com*

where T is the temperature, ρ, c, k are the density, specific heat and thermal conductivity of the tissue, the subscript b refers to the blood, w_b is the local blood perfusion rate in the tissue, q_{rad} represents additional heat sources activated by the irradiation.

The Pennes equation (1) has been used in a series of studies on the local heating or cooling of surface tissues, including the hyperthermia treatment of skin and underlying tissues by LED and LLL [7, 11]. The schema of the experiment and the layers of the surface tissues are shown in Figure 1. For the 1d case the analytical solutions of (1) can be easily obtained at different boundary conditions [12-15]. The corresponding solutions are given by quite cumbersome formulae and are not reproduced here for brevity. For the metabolic heat term the constant value corresponded to the average metabolic rate in the tissues $q_{met} = const$, $q_{rad} = 0$ have been mostly used. The typical surface temperature $T(t)$ curves for the heating and relaxation processes computed on the Pennes equation in this case can be found in Figure 2 that present the model validation curves. Similar dynamics have been detected for other types of the heat sources and boundary conditions [7]. Experimental studies on the infrared heating, LED irradiation or/and local cooling of the surface tissues produce dynamics [14,15] which differs from the one predicted by the Pennes equation. For example, the local heating (20 min) followed by thermal relaxation (30 min) exhibited complex dynamics of the temperature rise during the heating and clear two time decay relaxation process (Figure 2), while the theoretical estimations based on the Pennes bioheat equation posses simple monotonous temperature rise during the heating and exponential decay during the heat relaxation (see [7]).

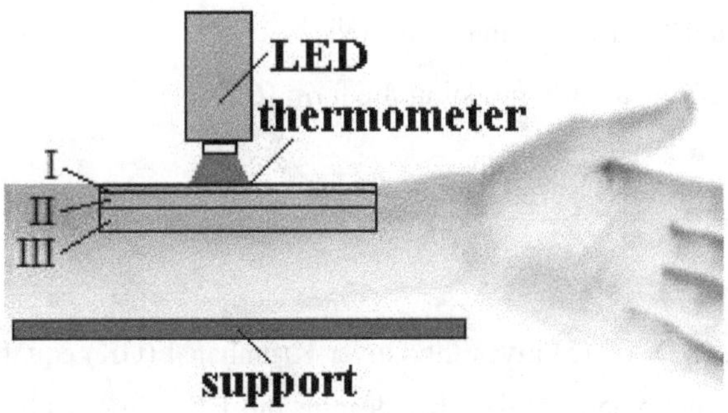

Figure 1. Schema of the surface heating experiment and three layer model of the surface tissues: dermis (I), hypodermis (II), muscle (III)

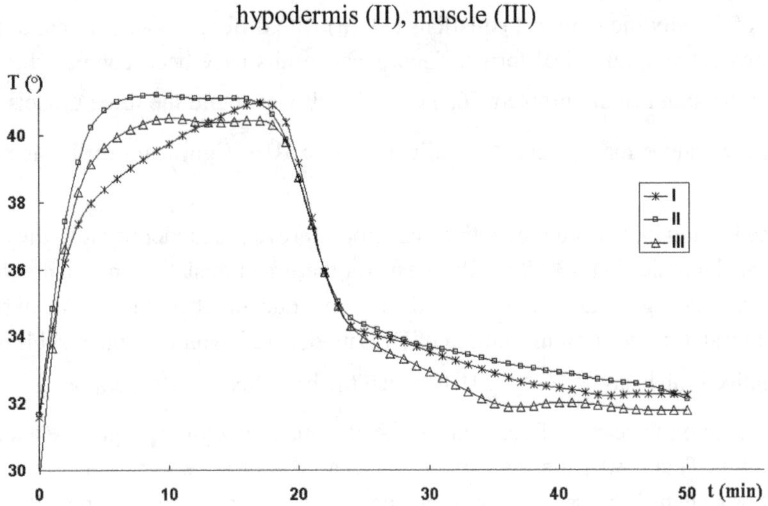

Figure 2. Temperature curves measured for the local infrared radiation (t=20 min) followed by thermal relaxation (30 min) over human skin (I), muscle (II) and bone (III) [15]

Different non-Fourier heat transport equations for the micro/nano scale systems, fast processes and complex biological systems with phase transitions at the interfaces have been proposed. The single phase lag (SFL) model introduces a time delay between the local heat flux q and the temperature gradient ∇T in the form [16,17]

$$q(t+\tau_q,\vec{r}) = -k\nabla T(t,\vec{r}) \tag{2}$$

where τ_q is the heat relaxation time.

Then the bioheat equation (1) has the extended form

$$\rho c\tau_q \frac{\partial^2 T}{\partial t^2} + \left(\rho c + \rho_b c_b w_b \tau_q\right)\frac{\partial T}{\partial t} + \rho_b c_b w_b T = \nabla\cdot\left(k\nabla T\right) + q_{met} + q_{rad} + \rho_b c_b w_b T_b \tag{3}$$

Unlike the parabolic equation (1), hyperbolic equation (3) describes the thermal wave expansion at a constant speed. When $w_b = 0, q_{met} = 0, q_{rad} = 0$, (3) gives the Cattaneo-Vernotte model [18,19] of the temperature balance equation.

Additional approach for the temperature gradient relaxation time τ_T has been proposed in the form of the dual phase lag (DFL) model

$$q(t+\tau_q,\vec{r}) = -k\nabla T(t+\tau_T,\vec{r}) \tag{4}$$

where the relaxation times τ_T and τ_q are not interconnected.

That approach gives the bioheat DFL model in the form [6,20]

$$\rho c\tau_q \frac{\partial^2 T}{\partial t^2} + \left(\rho c + \rho_b c_b w_b \tau_q\right)\frac{\partial T}{\partial t} + \rho_b c_b w_b T = k\left(\nabla^2 T + \tau_T \frac{\partial(\nabla^2 T)}{\partial t}\right) + q_{met} + q_{rad} + \rho_b c_b w_b T_b \tag{5}$$

When $w_b = 0, q_{met} = 0, q_{rad} = 0$, (5) gives the Guyer-Krumhansl (GK) equation that is widely used now in the heat transfer studies at the micro/nano scales, but despite the DFL, in the GK model the positive parameter $\kappa^2 = k\tau_T$ is related to the characteristic length scale κ and corresponds to the ballistic heat conduction mechanism.

The solutions of the modified bioheat equations (3), (5) with different boundary conditions have been found in analytical, semi-analytical and numerical forms. Theoretical results have been compared to the Pennes equation based computations for the skin heating problem [6, 11-13, 20]. It was found the three models predict similar $T(t)$ dynamics at relatively low values for τ_q and τ_T while at $\tau_{q,T} > 10\ s$ significant differences between the models have been detected.

The triple time lag and higher order equations can not improve the model because they are inconsistent with the second law of thermodynamics [21-23]. The SFL destroys the time translation invariance of the equation, while the DPL model cannot provide a general description of the heat conduction phenomena. The mathematical conditions to obtain an exponentially stable equilibrium solution of DPL model are beyond the physical limits [21,22,24]. Those conditions require negative time delay $\tau_q - \tau_T \le 0$ between the heat flux and temperature gradient (retarded effect), and the DPL theory is based on the cause-effect concept, i.e. the quantity with the higher relaxation time is the effect caused by another quantity. That is why the basic model for treatment of the experimental curves and analysis of the regularities in the heat propagation in tissues must be corrected based on a thermodynamically consistent model able to describe the experimentally observed two-time heat relaxation (Figure 2) in the tissues.

The heat conduction law must produce the corresponding heat transfer equation which is in agreement with the second law of thermodynamics and with space-time evolution. The correct thermodynamically consisted non-Fourier heat conduction law is the GK equation [24,25]

$$\tau\partial_t\vec{J}+\vec{J}=-k\nabla T+\kappa^2\Delta T \tag{6}$$

where τ is the relaxation time, κ is the dissipation parameter determined by the free path λ, $\kappa\sim\lambda$.

Despite the DFL model, all the parameters in the model (6) are nonnegative and have clear physical meaning. In this work the modified Pennes bioheat equation based on the GK law of heat propagation is studied in application to the experimental data measured on LED irradiation of human skin and correct estimation of the temperature related effects.

PROBLEM FORMULATION FOR THE PENNES BIOHEAT EQUATION WITH GUYER-KRUMHANSL LAW

Substitution of (6) into the general heat conduction equation gives the following modification of the Pennes bioheat equation

$$\begin{aligned}&\tau\frac{\partial^2 T}{\partial t^2}+\frac{\partial T}{\partial t}(1+\tau\tilde{\rho}\tilde{c}w_b)=(\alpha-\kappa^2\tilde{\rho}\tilde{c}w_b)\Delta T+\kappa^2\left(\frac{\partial}{\partial t}\Delta T\right)-\tilde{\rho}\tilde{c}w_b(T-T_b)+\\&+\tilde{q}_{met}+\tilde{q}_{rad}+\tau\frac{\partial}{\partial t}(\tilde{q}_{met}+\tilde{q}_{rad})-\kappa^2(\Delta\tilde{q}_{met}+\Delta\tilde{q}_{rad})\end{aligned} \tag{7}$$

where $\alpha=k/\rho c$ is the thermal diffusivity, $\tilde{\rho}=\rho/\rho_b$, $\tilde{c}=c/c_b$, $\tilde{q}_{met}=q_{met}/\rho c$, $\tilde{q}_{rad}=q_{rad}/\rho c$.

Since the heat sources $\tilde{q}_{met},\tilde{q}_{rad}$ include the radiation induced temperature dependent exothermal biochemical reactions, let us assume the tissues are uniform and $\tilde{q}_{met}=\tilde{q}_{met}(\mathrm{T})$, $\tilde{q}_{rad}=\tilde{q}_{rad}(T)$. Within a small interval of natural surface temperatures of human skin $\mathrm{T}=T_0(1\pm\varepsilon)$, $\varepsilon\le 1$ the following linear approximations can be assumed

$$\tilde{q}_{met}=\tilde{q}^0_{met}+\lambda_{met}(\mathrm{T}-\mathrm{T}_s),\ \tilde{q}_{rad}=\tilde{q}^0_{rad}+\lambda_{rad}(\mathrm{T}-\mathrm{T}_s) \tag{8}$$

where $\tilde{q}^s_{met}=\tilde{q}_{met}(\mathrm{T}_s)$, $\tilde{q}^s_{rad}=\tilde{q}_{rad}(T_s)$, $\lambda_{met},\lambda_{rad}=const$, T_s is the characteristic steady temperature [1]. In the considered experimental conditions (Figure 2) the measured temperature changed as $T=37\pm5^\circ C$, $\varepsilon=0.13$ and the approximations (8) are reasonable.

Then, assuming $T_b=const$ (7) can be presented in the form

$$\tau\frac{\partial^2 T}{\partial t^2}+\mathrm{a}_1\frac{\partial T}{\partial t}-\kappa^2\frac{\partial}{\partial t}\Delta T-\mathrm{a}_2\,\Delta T+\mathrm{a}_3\,T=\mathrm{a}_4 \tag{9}$$

where $\mathrm{a}_1=1-\tau(\tilde{\rho}\tilde{c}w_b+\lambda_{met}+\lambda_{rad})$, $\mathrm{a}_2=\alpha-\kappa^2\tilde{\rho}\tilde{c}w_b$, $\mathrm{a}_3=\tilde{\rho}\tilde{c}w_b-(\lambda_{met}+\lambda_{rad})$, $\mathrm{a}_4=\tilde{\rho}\tilde{c}w_bT_b+\tilde{q}^s_{met}+\tilde{q}^s_{rad}-\mathrm{T}_s(\lambda_{met}+\lambda_{rad})$, and the stroke sign means derivative with respect to temperature.

In general case (9) is the nonlinear PDE of the second order on time and space with mixed derivatives. Assuming (8) and $\tilde{q}'_{met}=\lambda_{met}$, $\tilde{q}'_{rad}=\lambda_{rad}$, $\tilde{q}''_{met}=0$, $\tilde{q}''_{rad}=0$, (9) can be rewritten in the form

$$\mathrm{a}_1\left(\frac{\partial T}{\partial t}-\frac{\mathrm{a}_2}{\mathrm{a}_1}\Delta T\right)+\tau\frac{\partial}{\partial t}\left(\frac{\partial T}{\partial t}-\frac{\kappa^2}{\tau}\Delta T\right)=\mathrm{a}_3\,T-\mathrm{a}_4 \tag{10}$$

When $\mathrm{a}_2/\mathrm{a}_1=\kappa^2/\tau$, (10) has the form

$$\left[\mathrm{a}_1+\tau\frac{\partial}{\partial t}\right]\left(\frac{\partial T}{\partial t}-\frac{\kappa^2}{\tau}\Delta T\right)=\mathrm{a}_3\,T-\mathrm{a}_4 \tag{11}$$

where the square brackets are related to the differential operator.

The standard bioheat equation (1) rewritten with new designations is

$$\frac{\partial T}{\partial t}-\alpha\Delta T=\mathrm{a}_3 T-\mathrm{a}_4 \tag{12}$$

When $\kappa^2/\tau=\alpha$, solutions of the classical GK (11) and Fourier heat conduction (12) equations are the same, and this case is called Fourier resonance condition [25]. In the case of the bioheat equation, the Fourier resonance condition $\mathrm{a}_2/\mathrm{a}_1=\alpha$ can be provided by the balance between the source terms in the form

$$\tilde{q}''_{met}-\tilde{q}'_{met}=-(\tilde{q}''_{rad}-\tilde{q}'_{rad}) \tag{13}$$

or, accounting for (8), $\lambda_{met}+\lambda_{rad}=0$. Since $\lambda_{met}>0$, the resonance condition is possible in the specific case when $\lambda_{met}=-\lambda_{rad}<0$, i.e. for the frequencies inducing the endothermal biochemical reactions [4].

When $\mathrm{a}_2/\mathrm{a}_1>\kappa^2/\tau$, solution of (11) is overdiffusive [25,26]. For the classical GK equation, it means the Peclet number $Pe=\kappa^2/\tau\alpha>1$ and convective heat conduction prevails over the molecular one. For the bioheat equation the condition is satisfied when $(\tilde{q}''_{met}-\tilde{q}'_{met})+(\tilde{q}''_{rad}-\tilde{q}'_{rad})<0$. The latter can be satisfied when, for instance, both metabolic and photoinduced reactions exhibit exponential growth with temperature.

ONE-DIMENSIONAL HEAT PROPAGATION PROBLEM FOR BIOLOGICAL MATERIAL

Let us consider one-dimensional heat propagation across the uniform layer of biological tissue $0<x<h$ (Figure 3). The material is irradiated by LLL, LED or other source of heat at the surface $x=0$ and the external heat flux $\vec{J}_1(t)$ is known. At the surface $x=h$ another heat flux $\vec{J}_2(t)$ is assumed to be given. Since in the biological tissues the surface $x=h$ is deep located and its temperature distribution cannot be measured directly, it is natural to assume $\vec{J}_2(t)=0$ and determine h as a penetration depth for the irradiation source applied. Here the solution will be found for the more general case $\vec{J}_2(t)\neq 0$

$$x=0:\ -\lambda\frac{\partial T}{\partial x}=J_1(t) \tag{14}$$

$$x=h:\ -\lambda\frac{\partial T}{\partial x}=J_2(t) \tag{15}$$

At $t=0$ the temperature distribution in the tissues is supposed to be correspondent to the temperature of blood

$$t=0:\ T=T_b \tag{16}$$

In the non-dimensional variables $T^{\circ}=(T-T_b)/T^*$, $x^{\circ}=x/h$, $t^{\circ}=t/t^*$ the equations (1), (3), (9) and boundary conditions (14)-(16) (the symbols $^{\circ}$ at the variables are omitted for simplicity) read

$$\frac{\partial T}{\partial t}=\alpha^0\frac{\partial^2 T}{\partial x^2}-\chi T+q \tag{17}$$

$$\tau^0 \frac{\partial^2 T}{\partial t^2} + \left(1 + \chi\tau^0\right)\frac{\partial T}{\partial t} = \alpha^0 \frac{\partial^2 T}{\partial x^2} - \chi T + q \tag{18}$$

$$\tau^0 \frac{\partial^2 T}{\partial t^2} + \left(1 + \tau^0(\chi - q')\right)\frac{\partial T}{\partial t} = Kn^2 \frac{\partial^3 T}{\partial t \partial x^2} + \left(\alpha^0 - Kn^2\chi\right)\frac{\partial^2 T}{\partial x^2} - (\chi - q')T + q \tag{19}$$

$$x = 0: \ \frac{\partial T}{\partial x} = J_1^0(t) \tag{20}$$

$$x = 1: \ \frac{\partial T}{\partial x} = J_2^0(t) \tag{21}$$

$$t = 0: \ T = 0 \tag{22}$$

where $\alpha^0 = \alpha t^* / h^2$, $w_b^0 = w_b t^*$, $\tau^0 = \tau / t^*$, $\chi = \tilde{\rho}\tilde{c}w_b^0$, $q = q_{met}^0 + q_{rad}^0 - \mathrm{T}_s^0(\lambda_{met}^0 + \lambda_{rad}^0)$, $q' = \lambda_{met}^0 + \lambda_{rad}^0$, $q_{met,rad}^0 = \tilde{q}_{met,rad}^s t^*$, $\mathrm{T}_s^\circ = \mathrm{T}_s / T^*$, $\lambda_{met,rad}^0 = \lambda_{met,rad} / \rho c$, $J_{1,2}^0(t) = J_{1,2}(t)\mathrm{h}/\lambda \mathrm{T}^*$, $Kn = \kappa / h$ is the Knudsen number. Generally speaking, the equivalent $\mathrm{T}_s = \mathrm{T}_b$ may be accepted, but in this study both temperatures are taken as different.

The terms $\sim Kn^2$ are essential for the nanostructural materials and nanofluid dynamics. In the biological tissues and other materials interacting with optical radiation, those terms determining additional sources of ballistic heat production become important [1].

Solution of (19) with the boundary and initial conditions (20)-(22) will be obtained and studied. The corresponding solutions of (18) and (17) with the same boundary conditions can be obtained then by assuming $Kn = 0$, $q' = 0$ in (18), and additionally $\tau^0 = 0$ in (17).

The heat equation with GK law has already been studied for solid materials [25], but the bioheat equations (1), (3), (9) with temperature dependent heat sources and boundary conditions (14)-(16) have not been properly studied yet. The analytical solution for the temperature independent heat sources has been studied in [13].

Solution of (19) can be found in the form

$$\mathrm{T}(t,x) = \Theta(t,x) + J_1^0(t)\mathrm{X}_1(x) + J_2^0(t)\mathrm{X}_2(x) \tag{23}$$

where $\Theta(t,x)$ satisfies the following PDE and boundary conditions

$$\tau^0 \frac{\partial^2 \Theta}{\partial t^2} + \mathrm{A}_1 \frac{\partial \Theta}{\partial t} - \mathrm{A}_2 \frac{\partial^3 \Theta}{\partial t \partial x^2} - \mathrm{A}_3 \frac{\partial^2 \Theta}{\partial x^2} + \mathrm{A}_4 T = q - F(t,x) \tag{24}$$

$$x = 0: \ \frac{\partial \Theta}{\partial x} = 0 \tag{25}$$

$$x = 1: \ \frac{\partial \Theta}{\partial x} = 0 \tag{26}$$

$$t = 0: \ \Theta = -J_1^0(0)X_1(x) - J_2^0(0)X_2(x) = 0 \tag{27}$$

where $F(t,x) = \sum_{k=1}^{2} \tau^0 J_k^{0//} X_k - A_3 J_k^0 X_k^{//} + A_4 J_k^0 X_k$, $\mathrm{A}_1 = 1 + \tau^0(\chi - q')$, $\mathrm{A}_2 = Kn^2$, $\mathrm{A}_3 = \alpha^0 - Kn^2\chi$, $\mathrm{A}_4 = \chi - q'$, stroke sign means the derivative of the functions with respect to their arguments.

The functions $X_{1,2}(x)$ satisfy the system of ODE and boundary conditions

$$X_{1,2}'' = \zeta X_{1,2} \tag{28}$$

$$x=0:\ X_1' = 1,\ \ X_2' = 0 \tag{29}$$

$$x=1:\ X_1' = 0,\ \ X_2' = 1 \tag{30}$$

where $\zeta = A_1 / A_2 = (1+\tau^0(\chi - q'))/\mathrm{Kn}^2$.

Solution of (28)-(30) can be easily found. Depending on sign of ζ two cases must be considered.

Case 1a. $\zeta > 0$ or $q' < \chi + 1/\tau^0$. Then,

$$X_1(x) = -\sqrt{\zeta}^{-1}(cth(\sqrt{\zeta})ch(\sqrt{\zeta}x) - sh(\sqrt{\zeta}x)),\quad X_2(x) = \frac{ch(\sqrt{\zeta}x)}{\sqrt{\zeta}sh(\sqrt{\zeta})} \tag{31}$$

Case 2a. $\zeta < 0$ or $q' > \chi + 1/\tau^0$. Then,

$$X_1(x) = \sqrt{\frac{1}{\zeta}}(ctg(\sqrt{\zeta})\cos(\sqrt{\zeta}x) + \sin(\sqrt{\zeta}x)),\quad X_2(x) = -\frac{\cos(\sqrt{\zeta}x)}{\sqrt{\zeta}\sin(\sqrt{\zeta})} \tag{32}$$

Case 3a. $\zeta = 0$ or $q' = \chi + 1/\tau^0$. Then (28) has no solutions satisfying (29), (30).

Now $F(t,x)$ can be computed and solution of (24)-(27) can be found for each of the cases (31)-(32). Due to the boundary conditions (25)-(26), the solution can be found as Fourier expansion

$$\Theta(t,x) = \sum_{n=0}^{\infty} \vartheta_n(\mathrm{t})\cos(\pi n\,\mathrm{x}) \tag{33}$$

Then, $\vartheta_k(\mathrm{t})$ can be found from the equation

$$\tau^0\vartheta_n'' + (A_1 - \pi^2 n^2 A_2)\vartheta_n' + A_4 - \pi^2 n^2 A_3 = \xi_n(t) \tag{34}$$

where $\xi_n(t) = \int_0^1 (q - F(t,x))\cos(\pi n x)dx$.

The fundamental solution of (34) is determined by the sign of the expression $D_n = (A_1 - \pi^2 n^2 A_2)^2 - 4\tau^0(A_4 - \pi^2 n^2 A_3)$, and different cases have to be studied.

Case 1b. $D_n > 0$. Then,

$$\theta_n(\mathrm{t}) = C_{1n}e^{-\lambda_{1n}t} + C_{2n}e^{-\lambda_{2n}t} + \Xi_n^1(\mathrm{t}) \tag{35}$$

where $\lambda_{1n,2n} = \dfrac{-(A_1 - \pi^2 n^2 A_2) \pm \sqrt{D_n}}{2\tau^0}$, $\Xi_n^1(t) = e^{\lambda_{1n}t}\int_0^t e^{-\lambda_{1n}\varsigma}\xi_n(\varsigma)d\varsigma - e^{-\lambda_{2}t}\int_0^t e^{\lambda_{2n}\varsigma}\xi_n(\varsigma)d\varsigma$.

Case 2b. $D_n < 0$. Then,

$$\theta_n(\mathrm{t}) = C_{3n}e^{-\lambda_{3n}t}\cos(\lambda_{4n}\,\mathrm{t}) + C_{4n}e^{-\lambda_{3n}t}\sin(\lambda_{4n}\,\mathrm{t}) + \Xi_n^2(\mathrm{t}) \tag{36}$$

where $\lambda_{3n} = \dfrac{\pi^2\,\mathrm{n}^2\mathrm{A}_2 - \mathrm{A}_1}{2\tau^0}$, $\lambda_{4n} = \dfrac{\sqrt{D_n}}{2\tau^0}$,

$$\Xi_n^2(t) = \frac{e^{-\lambda_{3n}t}}{\lambda_{4n}}\left[\sin(\lambda_{4n}\,\mathrm{t})\int_0^t e^{\lambda_{3n}\varsigma}\cos(\lambda_{4n}\varsigma)\xi_n(\varsigma)d\varsigma - \cos(\lambda_{4n}t)\int_0^t e^{\lambda_{2n}\varsigma}\sin(\lambda_{4n}\varsigma)\xi_n(\varsigma)d\varsigma\right].$$

Case 3b. $D_n = 0$. Then,

$$\theta_n(\mathrm{t}) = (C_{5n} + C_{6n}t)e^{-\lambda_{3n}t} + \Xi_n^3(\mathrm{t}) \tag{37}$$

where $\Xi_n^3(t) = te^{\lambda_{3n}t}\int_0^t e^{-\lambda_{3n}\varsigma}\xi_n(\varsigma)d\varsigma - e^{\lambda_3 t}\int_0^t \varsigma e^{\lambda_{3n}\varsigma}\xi_n(\varsigma)d\varsigma$.

The unknown coefficients in (35)-(37) can be found from the initial conditions

$$t = 0: \quad \theta_n = 0, \quad \theta_n^{/} = 0 \tag{38}$$

as

$$C_{1n} = \frac{\lambda_{2n}\Xi_n^1(0) + \Xi_n^{1/}(0)}{\lambda_{1n} - \lambda_{2n}}, \quad C_{2n} = -\frac{\lambda_{1n}\Xi_n^1(0) + \Xi_n^{1/}(0)}{\lambda_{1n} - \lambda_{2n}}, \quad C_{3n} = -\Xi_n^2(0)$$
$$C_{4n} = \frac{\lambda_{3n}\Xi_n^2(0) + \Xi_n^{2/}(0)}{\lambda_{3n}\lambda_{4n}}, \quad C_{5n} = -\Xi_n^3(0), \quad C_{6n} = -(\lambda_{3n}\Xi_n^3(0) + \Xi_n^{3/}(0)) \tag{39}$$

Due to the dependence $D_n(n)$ all the cases 1b – 3b can be found in the set of solutions $\{\theta_n(\mathrm{t})\}_{n=1}^{N}$.

RESULTS AND DISCUSSION

Numerical computations of the heat propagation across a layer of biological material with distributed positive (chemical reactions) and negative (carrying away by blood flow) sources have been carried out on (23), (31)-(33), (35)-(37). The following model parameters correspondent to human skin, muscle and bone tissues [1, 11, 13, 20, 27] have been used:

1) Common parameters: $T_b = 37^\circ C$, $\rho_b = 1160\,kg/m^3$, $c_b = 3650-3770\,J/\mathrm{kg}\cdot K$, $w_b = 0.5-3\ kg/\mathrm{m}^3\cdot s$.

2) Skin: $h = 0.25-2.5\,\mathrm{cm}$, $\rho = 1190\,kg/m^3$, $c = 3600-3768\,J/\mathrm{kg}\cdot K$, $\lambda = 0.235-0.42\,W/m\cdot K$, $q_{met} = 245-368\,W/m^3$, $\tau = 0.001\ s$.

3) Muscle: $h = 0.5-3.5\,\mathrm{cm}$, $\rho = 1190\,kg/m^3$, $c = 3600-3768\,J/\mathrm{kg}\cdot K$, $\lambda = 0.44-0.5\,W/m\cdot K$, $q_{met} = 368-684\,W/m^3$, $\tau = 0.4-0.5\ s$.

4) Bone: $h = 0.35-1\,\mathrm{cm}$, $\rho = 1190\,kg/m^3$, $c = 3600-3768\,J/\mathrm{kg}\cdot K$, $\lambda = 0.17-0.19\,W/m\cdot K$, $q_{met} = 220-300\,W/m^3$, $\tau = 1-1.5\ s$.

The physiological limits of the parameters are determined by individual variations. Other parameters are determined by the frequency of optical radiation and the chemical reactions induced by it. Therefore, the sets of non-dimensional values for q_{met}^0, q_{rad}^0, $\lambda_{met}^0, \lambda_{rad}^0 \in [0,1]$ have been tested. The numerical computations have been carried out for n=10 harmonics; that provides fast convergence of the series with accuracy $\sim 10^{-5}$. The typical dynamics of the temperature curves is similar to the skin, muscle and bone tissues with different rates of the temperature rise and relaxation determined by the thermal properties of the tissue. Since in the live tissues the noninvasive surface temperature measured over the skin surface is only available, the numerical results presented in Figure 3-5 have been computed for the skin parameters.

The computed temperature curves T(t) during the heating and thermal relaxation are depicted in Figure 3a-c in the dimensional form for different tissue models. The Fourier heat conduction law based bioheat equation (17) describes an exponential temperature rise during the heating and single time exponential temperature decay during the thermal relaxation at different depths of the layer (Figure 3a). The decay is quite slow so the depth of penetration of the irradiation is overestimated when the model (17) is used [4]. The modified model (18) better describes the complex dynamics of temperature rise with a plateau (Figure 3b), that closer corresponds to the experimental data (Figure 2) than the classical Pennes model (Figure 1). The model (19) correctly describes not only the temperature growth but also its thermal relaxation with two characteristic exponents corresponding to the solutions (35), (38) depending on the model parameters.

The temperature distributions in the same layer modeled by (18), (19) at different intensity of heating are presented in Figure 4.

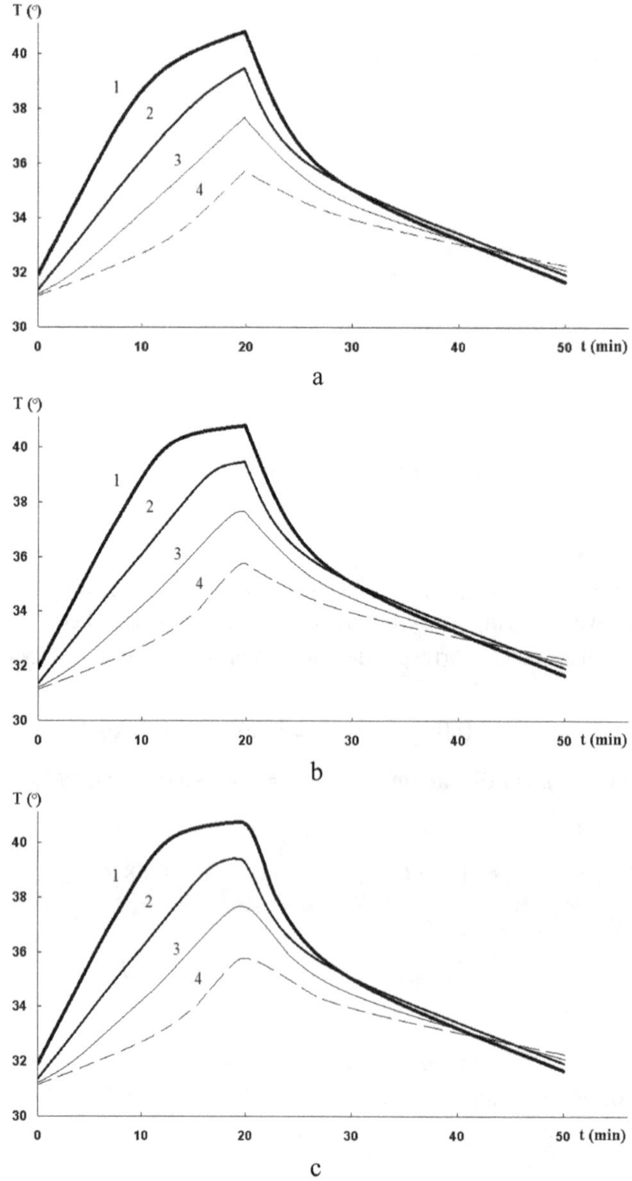

Figure 3. Computed curves T(t) on (a) the equation (17), (b) equation (18), (c) equation (19), where the curves 1-4 correspond to different depths h/L=0.25;0.5;0.75;0.95 of the layer

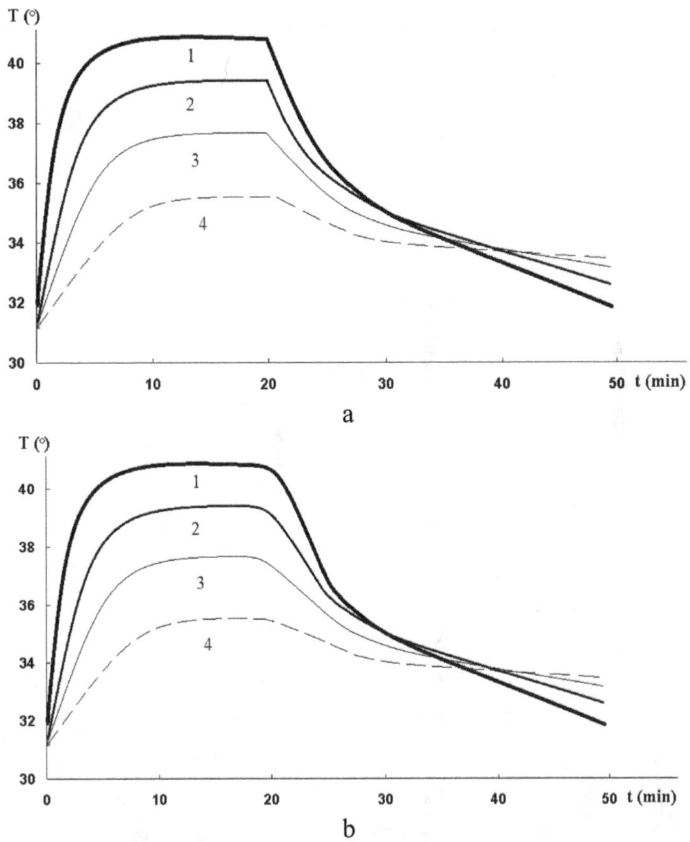

Figure 4. The temperature T(t) curves for 20 min heating followed by 30 min thermal relaxation computed on (a) the model (18), and (b) model (19), at $J_2^0 = 0$; the curves 1-4 correspond to $J_1^0 = 4;3;2;1$ accordingly

The maximal values of the temperatures at the plateau (Figure 3c, Figure 4b) are determined by individual thickness h, intensity of microcirculation w_b and metabolic heat production q_{met}. The thermal parameters of the tissues are more or less stable for different individuals and even in mammals [27]. Therefore, the computed solution allows identification of two individual parameters (w_b, q_{met}), while the thickness of skin and underlying tissues can be estimated by palpation [1]. Based on the individual (w_b, q_{met}) data, the patient specific safe dose of irradiation allowing the LLL or LED therapy without uncomforting or damaging temperatures in the treated tissues.

Comparative results on numerical computations of the temperature curves T(t) by the model with and without accounting for the distributed heat sources with corresponding experimental data obtained on the skin (Figure 5a), muscle (Figure 5b) and bone (Figure 5c) tissues with different w_b, q_{met} and the tissue parameters identified on the previous step on the averaged curves shows significant qualitative differences between two models. The most essential differences are marked in Figure 5 by arrows. The model without heat sources exhibited purely exponential temperature rise during the heating period and sudden exponential decay immediately after the heating had been stopped. It is single-time relaxation and the temperature decays exponentially and could reach low non-physiological values even at some realistic times t~1.5-2 h. When the hear sources activated by the photoinduced reactions are accounted for, the temperature rise (arrows 1 in Figure 5a-c) is not such acute as the exponential one and it is closer to the measurement data for all the surface tissues studied. The temperature decay clearly demonstrates two relaxation times: initial fast relaxation (arrows 2 in Figure 5a-c) provided by absence of the photoinduced heating due to the photoinduced exothermal biochemical reactions and activation of the microcirculation, but the transition to the relaxation is not sharp as in the exponential model because of the inertia of both biochemical processes and regulatory reactions of the activated microcirculation. The second phase of the relaxation is characterized by slow relaxation that produces high surface temperatures at t=50s than by the exponential relaxation due to presence of distributed metabolic heat sources, which maintains the surface temperatures on the physiological level close to the initial surface temperature before the experiment evet at t>2 h that correspond to the steady state of metabolism of the skin and other surface tissues.

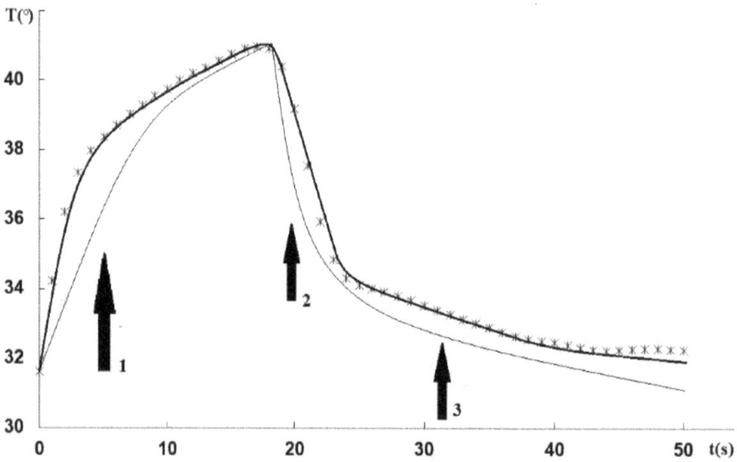

a

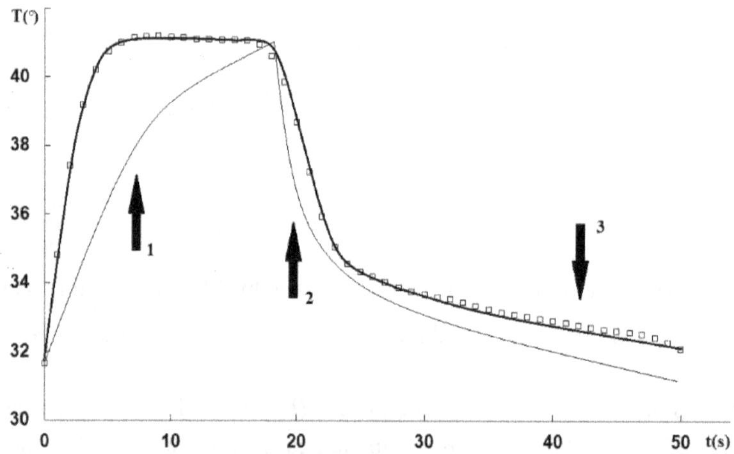

b

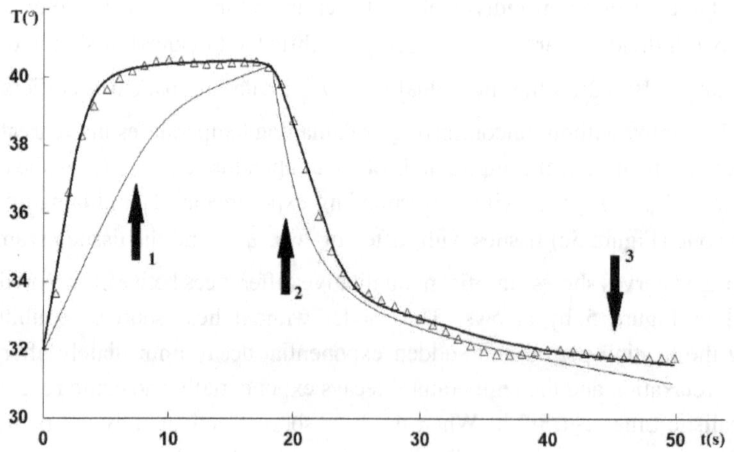

c

Figure 5. Temperature curves T(t) for (a) skin, (b) muscle and (c) bone tissues computed on the model with (thick lines) and without (thin lines) distributed heat sources, and measured on a volunteer (marked)

CONCLUSIONS

The heat conduction in the biological tissues is correctly described by the Pennes bioheat equation which accounts for both heat generation in the live tissues and heat removal by the circulating blood. Both phenomena are influenced by

the external heating of the tissues by LLL or LED with therapeutic purposes. Nevertheless, the basic heat propagating law included into the
Pennes bioheat equation influence significantly the dynamics of the heat rise and decay during the heating and thermal relaxation of the tissue accordingly. Comparative analysis of the theoretical computed curves to the measured ones reveals the better correspondence of the G-K heat propagation las in the biological tissues. When the heat relaxation time is included into the model, the complex two-time temperature growth is correctly described, but the relaxation decay is oversimplified. When the additional ballistic-type heat propagation is accounted for, the realistic two-time heat decay curves during the thermal relaxation are described correctly. The comparative analysis of the theoretical and measured curves is essential for identification of the individual tissue parameters. Those parameters can be used for quantitative estimations of the safe therapeutic dose of the irradiation based on the computations of the damaging parameter [13] that will be a subject of further studies.

REFERENCES

[1] Grossweiner, L. I. (2005). The Science of Phototherapy: An Introduction. Springer Science & Business Media.

[2] Gomer, Ch. J. (2010) Photodynamic Therapy: Methods and Protocols. Humana Press.

[3] Kizilova, N., & Korobov, A. (2018). On biomedical engineering techniques for efficient phototherapy. International Journal on Biosensors and Bioelectronics, 4(6), 289-295.

[4] Kizilova, N., & Korobov, A. (2016). Mechanisms of influence of the low-intense optical radiation on the microcirculatory system. A review. Photobiology and Photomedicine, 71(2), 79-93.

[5] Liu, K.-C. (2008). Thermal propagation analysis for living tissue with surface heating. International Journal of Thermal Sciences, 47, 507–513.

[6] Kumar, P., Kumar, D., Rai, K. N. (2015). A numerical study on dual-phase-lag model of bio-heat transfer during hyperthermia treatment. Journal of Thermal Biology, 49-50, 98–105.

[7] Ciesielski, M., Mochnacki, B., Piasecka-Belkhayat, A. (2016). Analysis of temperature distribution in the heated skin tissue under the assumption of thermal parameters uncertainty. 40th Solid Mechanics Conference, Book of Abstracts, Warsaw, IPPT, P048.

[8] Verma, S. K., Maheshwari, S., Singh, R. K., Chaudhari, P. K. (2012). Laser in dentistry: An innovative tool in modern dental practice. National Journal of Maxillofacial Surgery, 3(2), 124–132.

[9] Azadgoli, B., Baker, R. Y. (2016). Laser applications in surgery. Annals of Translational Medicine, 23(4), 452-458.

[10] Pennes, H. H. (1948). Analysis of tissue and arterial blood temperatures in the resting forearm. Journal f Applied Physiology, 1, 93–122.

[11] Xu, F., Lu, T. J., Seffen, K. A. (2008). Biothermomechanics of skin tissues. Journal of the Mechanics and Physics of Solids, 56, 1852–1884.

[12] Nóbrega, S., Coelho, P. J. (2017). A parametric study of thermal therapy of skin tissue. Journal of Thermal Biology, 63, 92–103.

[13] Lin, S.-M. (2013). Analytical solutions of bio-heat conduction on skin in Fourier and non-Fourier models. Journal on Mechanics in Medicine and Biology, 13 (4), 1350063.

[14] Kizilova, N., Korobov, A., Solovjova H. (2018). Heat transfer at local surface freezing of human tissues: experimental and theortical study. In: IX Internal Conference “Low temperature physics,” Abstracts book, Kharkiv, Ukraine, 148.

[15] Kizilova N., Korobov A. (2018). On bioheat equation and its modification. In: 3-rd International Conference “Differential equations and Control Theory,” Book of Abstracts, Kharkiv, Ukraine, 30-31.

[16] Tzou, D. Y. (1995). A unified field approach for heat conduction from macro- to microscales. Journal of Heat Transfer, 117(1), 8–16.

[17] Tzou, D. Y. (1997). Macro- to Microscale Heat Transfer: The Lagging Behavior, New York, Taylor and Francis.

[18] Cattaneo, C. (1948). Sulla conduzione del calore, Atti del Seminario Matematico e Fisico dell' Universita di Modena, 3, 83–101.

[19] Vernotte, P. (1958). Les paradoxes de la théorie continue de l’équation de la chaleur. Comptes rendus hebdomadaires des séances de l'Académie des sciences, 46, 3154–3155.

[20] Majchrzak, E., Turchan, Ł., Dziatkiewicz, J. (2015). Modeling of skin tissue heating using the generalized dual phase-lag equation. Archives of Mechanics, 67(6), 417-437.
[21] Rukolaine, S. A. (2014). Unphysical effects of the dual-phase-lag model of heat conduction. International Journal of Heat and Mass Transfer, 78, 58–63.
[22] Rukolaine, S. A. (2017). Unphysical effects of the dual-phase-lag model of heat conduction: higher-order approximations. International Journal of Thermal Sciences, 113, 83–88.
[23] Wang, M., Yang, N., Guo, Z.-Y. (2011). Non-Fourier heat conductions in nanomaterials. Journal of Applied Physics, 110(6), 064310.
[24] Kovács, R. (2017). Heat conduction beyond Fourier's Law: theoretical predictions and experimental validation, Budapest University of Technology and Economics, Ph.D. Thesis.
[25] Ván, P., Berezovski, A., Fülöp, T., Gróf, G., Kovács, R., Lovas, Á., & Verhás, J. (2017). Guyer-Krumhansl–type heat conduction at room temperature. Europhysics Letters, 118, 50005.
[26] Tang, D. W., Araki, N. (2000). Non-Fourier heat condution behavior in finite mediums under pulse surface heating. Material Science Engineering A, 292, 173.
[27] Berezovski, V. A., Kolotylov, N. N. (1990). Biophysical characteristics of human tissues: A reference book, Kiev, Naukova Dumka.

3

STATISTICAL ANALYSIS OF WIND SPEED DISTRIBUTION WITH SINOP-TURKEY APPLICATION

N. Balpetek[1], E. Kavak Akpinar[1,*]

ABSTRACT

In this study, the wind energy potential of the Sinop region was analyzed statistically by using the Turkish State Meteorological Station's hourly wind speed data between the years of 2005-2014. The two- parameter Weibull and one-parameter Rayleigh probability distribution functions were used to determine the wind energy potential of the region. The probability distribution functions were derived from the cumulative function and used to calculate the mean wind speed and the variance of the actual data. The best way of representing the performance of the Weibull and Rayleigh distributions is to use the statistical parameters such as the correlation coefficient (R^2), chi-square (χ^2) and the root mean square error analysis (RMSE). The results of the study showed that Sinop has a mean wind speed of 3.36 m/s with a maximum value of 4.28 m/s in February of 2011, and a minimum value of 2.41 m/s in March of 2013, while the corresponding mean wind power density is approximately 33.31 W/m^2 for the whole year. In general, it was determined the wind speed is higher during some winter and spring months, notably February and March, and is lower during the autumn months. The Weibull distribution function was found to be more appropriate than the Rayleigh distribution function.

Keywords: *Wind Energy, Statistical Analysis of Wind Energy Potential, Weibull Distribution, Rayleigh Distribution, Sinop-Turkey*

INTRODUCTION

Renewable energy technologies play a strategic role in achieving the goals of sustainable economical development and environmental protection. Wind energy is one of the most important energy sources used widely as an alternative source [1].

In wind energy studies, statistical properties of the wind speed provides remarkable information to researchers and designers for predicting the output energy of a wind energy conversion system. Wind speed at a site varies randomly and its variation in a certain region over a period of time can be represented by different probability distribution functions (PDFs). Two-parameter Weibull distribution is the most commonly used and accepted distribution as it is found to be a fairly accurate PDF for most of the wind regimes found in nature and also it is simple to use. However, it is not suitable for certain wind regimes, such as those having high frequencies of null winds and for short time horizons. Another distribution Rayleigh PDF has also been used in many applications, owing to its simplicity and ability of describing wind speed with sufficient accuracy when little detail is available about the wind characteristics of a site [2].

Statistical distributions have been analyzed in the literature for examining differences between the results for day and night, between various seasons, and for monthly and yearly data. The suitability of these distribution functions for a particular application depends upon a number of factors such as the type of wind regime, availability of data, and recording intervals Moreover, a number of methods for estimating the parameters of these distributions and model evaluation criteria are given in the literature [2-5]. In recent years, many studies have been conducted to evaluate wind energy potential in different regions of Turkey by using various probability distribution functions and the results have shown that the Weibull and Rayleigh distributions can describe wind speed probability distribution well [6-21]. Ulgen and Hepbasli [6] determined the Weibull density distribution function to be able to predict the energy output of wind energy systems for Izmir- Turkey. Celik [7] analyzed the acceptability of the Weibull and Rayleigh distributions for modeling local wind profiles based on 1-year measured hourly time-series wind speed data

This paper was recommended for publication in revised form by Regional Editor Tolga Taner
[1]Department of Mechanical Engineering, Firat University, Elazig, Turkey
**E-mail address: ebruakpinar@firat.edu.tr*

in the Iskenderun region of Hatay in Turkey. Karsli and Gecit [8] investigated the wind power potential of Nurdagı-Gaziantep in the southern of Turkey. Kose *et al.* [9] evaluated the wind energy potential in Kutahya-Turkey according to wind data from a wind observation station established at the Dumlupinar University Campus. Akpinar and Akpinar [10-11] carried out studies using Weibull density function to demonstrate wind energy potential of Maden-Elazıg [10] and Agın-Elazığ [11] in eastern of Turkey. Genc *et al.* [12] estimated wind power potential in Aksehir-Turkey using the Weibull wind speed distribution function. Akpinar [13] investigated wind energy potential with respect to wind speed data from four stations, which are located in Elazig and its close regions. In their study, wind data were fitted to the Weibull and Rayleigh probability distribution functions and compared with each other. Gokcek et al. [14] investigated wind power potential in Kırklareli-Turkey using Weibull and Rayleigh functions for the year 2004. Yılmaz and Celik [15] determined appropriate theoretical pdf distribution by comparing 10 pdf distributions (Beta, Erlang, Exponential, Gamma, Log-Logistic, Lognormal, Pearson V, Pearson VI, Uniform and Weibull) for the wind speed data measured in Gelibolu region. At the end of the comparison, Weibull was determined to be the best distribution representing wind data. Akdag and Güler [16] analyzed wind energy potential of Çanakkale region using Weibull distribution and real time series analysis, making use of hourly average wind data of the Turkish State Meteorological Service (TSMS) measured in 10 m height between 2001 and 2006. Bilgili and Sahin [17] investigated wind energy density in the southern and southwestern region of Turkey by using the Weibull and Rayleigh probability density functions. Mert and Karakus [18] determined the wind energy potential in Antakya through the use of a variety of statistical distributions, including the Burr, generalized gamma, and Weibull distributions. Dokur and Kurban [19] modelled wind energy potential based on two parameter Weibull distribution statistically for Bilecik, city in Turkey, by using the Turkish State Meteorological Service's hourly wind speed data between 2010 and 2014 measured. Kaplan and Aladag [20] analyzed statistically the wind energy potential in Hatay based on the data regarding the wind speed measured hourly between 2009 and 2013. Yanıktepe *et al.* [21] studied the wind energy potential of Osmaniye-Turkey using the Weibull and the Rayleigh models.

Wind data analysis and accurate wind energy potential assessment is critical for proper and efficient development of wind power application and is highly site-dependent. As a result, knowledge of the statistical properties of wind speed is essential for predicting the energy output of wind energy conversion system (WECS). Turkey has important wind energy potential especially in the coasts of western, southern and northern. The aim of this study is to investigate the wind energy potential of the Sinop province on Turkey's northern coast by using the wind data recorded at the Sinop Meteorological Station. An investigation of wind energy potential as a result of the distribution of wind speed in the Sinop province will draw attention to future wind energy projects.

THEORY

Wind speed data

The wind data used in this study were measured and recorded hourly at the Sinop station of the Turkish State Meteorological Service at 10 m above ground level between 2005 and 2014. The Sinop station (North Turkey), located in Mediterranean region (Figure 1) [22], is located at 42° 01′ N - 35° 11′ E, and its elevation is 32 m above sea level. The wind data were captured using a cup generator anemometer.

Theory of wind speed and wind power

The two-parameter Weibull distribution is the most widely used distribution in wind power applications. It is a versatile PDF, is simple to use, and is found to be accurate for most of the wind regimes encountered in nature. Rayleigh distribution is a special case of Weibull distribution in which the shape parameter is taken as $k = 2$. This function has also been used by many researchers due to its simplicity as there is only one parameter to be evaluated and is useful when not much information about the site is available [2].

Figure 1. Location of Sinop on Turkey Map [22]

Weibull and Rayleigh distribution of wind speed

The monthly mean wind speed values and the standard deviations are calculated from Equepments. (1) and (2) using measured data.

$$v_m = \frac{1}{N}\left[\sum_{i=1}^{N} v_i\right] \tag{1}$$

$$\sigma = \left[\frac{1}{N-1}\sum_{i=1}^{N}(v_i - v_m)^2\right]^{1/2} \tag{2}$$

The Weibull distribution function that is a special case of generalized gamma distribution for wind speed is expressed with Eq. (1).

$$f_w(v) = (\frac{k}{c})(\frac{v}{c})^{k-1}\exp(-(\frac{v}{c})^k) \tag{3}$$

where $f_w(v)$ is the probability of observing wind speed v, k the dimensionless Weibull shape parameter (or factor) and c the Weibull scale parameter, which has its reference value in the units of wind speed [6, 10-14].

The cumulative probability function of the Weibull distribution is calculated as below [6, 10-14]:

$$F_w(v) = 1 - \exp\left[-(\frac{v}{c})^k\right] \tag{4}$$

Determination of the parameters of the Weibull distribution requires a good fit of Eq. (4) to the recorded discrete cumulative frequency distribution. Taking the natural logarithm of both sides of Eq. (4) twice, gives

$$\ln\{-\ln[1-F(v)]\} = k\ln(v) - k\ln c \tag{5}$$

So, a plot of *ln{-ln[1-F(v)]}* versus $\ln v$ presents a straight line. The gradient of the line is k and the intercept with the y-axis is $-k\ln c$ [6, 10-11].

The k values range from 1.5 to 3.0 for most wind conditions. Another distribution function used in determination of the wind speed potential is Rayleigh distribution. This distribution is a special case of Weibull distribution and validate situation where the dimensionless shape parameter k of the Weibull distribution is assumed to be equal to 2. Probability density and cumulative function of the Rayleigh distribution are given by Eqs. (6) and (7), respectively [10-11],

$$f_R(v) = \frac{2v}{c^2} exp\left[-(\frac{v}{c})^2\right] \tag{6}$$

$$F_R(v) = 1 - exp\left[-(\frac{v}{c})^2\right] \tag{7}$$

The mean value of wind speed and its standard deviation may be computed using Eqs. (8) and (9), respectively,

$$v_m = c\Gamma(1+\frac{1}{k}) \tag{8}$$

$$\sigma = \sqrt{c^2\left\{\Gamma\left(1+\frac{2}{k}\right) - \left[\Gamma\left(1+\frac{1}{k}\right)\right]^2\right\}} \tag{9}$$

where $\Gamma()$ is the gamma function [10-12].

Based on the Weibull distribution, the wind speed with the largest frequency is calculated from Eq. (10) [12]:

$$V_{\text{mod}} = c\left(1-\frac{1}{k}\right)^{1/k} \tag{10}$$

The maximum wind speed can be determined by [12, 15]:

$$V_{\max E} = c\left(\frac{k+2}{k}\right)^{1/k} \tag{11}$$

When $k=2$ is taken in the above formulas, calculations are performed for Rayleigh distribution.

Calculations of wind power

The wind power per unit area in any windy site is of importance in assessing of the wind power projection for the power plants. The mean wind power density of the considered site per unit area based on any probability density function can be expressed as [16]:

$$P_m = \int_0^\infty P(v) f(v) dv \tag{12}$$

It is well known that the power of the wind that flows at speed v through a blade sweep area A increases as the cubic of its velocity and is given by,

$$P(v) = \frac{1}{2} \rho A v^3 \tag{13}$$

Where ρ is the air density for Sinop. The mean power density for the Weibull distribution is obtained from Eq. (14) as follows [7]

$$P_w = \frac{1}{2} \rho c^3 \Gamma(1 + \frac{3}{k}) \tag{14}$$

The mean power density for the Rayleigh distribution is determined by Eq. (15) [7]:

$$P_R = \frac{3}{k} \rho v^3 m \tag{15}$$

Statistical analysis of distributions

Modeling efficiency (R^2), chi-square (x^2) and root mean square error analysis (RMSE) and were used as the primary criterion to select the best distribution to account for the variation in the wind speed curves. Chi-square is the mean square of the deviations between the experimental and calculated values for the distributions and was used to determine the goodness of the fit. The lower are the values of chi-square, the better is the goodness of the fit. The RMSE gives the deviation between the predicted and experimental values, and it is required to approach zero. The R^2 also gives the ability of the model, and its highest value is 1. These statistical values can be calculated as follows:

$$R^2 = \frac{\sum_{i=l}^{N} (yi - zi)^2 - \sum_{i=l}^{N} (xi - yi)^2}{\sum_{i=l}^{N} (yi - zi)^2} \tag{16}$$

$$x^2 = \frac{\sum_{i=l}^{N} (yi - xi)^2}{N - n} \tag{17}$$

$$RMSE = \left[\frac{1}{N} \sum_{i=l}^{N} (yi - xi)^2 \right]^{1/2} \tag{18}$$

where yi is the ith experimental data, zi is the mean value of the experimental data, xi is the ith predicted data with the Weibull or Rayleigh distribution, N is the number of observations and n is the number of constants [10, 11,-13].

RESULTS AND DISCUSSION

The measured hourly wind data between the years of 2004-2015 have been used to evaluate the wind energy potential of the Sinop province. The calculations were made to obtain the Weibull and Rayleigh distribution parameters, the mean power density and the mean wind speed. The main results obtained from the present study can be summarized as follows:

The monthly mean wind speed variations in Sinop are illustrated in Figure 2. It is also clear from Figure 2 that the wind speed for the investigated years had the lowest value in March of 2013 and the highest in February of 2011, ranging from 2.41 to 4.28 m/s. The annual mean value of the wind speed was calculated as 3.36 m/s.

In Figure 3, the diurnal variation of the mean wind speed is plotted. According to the yearly average results, the lowest wind speed was 2.15 m/s between the hours of 05:00 and 06:00 in 2006, and the highest wind speed was 3.89 m/s between the hours of 13:00 and 14:00 in 2007. There is a similar trend in the mean wind speed of different daily hours during different years. The diurnal wind speed has its minimum between the hours of 0:00 and 18:00 and its maximum between the hours of 6:00 and 16:00. Figure 3 clearly shows that daily variations in wind speed from morning to midnight in Sinop follows a parabolic curve, where the maximum point occurs in the afternoon. In most locations around the world, the wind is more turbulent and conditions become windier during the daytime relative to the night time, mostly due to temperature differences. Therefore, the aforementioned situation is advantageous because the wind energy is mostly produced during the daytime, since the most electricity consumption is at night time.

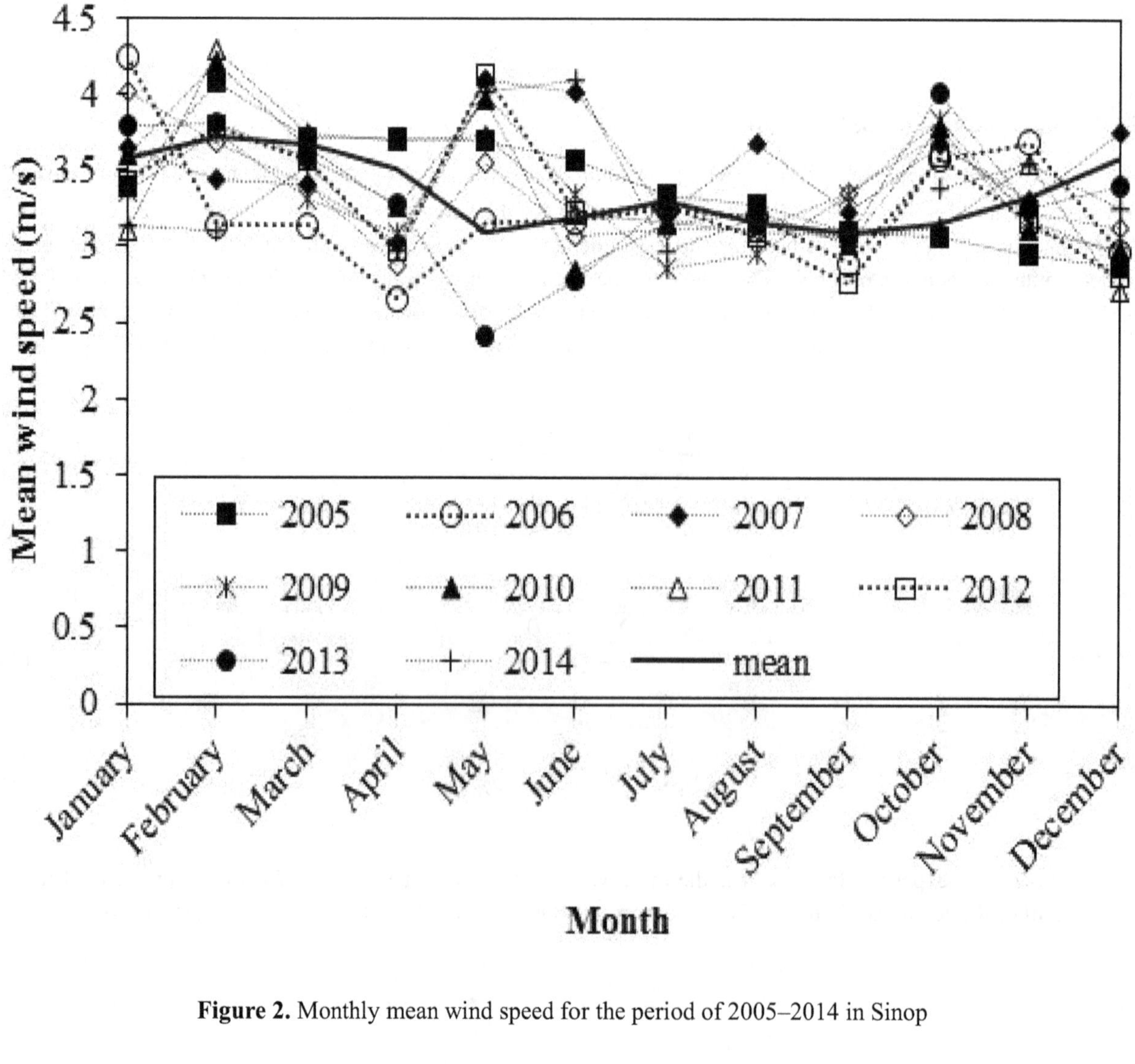

Figure 2. Monthly mean wind speed for the period of 2005–2014 in Sinop

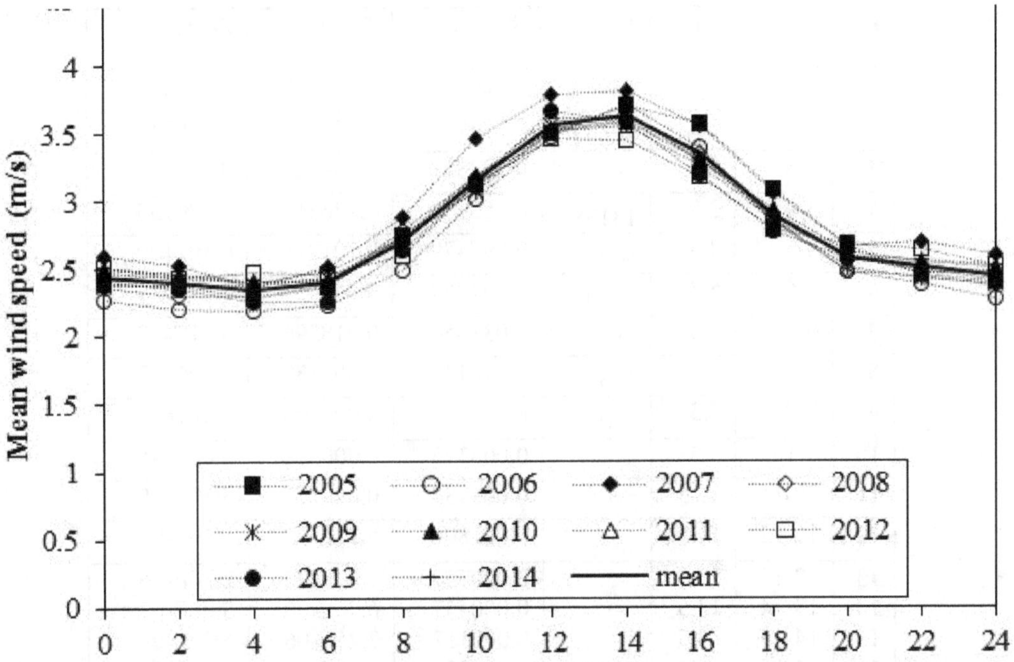

Figure 3. Diurnal variation of mean wind speed for the period of 2005–2014 in Sinop

The wind speed data in time-series format is usually arranged in the frequency distribution format since it is more convenient for statistical analysis. Therefore, the available time-series data were arranged as frequency distribution format for the period of 2005-2014 in Table 1. The wind speed is grouped into classes in the second column of Table 1. The mean wind speeds are calculated for each speed class intervals in the third column using the arithmetic mean. The fourth column shows the frequency of occurrence of each speed class. The probability density distributions calculated from the actual, Weibull and Rayleigh functions are presented in the fifth, sixth and seventh columns, respectively. It can be seen from Table 1 that the maximum probability density values are in the range of 1-2 and 2-3 m/s.

The yearly probability density and the cumulative distributions derived from the time-series data from Sinop are presented in Figures 4 and 5, respectively. The yearly probability density and the cumulative distributions are illustrated in Figure 6 for all the investigated years. All the curves show a similar trend for wind speeds' probability and cumulative distribution.

The monthly mean wind speed values and the standard deviations were calculated from Eqs. (1) and (2) using measured data. The monthly mean wind speed values (v_m) and standard deviations (σ) are given in Table 2 for Weibull distribution. Most of the monthly mean wind speed values are between 2.0 and 4.0 m/s. February of 2010 has the highest monthly mean wind speed value with 4.07 m/s, March of 2013 has the lowest monthly mean wind speed value with 2.39 m/s. Generally, it is determined that the highest mean wind speed values are in February and March, and the lowest mean wind speed values are in September and October. Having analyzed the 118 months of wind speed data, it can be concluded that the wind speed distribution differs remarkably from one month to the next. The monthly and yearly standard deviation values are mostly between 1.0 and 2.0 m/s.

Table 1. The measured hourly time-series data in frequency distribution format for the period of 2005-2014 and the probability density distributions calculated from the Weibull, $(f_W(vj))$, and Rayleigh $(f_R(vj))$

j	v_j	$v_{m,j}$	f_j	$f(v_j)$	$f_w(v_j)$	$f_R(v_j)$
1	0-1	0.5	10262	0.120892	0.153598	0.155466
2	1-2	1.5	22848	0.274176	0.239059	0.24075
3	2-3	2.5	18635	0.222812	0.238165	0.236103
4	3-4	3.5	13138	0.158936	0.176315	0.174135
5	4-5	4.5	8029	0.097479	0.101911	0.102129
6	5-6	5.5	4730	0.057069	0.047457	0.048914
7	6-7	6.5	2571	0.031054	0.018396	0.019433
8	7-8	7.5	1364	0.016134	0.006189	0.006472
9	8-9	8.5	810	0.009546	0.001906	0.001821
10	9-10	9.5	456	0.005325	0.000572	0.000435
11	10-11	10.5	279	0.003284	0.000177	8.89E-05
12	11-12	11.5	153	0.001806	5.77E-05	1.56E-05
13	12-13	12.5	65	0.000757	1.97E-05	2.36E-06
14	13-14	13.5	36	0.000437	6.76E-06	3.08E-07
15	14-15	14.5	10	0.000117	2.25E-06	6.23E-09
16	15-16	15.5	7	7.8E-05	7.59E-07	9.45E-11
17	16-17	16.5	5	0.000056	2.47E-07	6.01E-12
18	17-18	17.5	2	2.24E-05	7.71E-08	3.23E-13
19	18-19	18.5	1	1.12E-05	2.32E-08	1.46E-14
20	19-20	19.5	1	1.12E-05	6.75E-09	3.97E-19

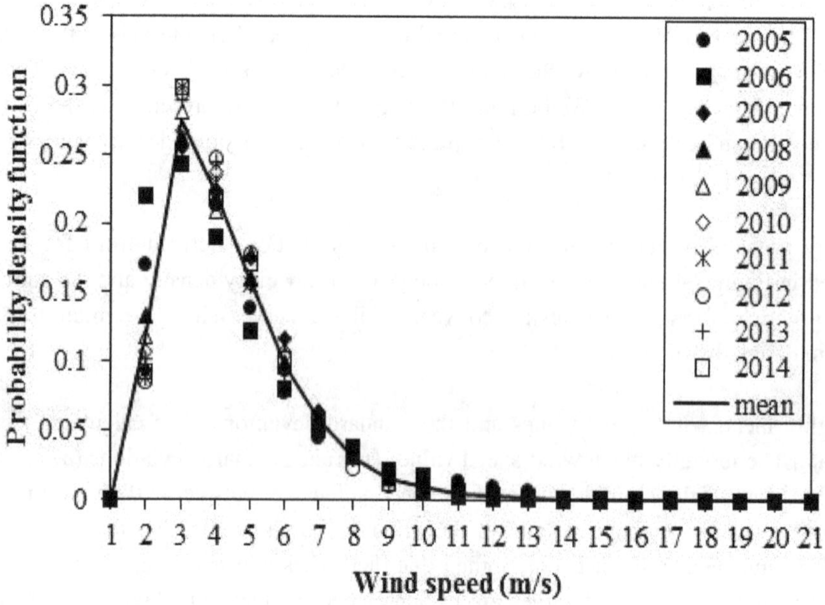

Figure 4. Yearly wind speed probability density distributions, derived from the measured. hourly time series

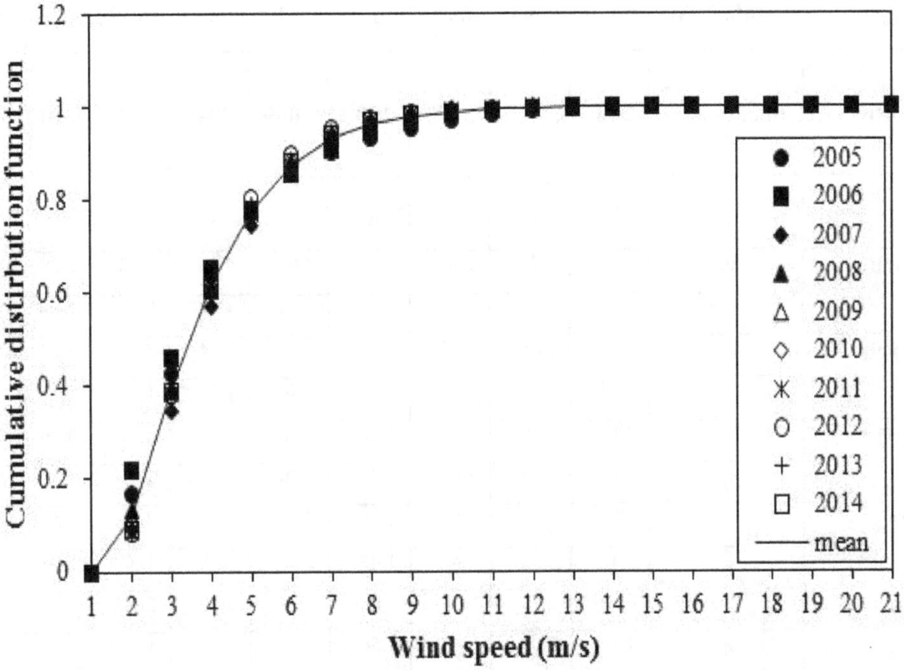

Figure 5. Yearly wind speed cumulative probability distributions, derived from the measured hourly time series

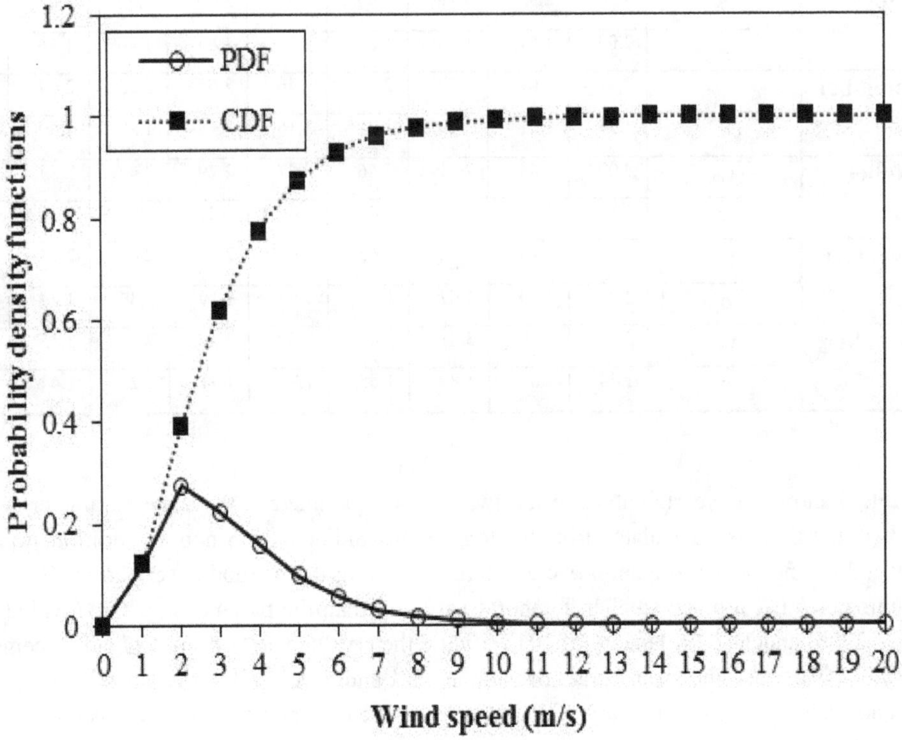

Figure 6. The wind speed probability density and cumulative probability distributions for whole years, derived from the measured hourly time-series

Table 2. Monthly mean wind speed and standard deviations to Weibull distribution

Month	Parameters	2005	2006	2007	2008	2009	2010	2011	2012	2013	2014
January	v_m	2.67	3.61	3.10	3.43	2.97	3.27	2.68	3.09	3.19	2.83
	σ	1.49	2.19	1.42	2.04	1.50	1.59	1.32	1.38	1.03	1.55
February	v_m	3.21	2.47	3.22	3.25	3.61	4.07	3.82	3.29	3.78	2.71
	σ	1.67	1.91	1.59	1.66	1.63	1.92	2.01	1.62	1.79	1.18
March	v_m	3.01	2.88	3.73	3.26	3.97	3.54	3.25	3.63	2.39	3.90
	σ	1.58	1.20	1.91	1.43	1.96	1.92	1.71	1.71	1.08	1.71
April	v_m	3.18	2.99	3.38	3.30	3.06	3.59	3.61	3.36	3.42	3.41
	σ	1.64	1.85	1.62	1.45	1.60	1.97	1.55	1.45	1.70	1.43
May	v_m	2.71	2.51	3.04	3.16	3.11	2.93	2.74	2.53	2.86	-
	σ	1.47	1.34	1.43	1.49	1.36	1.38	1.07	1.04	1.26	-
June	v_m	2.89	2.90	3.54	3.05	2.89	3.09	2.92	2.94	2.92	3.02
	σ	1.78	2.00	1.73	1.56	1.40	1.50	1.28	1.30	1.29	1.41
July	v_m	2.58	3.47	3.15	3.11	2.91	2.78	3.59	2.87	3.01	3.47
	σ	1.33	3.00	1.46	1.66	1.51	1.35	1.73	1.34	1.61	1.63
August	v_m	3.10	2.93	3.11	2.96	2.75	3.00	2.98	3.02	3.02	2.71
	σ	2.09	1.57	1.55	1.63	1.43	1.64	1.46	1.46	1.55	1.32
September	v_m	2.81	2.61	3.57	2.91	2.46	2.65	2.63	2.69	3.03	2.86
	σ	1.74	1.91	2.19	1.71	1.16	1.15	1.29	1.30	1.34	1.39
October	v_m	3.09	2.43	2.81	2.46	2.77	2.79	3.47	2.84	2.87	2.52
	σ	1.83	1.54	1.41	1.45	1.45	1.25	1.75	1.30	1.55	1.07
November	v_m	3.22	2.46	3.53	2.62	2.94	2.58	2.66	2.74	2.61	3.20
	σ	2.02	1.80	1.67	1.60	1.67	1.19	1.37	1.27	1.20	1.85
December	v_m	2.54	2.29	3.41	3.49	3.60	3.40	2.76	3.35	3.08	3.14
	σ	1.44	2.09	1.88	1.88	1.76	1.71	1.25	1.44	1.71	1.61

Table 3 shows the yearly values of the two Weibull parameters, the scale parameter c (m/s) and the shape parameter k (dimensionless), calculated from the long term wind data for Sinop. In addition, Table 3 shows monthly values of k and c. The values of c and k were determined using the method described in *Theory section.* It is clear that the parameter k has a much smaller, temporal variation than the parameter c. The k value is between 1.09 (in December of 2006) and 3.42 (in January of 2013), while the c value varies from 2.37 (in December of 2006) to 4.60 (in February of 2010) m/s. The mean values of parameters k and c were calculated as 2.02 and 3.43 m/s, respectively. The shape and scale parameters for May of 2014 could not be obtained for various reasons provided by the Turkish State Meteorological Service. The mean wind intensity and standard deviation values are important in predicting shape and scale parameters.

Table 3. Monthly shape parameter (k) and scale parameter (c) values according to the Weibull distribution for period of 2005-2014

Month	Parameters	2005	2006	2007	2008	2009	2010	2011	2012	2013	2014
January	**k**	1.85	1.69	2.31	1.72	2.06	2.17	2.12	2.39	3.42	1.90
	c	3.01	4.04	3.50	3.84	3.35	3.70	3.03	3.49	3.55	3.19
February	**k**	2.01	1.30	2.12	2.05	2.35	2.23	1.97	2.13	2.22	2.44
	c	3.62	2.68	3.64	3.67	4.08	4.60	4.31	3.72	4.27	3.05
March	**k**	1.98	2.57	2.03	2.42	2.13	1.91	1.98	2.24	2.34	2.43
	c	3.39	3.24	4.21	3.68	4.48	3.99	3.66	4.10	2.69	4.39
April	**k**	2.02	1.66	2.20	2.42	1.99	1.89	2.48	2.47	2.10	2.55
	c	3.59	3.35	3.82	3.72	3.45	4.05	4.07	3.79	3.86	3.84
May	**k**	1.92	1.94	2.24	2.24	2.43	2.23	2.75	2.59	2.41	-
	c	3.06	2.83	3.43	3.57	3.51	3.30	3.08	2.85	3.23	-
June	**k**	1.67	1.47	2.15	2.04	2.16	2.16	2.42	2.41	2.40	2.25
	c	3.24	3.20	4.00	3.45	3.26	3.48	3.30	3.32	3.29	3.40
July	**k**	2.02	1.15	2.28	1.95	2.00	2.17	2.18	2.27	1.94	2.24
	c	2.91	3.65	3.56	3.51	3.28	3.14	4.06	3.24	3.39	3.92
August	**k**	1.50	1.94	2.10	1.89	2.00	1.89	2.15	2.16	2.03	2.16
	c	3.43	3.30	3.51	3.34	3.10	3.38	3.37	3.41	3.41	3.07
September	**k**	1.65	1.38	1.67	1.75	2.24	2.46	2.13	2.16	2.40	2.16
	c	3.14	2.86	4.00	3.27	2.78	2.99	2.97	3.03	3.41	3.23
October	**k**	1.73	1.61	2.09	1.73	1.99	2.36	2.07	2.30	1.91	2.50
	c	3.47	2.72	3.17	2.76	3.12	3.14	3.92	3.21	3.23	2.84
November	**k**	1.63	1.38	2.22	1.68	1.81	2.29	2.03	2.28	2.30	1.78
	c	3.60	2.70	3.98	2.94	3.30	2.91	3.01	3.09	2.95	3.60
December	**k**	1.83	1.09	1.88	1.93	2.15	2.07	2.34	2.49	1.87	2.03
	c	2.86	2.37	3.84	3.94	4.07	3.83	3.11	3.78	3.47	3.54

The Weibull and Rayleigh approximations of the actual probability distribution of wind speeds are shown in Fig. 7, while a comparison of the two approximations is given in Table 4. In Fig. 7, the probability distribution of the actual data, the Weibull probability distribution, and the Rayleigh probability distribution are plotted versus the wind speed according to the average of ten years data. The probability ratio of the Weibull distribution is higher than the Rayleigh distribution. It can be seen from Fig. 7 that the Weibull distribution fit to the actual distribution data is closer than the Rayleigh distribution.

Table 4. R^2, RMSE and χ^2 values obtained from Weibull and Rayleigh distributions

Years	Weibull Distribution			Rayleigh Distribution		
	R^2	RMSE	χ^2	R^2	RMSE	χ^2
2005	0.982	0.000060	0.000140	0.969	0.000107	0.000231
2006	0.994	0.000016	0.000036	0.930	0.000209	0.000439
2007	0.969	0.000110	0.000256	0.966	0.000120	0.000257
2008	0.969	0.000066	0.000155	0.981	0.000069	0.000149
2009	0.962	0.000146	0.000340	0.962	0.000147	0.000316
2010	0.966	0.000128	0.000298	0.965	0.000135	0.000290
2011	0.937	0.000258	0.000597	0.931	0.000284	0.000608
2012	0.957	0.000191	0.000446	0.937	0.000284	0.000612
2013	0.961	0.000140	0.000314	0.955	0.000164	0.000345
2014	0.941	0.000237	0.000547	0.936	0.000258	0.000552

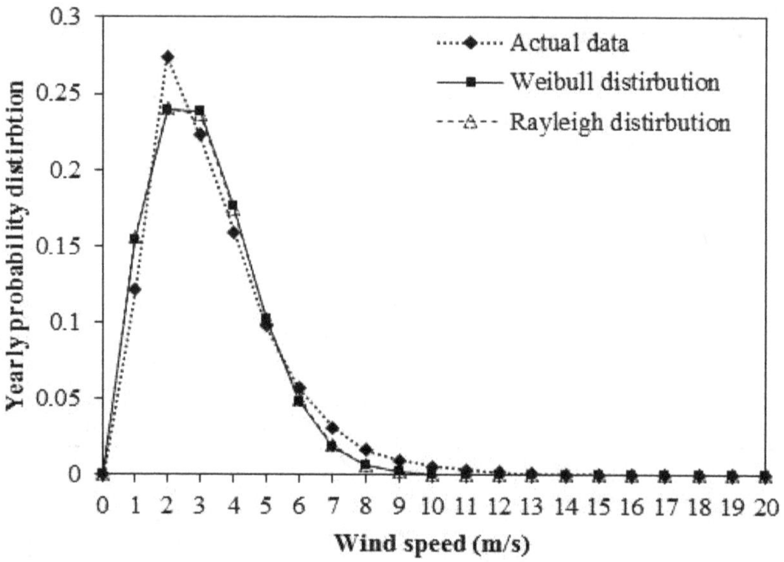

Figure 7. The comparison of Weibull and Rayleigh approximations with the actual probability distribution of wind speeds

The values of the statistical parameters (R^2, *RMSE*, and χ^2) obtained from the Weibull and Rayleigh distributions are given in Table 4. The R^2 value was found to be between 0.937-0.994 in the Weibull distribution and 0.930-0.981 in the Rayleigh distribution. The RMSE and χ^2 values range from 1.6x10^{-6} to 2.58x10^{-5} and 3.6x10^{-6} to 5.97x10^{-5} for the Weibull distribution, while they vary between 6.9x10^{-6}-2.84x10^{-5} and 1.49x10^{-5}- 6.12x10^{-5} for the Rayleigh distributions, respectively. As can be seen in Table 4, the highest R^2 value was obtained by using the Weibull distribution. However, the results have shown that the *RMSE* and χ^2 values of the Weibull distribution were lower than the values obtained for the Rayleigh distribution. As a result, the Weibull approximation was found to be the most accurate distribution according to the highest value of R^2 and the lowest values of *RMSE* and χ^2.

Fig. 8 shows the monthly change in the R^2 obtained from the Weibull and Rayleigh distributions for Sinop using the ten-year data. The range of R^2 values change from 0.89 to 0.95 in the Weibull distribution and 0.81 to 0.92 in the Rayleigh distribution. Because the R^2 value is closer to 1 in the Weibull distribution, it is understood that the Weibull distribution is more suitable for modeling the wind data for Sinop.

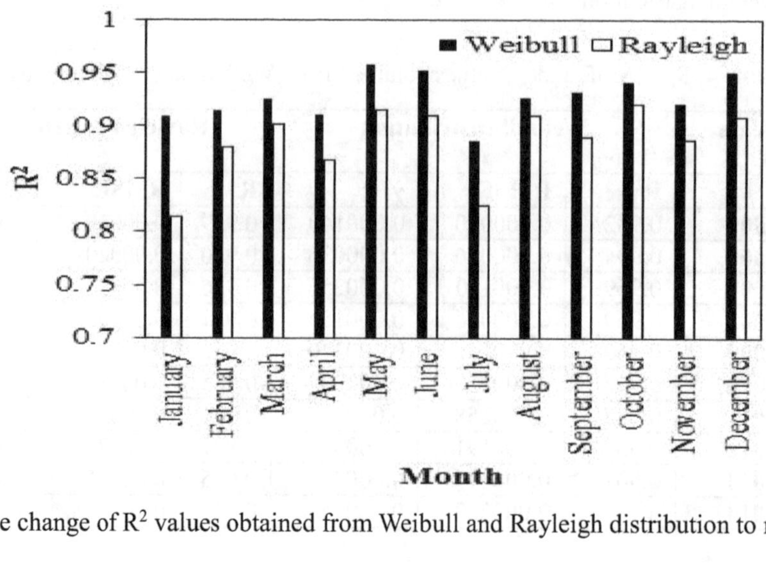

Figure 8. The change of R^2 values obtained from Weibull and Rayleigh distribution to months

The Weibull distribution parameters and the Rayleigh distribution parameters are given in Table 5 and Table 6 according to the investigated years, respectively. The maximum mean velocity (V_{max}) in the Weibull distribution ranged from 4.48 to 5.46 m/s and the mean power density (P_w) ranged from 28.29 to 39.20 W/m^2. In the Rayleigh distribution, V_{max} varied from 4.39 to 5.25 m/s and P_R varied from 24.27 to 41.65 W/m^2.

Table 5. Weibull distribution parameters

Year	k	c (m/s)	V_m (m/s)	σ (m/s)	V_{mod} (m/s)	V_{maxE} (m/s)	P_W (W/m^2)
2005	1.79	3.29	2.93	1.69	2.08	5.00	33.04
2006	1.52	3.15	2.84	1.90	1.57	5.09	37.21
2007	2.09	3.69	3.27	1.64	2.71	5.46	39.20
2008	1.94	3.49	3.10	1.66	2.41	5.02	35.74
2009	2.02	3.48	3.08	1.59	2.49	4.88	33.85
2010	2.07	3.51	3.11	1.57	2.56	4.86	33.97
2011	2.15	3.46	3.06	1.50	2.58	4.69	31.30
2012	2.28	3.40	3.01	1.39	2.64	4.48	28.29
2013	2.16	3.38	2.99	1.45	2.54	4.57	29.04
2014	2.13	3.46	2.06	1.50	2.57	4.71	31.48

Table 6. Rayleigh distribution parameters

Year	c (m/s)	V_m (m/s)	σ (m/s)	V_{mod} (m/s)	V_{maxE} (m/s)	P_R (W/m^2)
2005	3.25	2.88	1.50	2.29	4.59	27.89
2006	3.10	2.75	1.43	2.19	4.39	24.27
2007	3.71	3.29	1.72	2.62	5.25	41.65
2008	3.49	3.09	1.61	2.46	4.93	34.47
2009	3.49	3.09	1.61	2.46	4.93	34.48
2010	3.53	3.13	1.63	2.50	5.00	35.89
2011	3.49	3.10	1.62	2.47	4.94	34.75
2012	3.46	3.07	1.60	2.45	4.90	33.79
2013	3.42	3.03	1.58	2.42	4.84	32.64
2014	3.49	3.09	1.61	2.46	4.93	34.54

The variation of mean wind speed with Weibull power density and Rayleigh power density are shown in Fig. 9 and in Fig. 10 with respect to the investigated years, respectively. The mean wind velocity was determined to be 2.95 m/s for the Weibull distribution and 3.05 m/s for the Rayleigh distribution. In the Weibull distribution, the highest power density was 39.20 W/m^2 in 2007, the lowest power density was 28.30 W/m^2 in 2012, and the mean power density was 33.31 W/m^2. In the Rayleigh distribution, the highest and the lowest power densities were determined to be 41.66 W/m^2 in 2007 and 24.27 W/m^2 in 2006, respectively, and the obtained mean power density was 33.44 W/m^2.

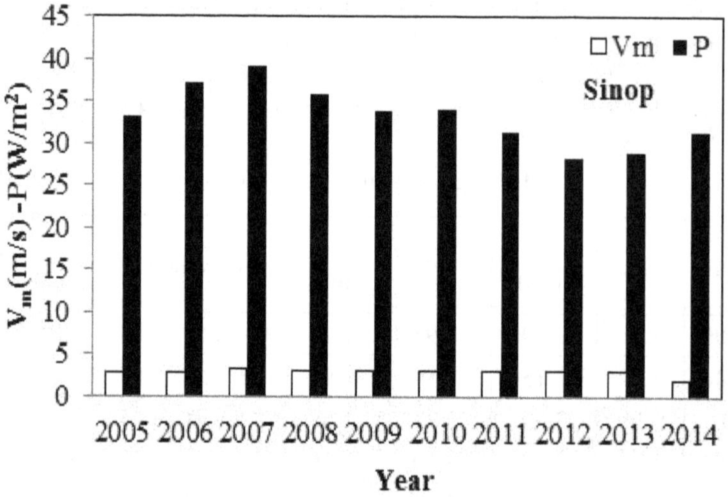

Figure 9. Yearly mean power density and wind speed according to Weibull distribution

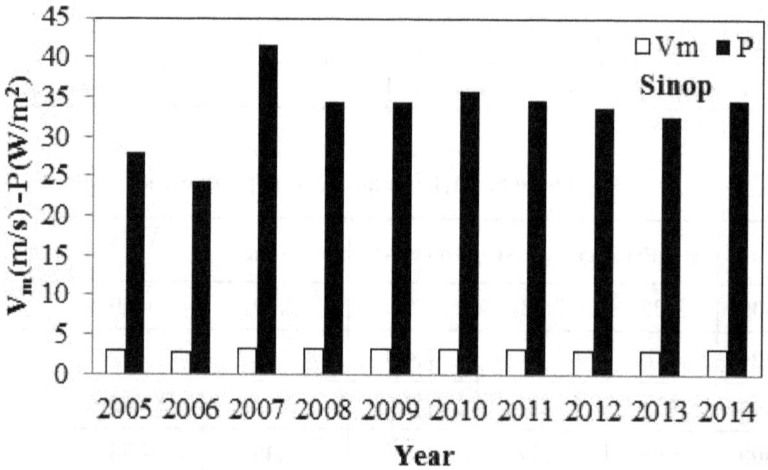

Figure 10. Yearly mean power density and wind speed according to Rayleigh distribution

CONCLUSION

In this study, the wind power density and wind speed distribution parameters of Sinop were statistically analyzed during the years of 2005-2014. Two probability density functions were fitted to the measured probability distributions on a yearly and monthly basis. Performances of the probability models were compared to the measured monthly and yearly wind speed values. Weibull and Rayleigh distributions were used for modeling and the success of this modeling process was evaluated according to R^2, RMSE and χ^2 parameters. The results can be summarized as follows:

- The mean wind speed was found as 3.36 m/s for all years between 2004 and 2015.
- The mean power densities were calculated as 33.31 and 33.44 W/m^2, respectively, for Weibull and Rayleigh distribution for the whole year.

- The Weibull model is generally better in fitting the measured yearly probability density distributions than the Rayleigh model, to the statistical criteria such as R^2, RMSE and χ^2. Therefore, it was concluded that it would be more appropriate to use the Weibull distribution for the analysis of Sinop's wind data.

- The values of the shape parameter (k) and scale parameter (c) at Sinop were examined. The yearly mean values of k and c for thevWeibull distribution were determined to be 2.02 and 3.44 m/s over the 10 year period of 2005-2014. The yearly mean c value for the Rayleigh distribution was found to be 3.45 m/s. The Weibull shape parameter follows very closer to the Raleigh function, k=2, for Sinop which means that the characteristics for the wind wave are regular and uniform.

- As a result, monthly average power and annual average power densities in Sinop are as small as 100 W/m^2. Therefore, it is not possible to support the network directly by wind energy systems; however it can be used in applications where there is no network access or in rural areas where low-power density is required. It has been decided that electricity generation from wind energy is suitable because the average speed on a daily and monthly basis usually is 3 m/s.

REFERENCES

[1] Pishgar-Komleh, S.H., Keyhani, A., Sefeedpari, P. (2015). Wind speed and power density analysis based on Weibull and Rayleigh distributions (a case study: Firouzkooh county of Iran). Renewable and Sustainable Energy Reviews, 42, 313–322.

[2] Sohoni, V., Gupta, S., Nema, R. (2016). A comparative analysis of wind speed probability distributions for wind power assessment of four sites. Turk J Elec Eng & Comp Sci, 24, 4724-4735.

[3] Morgan, V.T. (1995). Statistical distributions of wind parameters at Sydney, Australia. Renew Energy, 6, 39-47.

[4] Seguro, J.V., Lambert, T.W. (2000). Modern estimation of the parameters of the Weibull wind speed distribution for wind energy analysis. J Wind Eng Ind Aerod, 85, 75-84.

[5] Costa Rocha, P.A., De Sousa, R.C., De Andrade, C.F., Da Silva, M.E.V. (2012). Comparison of seven numerical methods for determining Weibull parameters for wind energy generation in the northeast region of Brazil. Appl Energ, 89, 395-400.

[6] Ulgen, K., Hepbasli, A. (2002). Determination of Weibull parameters for wind energy analysis of Izmir, Turkey. Int J Energy Res, 26, 494–506.

[7] Celik, A.N. (2003). A statistical analysis of wind power density based on the Weibull and Rayleigh models at the southern region of Turkey. Renewable Energy, 29(4), 593–604.

[8] Karsli, V.M., Gecit, C. (2003). An investigation on wind power potential of Nurdagı- Gaziantep, Turkey. Renew Energy, 28, 823–830.

[9] Kose, R., Ozgur, M. A., Erbas, O., Tugcu, A. (2004). The analysis of wind data and energy potential in Kutahya, Turkey. Renew Sustain Energy Rev, 8, 277–288.

[10] Akpinar, E.K., Akpinar, S. (2004). Determination of the wind energy potential for Maden-Elazığ, Turkey. Energy Conversion and Management, 45, 2901-2914.

[11] Akpinar, E.K., Akpinar, S. (2004). Statistical Analysis of wind energy potential on the basis of the Weibull and Rayleigh distribution for Ağın-Elazığ, Turkey. J.Power Energy, 218, 557-565.

[12] Genc, A., Erisoglu, M., Pekgor, A., Oturanc, G., Hepbasli, A., Ulgen, K. (2005). Estimation of wind power potential using Weibull distribution. Energ Source, 27, 809-822.

[13] Akpinar, E.K. (2006). A statistical investigation of wind energy potential. Energy Sources, Part A, 28, 807–820.

[14] Gökcek, M., Bayülken, A., Bekdemir, Ş. (2007). Investigation of wind characteristics and wind energy potential in Kirklareli, Turkey. Renewable Energy, 32, 1739-1752.

[15] Yilmaz, V., Çelik, H.E. (2008). A statistical approach to estimate the wind speed distribution: the case of Gelibolu region. Doğuş Üniversitesi Dergisi, 9 (1), 122-132.

[16] Akdag, S.A., Güler, Ö. (2009). Calculation of wind energy potential and economic analysis by using Weibull Distribution—A case study from Turkey. Part 1: Determination of Weibull parameters. Energy Sources, Part B, 4, 1–8.

[17] Bilgili, M. and Şahin, B. (2009). Investigation of wind energy density in the Southern and Southwestern region of Turkey. Journal of Energy Engineering, 135, 1(12), 12-20.

[18] Mert, I., Karakus, C. (2015). A statistical analysis of wind speed data using Burr, generalized gamma, and Weibull distributions in Antakya, Turkey. Turk J Elec Eng & Comp Sci, 23, 1571 -1586.

[19] Dokur, E., Kurban, M. (2015). Wind speed potential analysis based on Weibull distribution. Balkan Journal of Electrical & Computer Engineering, 3(4), 231-235.

[20] Kaplan, Y.A., Aladağ, C. (2016). Comparison of different methods in estimating Weibull distribution parameters for wind power application. International Journal of Innovative Research in Science, Engineering and Technology. 5(12), 232 242.

[21] Yanıktepe, B., Özalp, C., Kaşka, Ö., Köroğlu T. (2011). An assessment of wind power potential in Osmaniye, Turkey. 6th International Advanced Technologies Symposium (IATS'11), 16-18 May 2011, Elazığ, Turkey, 82-88.

[22] https://tr.depositphotos.com/vector-images/marmara.html?qview=53816051.

4

EVALUATING INTEGRATION OF BIOMASS GASIFICATION PROCESS WITH SOLID OXIDE FUEL CELL AND TORREFACTION PROCESS

H. Abbasi[1], F. Rahimpour[2], F. Pourfayaz[1], A. Kasaeian[1,*]

ABSTRACT

In the present study, the integration of a solid oxide fuel cell with the biomass gasification process in which the torrefied biomass produced in a torrefaction process is used as the feedstock has been investigated. This novel design gives the power generation system the advantage of eliminating the filtration of the fuel cell inlet gas. This is because of the absence of sulfur and its derivatives in the synthesized gas owing to the terrified biomass, as the feedstock of the gasification process. Moreover, the integration of the processes makes it possible to employ the heat recovery methods. Therefore, by using the high-grade thermal energies for preheating process flows, the presented design considers the maximum available heat recovery and minimum heat and mass losses. The optimum design is determined by the sensitivity analysis and then simulated using the ASPEN PLUS software and its performance has been studied. It was specified that in the optimum operation state, the gasifier outlet temperature and pressure are 950 °C and 5 bar, respectively. Also, the oxygen flow rate in the anode of SOFC and the combustion chamber are 3.03 and 0.81 kmol/h, respectively. Moreover, the results showed that the presented design causes an improvement in the performance of the fuel cell. The electrical efficiency and the overall efficiency of the system are determined to be in the range of 63-69% and 80-85%, respectively. Also, it was revealed that the presented design has the power generation capacity of 100 to 997 kW.

Keywords: *Solid Oxide Fuel Cell, Biomass, Gasification, Renewable Energy*

INTRODUCTION

The drawbacks of fossil fuels especially their pollution has led the worldwide attention to the clean energy sources [1]. The energy sources known as the renewable energies are of the most developed alternatives for the fossil fuels [2]. Different technologies and system designs based on the renewable energies have been developed during past decades [3]. The gasification of the biowastes plays an important role in providing the world with the renewable energies [4]. Biomass gasification is a chemical conversion technology in which the fuel is converted to a gas mixture called as the Syngas or the synthesis gas [5]. The gas mixture composition is dependent on different parameters including the fuel, solid wastes, gasifier type, fluid bed, oxygen, steam, and other operating parameters such as the temperature and pressure [6]. Considering the local availability of the biomass wastes, their employment in the CHP power generation systems is an appropriate application for them [7]. Currently, most of the operating bio waste gasifiers are combined with the gas engines or the gas turbine combined cycles to produce power. By utilizing efficient power generation systems like the solid oxide fuel cells (SOFC), the performance of a biomass gasification process can be highly improved [8].

SOFC is a high-temperature fuel cell which can operate in the temperature range of 500 to 1000 °C, according to the way it has been constructed [9]. The advantages of this type over the low-temperature types are the simplicity of the design because there will be no liquid phases inside it, flexibility on the type of fuel, the internal reforming of the gasses, and the ability to be combined with other systems like the gas turbines and the gasifiers [10]. On the other hand, the disadvantages of SOFC are the challenging methods of making their structure stable in the high temperatures, and the problems that can be caused by the carbon and sulfur inside the fuel such as the sedimentation of carbon or the poisoning effects of sulfur [11].

The torrefaction methods are being used as a pre-filtration stage to remove the sulfur derivatives and the solid particles from the biomass slurry. The torrefaction process lasts for 30 to 60 minutes and operates in the temperature

This paper was recommended for publication in revised form by Regional Editor Omid Mahian
[1]Faculty of New Sciences and Technologies, University of Tehran, Tehran, Iran
[2]School of Chemical Engineering, College of Engineering, University of Tehran, Tehran, Iran
**E-mail address: akasa@ut.ac.ir*

range of 200 to 300 °C in the atmospheric pressure [12]. This method not only removes the sulfur and other redundant materials but also improves the quality of the produced biomass [13]. Among the specifications of the torrefaction process, these ones can be pointed out: the significant decrease in the humidity of the biomass, the increase in its heating value and consequently reduction of the Oxygen to carbon ratio and the increase in the energy density of biomass while being compressed [14].

Considering the advantages of using the torrefied biomass in the gasification processes, many researches have been conducted on this subject, recently. Yue et al. [15] tested the torrefaction process on sorghum and sweet sorghum as the biomass to examine the product yield. They have reported that the energy density of the biomass was increased by an amount of about 1.5 folds. Pinto et al. [16] compared the performance of gasification process with raw and torrefied Eucalyptus globulus stumps. The best performance of the process was achieved by consuming the torrefied stumps with the gas yield value of 1.22 Nl/g daf. In another study, Woytiuk et al. [17] have investigated the effects of using torrefied short rotation coppice in a fluidized bed gasification process. It has been reported that the torrefaction process can increase the gas yield of the gasification process from 2.02 to 2.47 m^3/kg.

On the other hand, there are many recent studies carried out on the processes including the combination of the gasification process and the fuel cells. Gadsboll et al. [18] have combined the biomass gasification process with a SOFC. They have reached the maximum biomass to electricity efficiency of 43%, experimentally. However, by the theoretical methods, they have claimed that this efficiency can be brought up to 62%. Borji et al. [19] have analyzed the performance of CHP plant including a SOFC and auto-thermal gasification system, numerically. They have investigated the effects of different air to steam ratios on the performance of the plant. It is concluded that increasing the air to steam ratio will lead to lower efficiency of the plant, lower heating value of the product, higher steam to carbon ration, and lower CH_4 content of the product. Ozgoli et al. [20] have presented an economic analysis of a hybrid cycle, composed of a biomass gasification process, a SOFC, and a gas turbine. The presented design has the capacity of 1.7 MW and they have claimed that Europe is the best choice for the construction of such power plant, economically.

In this study, the feedstock of the gasification process is considered to be the torrefied biomass. Furthermore, the effects of employing the torrefied biomass on the performance of BGFC are investigated. The employment of the torrefaction process eliminates the need for filtration of the SOFC inlet stream. Moreover, considering the thermal requirements of the three integrated systems, the heat recovery methods can be applied. These optimum operating conditions have been determined by the sensitivity analysis and the performance of the whole process are studied. The integration of the mentioned systems and also the optimum design is not studied in the literature. For this purpose, the whole process is simulated using the ASPEN PLUS software. There are several simulation software developed for different applications, but for the processes containing physical and chemical reactions, ASPEN PLUS and HYSYS are among the most accurate ones. The reason for choosing ASPEN PLUS is the availability of a wide range of thermodynamic relations and also a wide variety of operation units.

DESIGN OF THE SYSTEM

Different parts of the system are under the influence of a variety of considerations. Here, six main characteristics of the system are explained.

The Gasifier Design

The gasification process considered in this study consists of two integrated fluid beds inside the combustion chamber of the biomass (char). The biomass begins to burn and react in the presence of the air and the gasifier steam and then the volatile biomass converts to gas or tar. The gasification is of the allothermal type in which the direct contact of the gasification and the combustion processes are prevented. The heat required by the gasification process is provided by the flow between the beds. This system is composed of a continuous ring of a moving fluid bed as the combustion chamber, a cyclone for solid particles separation, and a bubble-type fluid bed as the gasifier. This design prevents the dilution of the produced gas with Nitrogen, so there is no need for an Oxygen production unit to provide the gasifier with a pure Oxygen. The process operates at a temperature about 800 °C to produce a gas mixture, rich with H_2, CO, and CH_4 with a medium-grade heating value. These substances are appropriate fuels for the SOFCs. The fluid bed is a suitable choice for a perfect mixing and a proper contact of the gas and solid phases. This design has some advantages such as the simple structure, flexibility on the size range of the bed particles, an appropriate control on the temperature, high rate of reactions, high level of carbon conversion, high value of the specific heat, high conversion efficiency, less limitations on the process, and the ability of improving its potential.

The Out Coming Steam from the SOFC

A high-temperature steam is produced as a result of the electrochemical reactions of the oxidation of the hydrogen existing in the synthesized gas. This steam can be used for the gasification process. According to the higher amounts of the steam rather than the amount of humidity inside the biomass, the concentration of hydrogen in the inlet of the SOFC will be increased. Usage of this steam is necessary for refinement of the gas and reduction of tar. A tar, rich with phenol, can be upgraded by the catalytic steam gasification process. Hence, a part of the outlet steam from the anode of the SOFC can be returned to the steam gasifier unit [21].

Steam Production from the Outlet Gas Cooler of SOFC

Some parts of the high-temperature gas in the outlet of SOFC cannot be returned to the gasifier as a heat source. Instead, it can be utilized in the boiler tank. The thermal energy of the gas can be extracted into the boiler tank to produce superheated steam.

Air Supply for the Gasification Process and the SOFC

The SOFC and the gasification process need Oxygen (air flow) to operate. In this design, the required Oxygen is added selectively in different stages of the gasification process. The stages include the secondary zones in a pyrolysis reactor for a selective oxidation of the char. The primary role of Oxygen is to supply the thermal energy needed for the gasifier, by combustion of coal (tar) in the combustion chamber. In an overall energy balance of the BGFC system, the thermal energy of the combustion chamber after being integrated with the SOFC outlet gasses must fulfill the thermal requirements of the gasifier. With this regard, the discharge air flow from the cathode of the SOFC can be utilized as an Oxygen supply in the tar combustion chamber.

Feedstock preheating

The air, as a feedstock of the SOFC, needs to be warmed up before it reaches the SOFC cathode. The aim of air preheating is first to prevent the thermal shocks to the ceramic components, and also increasing the amount of the electricity generated in the electrochemical processes. The hot outlet gas flows from the SOFC, and the tar combustion chamber can be utilized to preheat the air flow. For the purpose of thermal integration, in the presented design, a heat exchanger is considered to transfer the thermal energy from the outlet gas flow of the tar combustion chamber to the air flow. The hot outlet gas from the gasifier cannot be directly moved to the SOFC, and it must be cooled down to the temperatures in the range of the operating temperature of the anode [21].

Excess Steam

The excess heat of the gasifier outlet gas flow can be used by a superheater to produce superheated steam. The resulting superheated steam can be sent to a steam turbine along with some parts of the produced steam from other boiling tanks.

THERMAL INTEGRATION OF THE BGFC

After determination of the temperature, pressure, and composition of each of the flows in the primary process units, the heat exchangers are placed on the hot and cold flows to achieve the desired temperatures. Three main strategies are defined for this aim: integration of the gasifier synthesized gas with the SOFC, integration of the outlet gas (rich in steam and unreacted gasses such as hydrogen and carbon monoxide) and exhaust air from the SOFC with the gasifier, and all other types of integrations between the coolers and boilers.

BGFC SIMULATION USING THE ASPEN PLUS SOFTWARE

Figure 1 shows the schematics of the designed process.

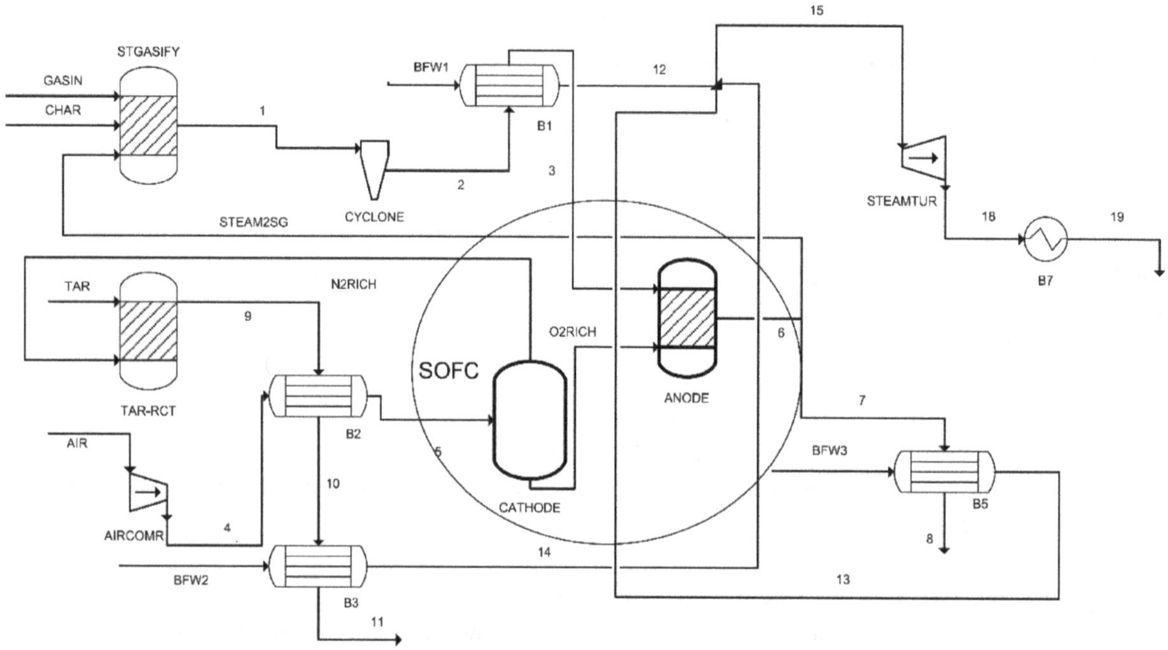

Figure 1. Schematics of the integrated BGFC process

This process is simulated in the ASPEN Plus software. In this simulation, the analysis of the feedstock, which is the outlet of the torrefied biomass unit, is based on the Tables 1 and 2.

Table 1. Analysis of the biomass resulted from the torrefaction process

Char	H_2O	Tar	CO	CH_4	CO_2
0.20	0.20	0.40	0.019	0.001	0.18

Table 2. LHV of the compositions in the feedstock

Feedstock	LHV (MJ/kmol)
methane	802.7
carbon	393.5
benzene	3171
phenol	2921
p-xylene	4377
m-xylene	4377
o-xylene	4377
e-benzene	4379
carbon monoxide	283

In order to generate power in the range mentioned before, the SOFC needs to operate under the pressure of 5 bar. The same or a little less pressure can be put on the gasifier. Furthermore, it has been assumed that the ambient temperature is equal to 25 °C. Also, as it is shown in Figure 1, two continuous fluid bed reactors along with the Gibbs reactors, one of which consumes gas and char named as STGASIFY, and the other uses tar and ash and named as TAR-RCT, are simulated.

Several studies have defined phenol, benzene, and xylene as the primary component of char and tar and ash are considered as uncommon elements. In the presented simulation, the NRTL thermodynamic model is selected for calculation of the material properties. The SOFC is consisted of a cathode and an anode and is modeled with a 95% efficient separator and a Gibbs reactor. More specifications of the model are given in Table 3.

Table 3. More specifications of the BGFC system simulated in the ASPEN Plus software

Unit name	Aspen Model	Outlet Temperature (°C)	Pressure (bar)	Efficiency (%)
STGASIFY	RGibbs	950	5	---
CYCLONE	SSpilit	---	---	99
TAR-RCT	RGibbs	950	5	---
AIRCOMPR	compressor	---	5	75
CATHODE	Sep2	---	5	95
ANODE	RGibbs	800	5	---
STEAMTUR	turbine	---	1	75
B1	heatX	800	5	---
B2	heatX	800	5	---
B3	heatX	Minimum approach=15	5	---
B4	Spilit	---	---	Steam2SG=26%
B5	heatX	Minimum approach=15	5	---
B6	Mixer	---	---	---
B7	Cooler	25	1	---

A BGFC power plant can be a good heat source because both the gasification process and the SOFC operate at high temperatures. It has been observed that by a rise in the pressure, will increase the amount of the generated power by the BGFC system. On the other hand, in the pressures lower than that of the gasifier, by increasing the temperature, the concentration of hydrogen in the synthesized gasses rises. This phenomenon increases the heating value of the synthesized gasses. However, in the temperatures more than 1000 °C, there might be some operational failures, and more maintenances will be needed. The SOFC operates at the high temperatures in the range of 500 to 1000 °C, and hence it can be considered as a proper heat source. According to case studies reported in the literature, the gasifier and the SOFC operate in 950 and 800 °C, respectively. The pressure in the SOFC varies between 1 to 5 bar to generate 1 kW to 1 MW of power.

RESULTS AND DISCUSSION

In this section, firstly the results of the sensitivity analysis are discussed and then the simulation results of the presented design will be reported.

Sensitivity Analysis

The BGFC system is composed of three main operating zones, and it is necessary to optimize the mass and energy transfers to achieve the maximum efficiency of the system. These zones are described separately in the following sections.

The Operating Pressure and Temperature

Figure 2 shows the relation between the steam gasifier operating temperature and the concentration of the produced hydrogen.

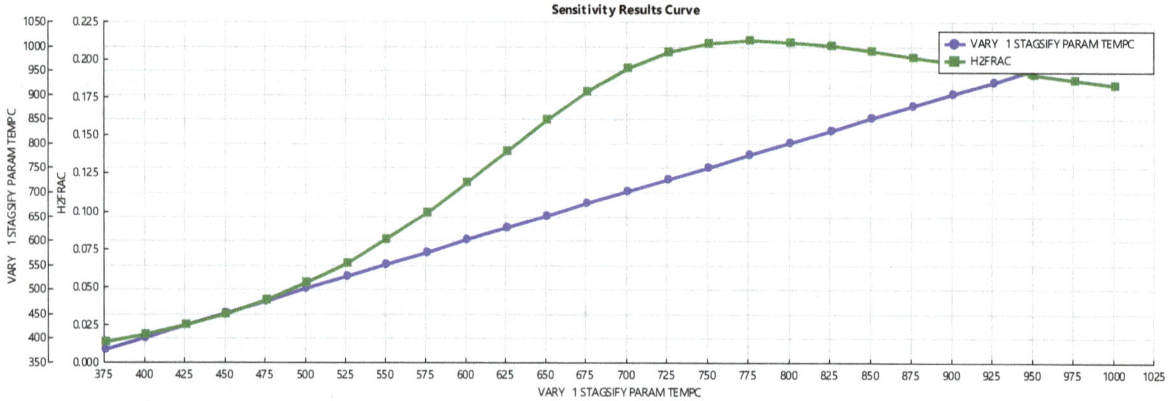

Figure 2. The variation of the produced hydrogen concentration with the changes in the gasifier temperature

According to Figure 2, the optimum gasifier operating temperature is approximately 800 °C. But, because of the thermal requirements of the whole system and the temperature of the tar combustion chamber (between 900 and 1000 °C), the gasifier temperature can be brought up to 950 °C. As it is shown in Figure 2, there will not be a significant change in the concentration of the produced hydrogen. The selected temperature of 950 °C can also create the required temperature difference between the gasifier and SOFC which makes the designed heat exchangers able to transfer the thermal energies efficiently.

Figure 3 shows the changes in the concentration of the produced hydrogen with the changes in the pressure of the gasifier unit. It can be seen from this figure that with the increase of the gasifier pressure up to 10 bar, the rate of biomass conversion to hydrogen reduces. But the reduction is negligible, and the pressure of 5 bar is selected because of the operational requirements of the other units such as the steam turbine. Moreover, it must be noted that the decrease in the amount of the produced hydrogen can be compensated with the increased superheated steam. The higher pressures lead to the higher amount of the superheated steam which is the outlet of the SOFC anode and enters the gasifier as the inlet steam.

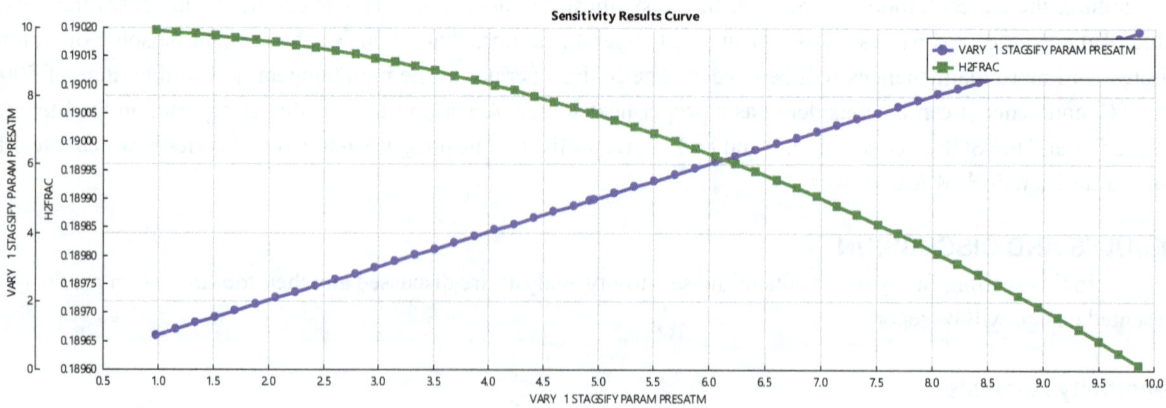

Figure 3. The variation of the produced hydrogen concentration with the changes in the gasifier pressure

The Oxygen Required by the SOFC

Figure 4 shows the concentration of the produced hydrogen in the SOFC based on the inlet Oxygen flow rate.

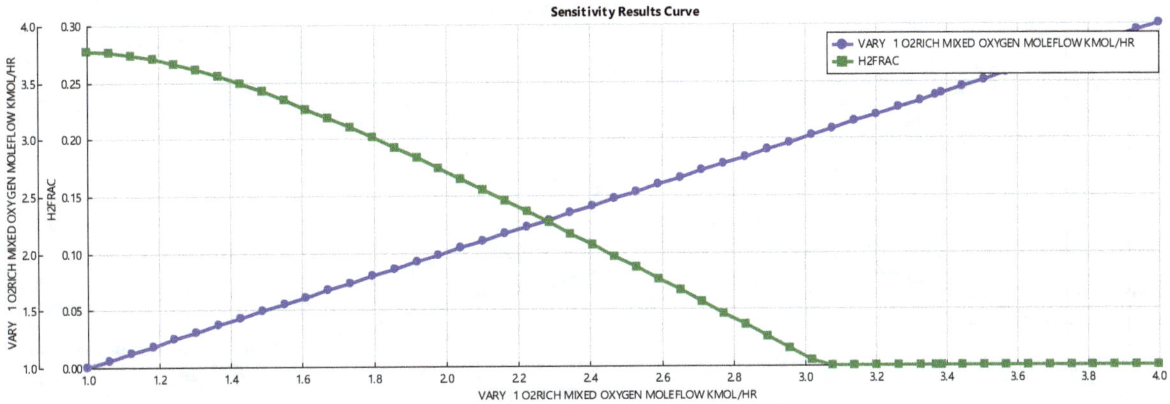

Figure 4. The variation of hydrogen conversion in the SOFC with the changes in the inlet Oxygen flow rate

According to this figure, in the inlet flow rate of 3.03 kmol O_2/h into the anode of the SOFC, the synthesized gas (hydrogen) is completely consumed, and the maximum efficiency of the fuel consumption is achievable. For the flow rates more than that, the same results will be obtained for the conversion rate of hydrogen. But, the Oxygen flow rate also has some effects on the amount of the power generated by the SOFC. To take those effects into account, Figure 5 shows the relation between the Oxygen flow rate into the SOFC and the power generation status. It can be seen from this figure that in the flow rate of 3.03 kmol O_2/h, the maximum output power can be obtained. In the higher or lower rates, less power will be generated by the SOFC. For the higher amounts, more heat is required to warm up the excess oxygen, and this will result in the lower amounts of generated power. Moreover, higher flow rates of the inlet Oxygen causes more power consumption by the compressor and more heat losses from the tar combustion chamber to preheat the excess Oxygen.

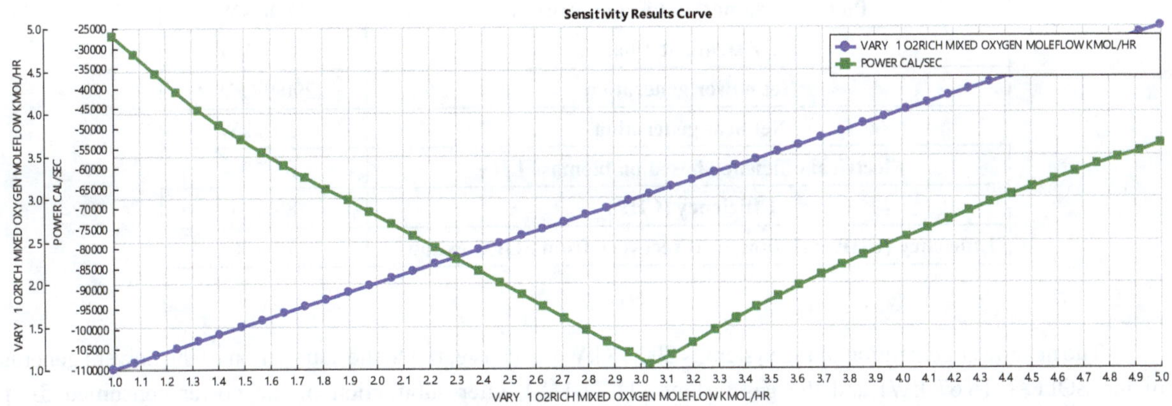

Figure 5. The variation of SOFC power generation with the changes in the inlet Oxygen flow rate

The Oxygen Required by the Tar Combustion Chamber

The flow rate of the inlet Oxygen to the tar combustion chamber can be determined in the same way that has been used for the SOFC anode. Figure 6 shows the changes of the heat generated by the tar combustion chamber with the changes in its inlet Oxygen flow rate.

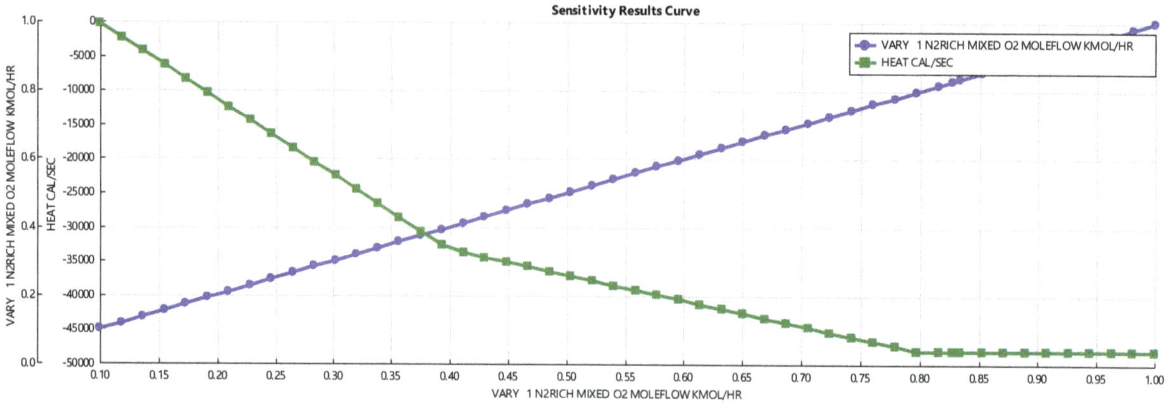

Figure 6. The variation of heat generation of tar combustion chamber with changes in the inlet Oxygen flow rate

Based on the results obtained for this parameters, the value of 0.81 kmol O_2/h has been selected for the Oxygen flow rate, in which the maximum thermal energy will be produced by the tar combustion chamber. It must be noted that the cathode of the SOFC is designed based on the Oxygen flow rates into the anode and the tar combustion chamber.

The Simulation Results

The results given in Table 4 shows the analysis of the integrated BGFC system presented in this paper.

Table 4. Results of the simulation

Parameter	Results
Power generation from SOFC based on 85 % fuel efficiency	316.67 kW
Power generation from steam turbines	11.87 kW
Power consumption by compressors	37.46 kW
LP steam at 1 bar	100.5 kW
Net power generation	290.98 kW
Net heat generation	100.5 kW
Electrical efficiency based on biomass LHV	65.7 %
Efficiency (CHP)	88.5 %
Efficiency (CHP excluding heat recovery from SOFC exhaust)	73.3 %

The net power generation of the system is 290.98 kW which is equal to the summation of the powers generated from the SOFC (316.67 kW) and the gas turbine (11.77 kW), after subtraction of the power consumed for the compressor (37.76 kW). The electrical efficiency of the designed BGFC system is 65.7%, and its CHP efficiency equals to 88.5%. In the calculation of the CHP efficiency, the heat recovery of the SOFC exhaust gasses, the heat gain from the cooling of the synthesized gas, and other heat recoveries are included. Even without considering the heat recovery from the SOFC outlet gasses, the CHP efficiency will be 73.3%. Therefore, an integrated BGFC system has a much higher efficiency rather than the normal combined cycles of biomass gasification with the gas turbines. Also, it must be noted that the hot water produced in the SOFC unit is almost pure and can be utilized in different other applications without the need of purification.

CONCLUSION

In this paper, an integration of a solid oxide fuel cell with a biomass gasifier (BGFC) is presented and simulated to investigate its performance. The feedstock biomass is taken from the outlet of a torrefaction process and according to the absence of sulfur derivatives, there is no need for any filtration or purification process to be done on the biomass. The results indicated that the upgraded biomass would improve the performance of the whole BGFC system. The integration of the different operation units is based on the maximum achievable heat recovery to reduce the heat losses of the system. With this regard, the optimum system design is determined and is simulated using the ASPEN Plus software. The main findings of this research can be summarized as followings:

The optimum design parameters:

- The gasifier outlet temperature: 950 °C
- The gasifier pressure: 5 bar
- The SOFC anode inlet Oxygen flow rate: 3.03 kmol/h
- The tar combustion chamber Oxygen flow rate: 0.81 kmol/h

The performance of the system in the optimum design:

- SOFC power generation: 316.67 kW
- Steam turbines power generation: 11.87 kW
- Net power generation: 290.98 kW
- Net heat generation: 110.5 kW

Moreover, the performance of the system without the heat recoveries has also simulated. The results show that in this case, the overall efficiency of the system will be 73.3%, whilst by employing the heat recovery methods presented in this paper, the overall efficiency will have the value of 88.5%. In other words, by applying the heat recovery methods, more than 15% increase is expected for the efficiency of the system.

NOMENCLATURE

AIRCOMR	Air Compressor
B	Boiler
BFW	Boiler Feed Water
H2FRAC	Hydrogen Fraction
heatX	Heat Exchanger
GASIN	Inlet Gass
RGibbs	Gibbs Reactor
Sep	Separator
SOFC	Solid Oxide Fuel Cell
STEMTUR	Steam Turbine
STGASIFY	Gasifier unit
TAR-RCT	Tar Reactor
TEMPK	Temperature in Kelvin

REFERENCES

[1] Nazari, M.A., Aslani, A., Ghasempour, R. (2018). Analysis of Solar Farm Site Selection Based on TOPSIS Approach. International Journal of Social Ecology and Sustainable Development, 9, 12-25.

[2] Aramesh, M., Kasaeian, A., Pourfayaz, F., Wen, D. (2017). Energy analysis and shadow modeling of a rectangular type salt gradient solar pond. Solar Energy, 146, 161-71.

[3] Nazari, M.A., Ghasempour, R., Ahmadi, M.H., Heydarian, G., Shafii, M.B.(2018). Experimental investigation of graphene oxide nanofluid on heat transfer enhancement of pulsating heat pipe. International Communications in Heat and Mass Transfer, 91, 90-4.

[4] Uhlenhut, F., Schlüter, K., Gallert, C. (2018). Wet biowaste digestion: ADM1 model improvement by implementation of known genera and activity of propionate oxidizing bacteria. Water research, 129, 384-93.

[5] Zou, X., Ma, Z., Liu, H., Chen, D., Wang, C., Zhang, P., et al. (2018). Green synthesis of Ni supported hematite catalysts for syngas production from catalytic cracking of toluene as a model compound of biomass tar. Fuel, 217, 343-51.

[6] Singhal, S.C. (2002). Solid oxide fuel cells for stationary, mobile, and military applications. Solid State Ionics, 152, 405-10.

[7] Salem, A.M., Kumar, U., Izaharuddin, A.N., Dhami, H., Sutardi, T., Paul, M.C. (2018). Advanced numerical methods for the assessment of integrated gasification and CHP generation technologies. Coal and Biomass Gasification. Springer, Singapore, pp. 307-30.

[8] Götz, T., Saurat, M., Tholen, L., Adisorn, T., Obernberger, I., Brunner, T., et al. (2018). Green on-site power generation: environmental considerations on small-scale biomass gasifier fuel-cell CHP systems for the residential sector, 909-21.

[9] Santos, T.H., Grilo, J.P., Loureiro, F.J., Fagg, D.P., Fonseca, F.C., Macedo, D.A. (2018). Structure, densification and electrical properties of Gd3+ and Cu2+ co-doped ceria solid electrolytes for SOFC applications: Effects of Gd2O3 content. Ceramics International, 44, 2745-51.

[10] Sadaka, S., Negi, S. (2009). Improvements of biomass physical and thermochemical characteristics via torrefaction process. Environmental Progress & Sustainable Energy, 28, 427-34.

[11] Arias, B., Pevida, C., Fermoso, J., Plaza, M.G., Rubiera, F., Pis, J. (2008). Influence of torrefaction on the grindability and reactivity of woody biomass. Fuel Processing Technology, 89, 169-75.

[12] Tapasvi, D., Kempegowda, R.S., Tran, K.Q., Skreiberg, Ø., Grønli, M. (2015). A simulation study on the torrefied biomass gasification. Energy Conversion and Management, 90, 446-57.

[13] Zhang, Y., Geng, P., Liu, R. (2017). Synergistic combination of biomass torrefaction and co-gasification: Reactivity studies. Bioresource Technology, 245, 225-33.

[14] Sansaniwal, S.K., Pal, K., Rosen, M.A., Tyagi, S.K. (2017). Recent advances in the development of biomass gasification technology: A comprehensive review. Renewable and Sustainable Energy Reviews, 72, 363-84.

[15] Yue, Y., Singh, H., Singh, B., Mani, S. (2017). Torrefaction of sorghum biomass to improve fuel properties. Bioresource Technology, 232, 372-9.

[16] Pinto, F., Gominho, J., André, R.N., Gonçalves, D., Miranda, M., Varela, F., et al. (2017). Improvement of gasification performance of Eucalyptus globulus stumps with torrefaction and densification pre-treatments. Fuel, 206, 289-99.

[17] Woytiuk, K., Campbell, W., Gerspacher, R., Evitts, R.W., Phoenix, A. (2017). The effect of torrefaction on syngas quality metrics from fluidized bed gasification of SRC willow. Renewable Energy, 101, 409-16.

[18] Gadsbøll, R.Ø., Thomsen, J., Bang-Møller, C., Ahrenfeldt, J., Henriksen, U.B. (2017). Solid oxide fuel cells powered by biomass gasification for high efficiency power generation. Energy, 131, 198-206.

[19] Borji, M., Atashkari, K., Ghorbani, S., Nariman-Zadeh, N. (2016). Model-based evaluation of an integrated autothermal biomass gasification and solid oxide fuel cell combined heat and power system. Proceedings of the Institution of Mechanical Engineers, Part C: Journal of Mechanical Engineering Science, 231, 672-94.

[20] Ozgoli, H.A., Ghadamian, H., Pazouki, M. (2017). Economic Analysis of Biomass Gasification-Solid Oxide Fuel Cell-Gas Turbine Hybrid Cycle. International Journal of Renewable Energy Research (IJRER), 7, 1007-18.

[21] Veyo, S.E. (1996). The Westinghouse solid oxide fuel cell program-a status report. Energy Conversion Engineering Conference, 1996 IECEC 96, Proceedings of the 31st Intersociety. IEEE, pp. 1138-43.

5

SECOND LAW ANALYSIS OF MIXED CONVECTION OF MAGNETOHYDRODYNAMIC FLOW IN AN INCLINED SQUARE LID-DRIVEN ENCLOSURE

E. B. Öğüt[1,*]

ABSTRACT

Second law of thermodynamics analysis is formulated for the case of laminar mixed convection in an inclined square lid-driven enclosure in the presence of magnetic field. Vertical sides of the enclosure moves upward when inclination angle is zero. Governing equations of flow and temperature in the form of stream function-vorticity formulation were solved numerically using the differential quadrature method. Governing parameters are: Richardson number (from 0.01 to 100), Prandtl number (from 0.1 to 1.0), inclination angle of enclosure (from 0° to 180°), Hartmann number (from 0 to 100) and magnetic field direction (0°). It is found that the inclination angle of enclosure is effective parameter on entropy generation especially for higher Richardson number (Ri > 1) due to domination of natural convection. However, Hartmann number is effective on both heat transfer and entropy generation for all values of Richardson and Prandtl numbers and it decreases the convective fluid flow and entropy generation.

Keywords: *Lid-driven Enclosure, Mixed Convection, Entropy Generation, Magnetic Field*

INTRODUCTION

Flow fields and temperature distributions in lid-driven cavities are important for both thermal sciences and fluid mechanics due to their importance in areas such as cooling of electronic devices, geothermal applications, furnaces, air-solar collectors etc. Shankar and Despande [1] showed an example for the lid-driven configuration. They indicated that in the short-dwell coater used to produce high-grade paper and photographic film, the structure of the field in the liquid pond can greatly influence the quality of the coating on the roll and the cavity is a rectangular parallelepiped in which the lid generates the motion. This phenomenon contains pure fluid motion. Ghia et al. [2] performed a study to show the flow fields inside the lid-driven enclosure as a benchmark study. When a temperature gradient is imposed such that the shear driven and buoyancy effects are of comparable magnitude then the resulting flow falls under mixed convection regime as indicated by Sharif [3]. In this case, the analysis becomes more complex. Investigation of opposing or aiding mechanism of the mixed convection is performed by Torrance et al. [4], Iwatsu et al. [5] and Mohammad and Viskanta [6]. Control of heat transfer is very important issue in both industrial heat transfer and other applications from the energy saving point of view. One way of this control parameter is application of magnetic field at different directions. It is important especially in crystal growth processes. In earlier studies, effects of magnetic field on natural convection heat transfer in enclosures with different configurations have been studied by Buyuk [7] and Ece and Buyuk [8]. Magnetohydrodynamic mixed convection also important for channel flows as studied by Chamkha [9]. However, Chamkha [10] made a numerical analysis to obtain flow and temperature field of hydromagnetic combined convection flow in a lid-driven enclosure with left side moves up/down with internal heat generation or absorption. They solved the governing equations based on finite-volume approach along with the ADI scheme. They found that flow behavior and the heat transfer characteristics inside the cavity are strongly affected by the presence of the magnetic field. Significant reductions in the mean Nusselt number were produced for both aiding and opposing flow situations as the strength of the applied magnetic field was increased.

The best way of analysis of the thermofluid problems is second-law analysis of thermodynamics and its design-related concept of entropy generation and minimization. Entropy generation minimization is the method of modeling and optimization of real devices that owe their thermodynamic imperfection to heat transfer, mass transfer, and fluid flow irreversibilities [11]. Thus, the method combines the three scientific areas as thermodynamics, heat

This paper was recommended for publication in revised form by Regional Editor Hakan Demir
[1]Department of Vocational School, Kocaeli University, Kocaeli, TURKEY
**E-mail address: elif.ogut@kocaeli.edu.tr*

transfer and fluid mechanics. The existence literature showed that there are many studies on entropy generation due to natural [12-14] or mixed convection [15-17] in enclosures and ducts. Mansour et al. [15] made an analytical study to investigate the dissipation and entropy generation in fully developed forced and mixed laminar convection in a vertical tube. They used power series expansion of the axial velocity to solve the coupled partial differential equations. Tasnim and Mahmud [16] examined the entropy generation for mixed convection flow in a vertical annular space. Mahmud and Fraser [17] investigated the mixed convection including Radiation and magnetic forces analytically. They analysed the entropy generation. They observed that entropy generation number is characterized by the concave shaped profile and is symmetric about the channel centerline for symmetrical temperature boundary conditions.

To the best knowledge of the present author no attention has been paid to analyze of entropy generation due to mixed convection in an inclined lid-driven enclosure under magnetic field. Mainly, the study is an extended study by adding the second law analysis to the study performed by Buyuk [7]. The effects of key governing parameters effective on entropy generation, heat transfer and flow field are investigated. These parameters are Richardson number, inclination angle of the enclosure, Prandtl number.

PHYSICAL MODEL

Considered physical model is shown in Figure 1 with coordinate system and boundary conditions, schematically. Inclination angle of the enclosure with respect to coordinate system is also shown in Figure. It is a square inclined lid-driven enclosure. Moving sides of the enclosure are isothermal and they have different temperature. They have constant speed as U. In the figure, $T_1 > T_0$. Other two sides are kept adiabatic. Magnetic field of strength B is applied at an angle φ according to coordinate system.

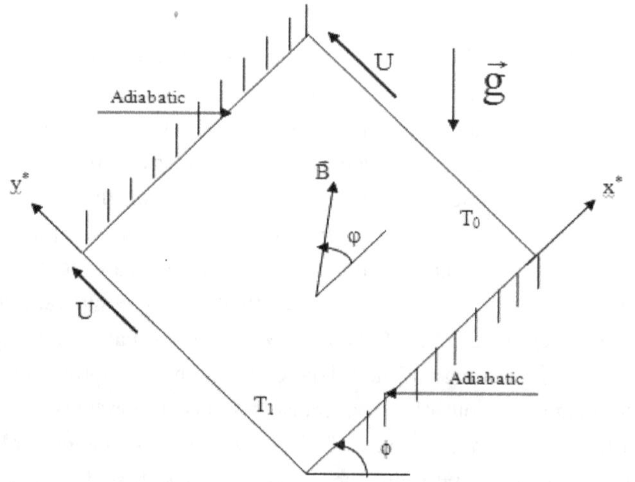

Figure 1. Schematic diagram of the physical system

COMPUTATIONAL DETAILS

Steady, laminar, mixed convection flow under the presence of a magnetic field in an inclined square enclosure with isothermally heated and moving side walls, and stationary and adiabatic horizontal walls was considered. Dimensional coordinates with the x*-axis measuring along the bottom wall and y*-axis being normal to it along the left wall were used. The geometry and the coordinate system are schematically shown in Fiure 1. The angle of inclination of the enclosure is denoted by ϕ. Magnetic field of strength B is applied at an angle φ. The top and the bottom walls are insulated and the fluid is isothermally heated and cooled by the left and the right side walls at uniform temperatures of T_1 and T_0, respectively. The vertical walls are also allowed to move with equal and constant velocity in the same or opposite directions.

Dimensionless variables used in the analysis are:

$$x = \frac{x^*}{L} \quad , \quad y = \frac{y^*}{L} \tag{1}$$

$$u = \frac{u^*}{U} \quad , \quad v = \frac{v^*}{U} \quad , \quad p = \frac{\left[p^* + \rho_0 g \left(x^* \sin\phi + y^* \cos\phi\right)\right]}{\rho_0 U^2} \quad , \quad \theta = \frac{T - T_0}{T_1 - T_0} \quad . \tag{2}$$

Here g is the gravitational acceleration, L is the dimensional length of the square enclosure, U is the velocity of the side walls of the enclosure, u* and v* are the dimensional velocity components in the x* and y* directions respectively, p* is the dimensional pressure and ρ_0 is the density of the fluid at temperature T_0.
A dimensionless stream function and vorticity were defined as follows,

$$u = \frac{\partial \psi}{\partial y} \quad , \quad v = -\frac{\partial \psi}{\partial x} \tag{3}$$

$$\omega = \frac{\partial v}{\partial x} - \frac{\partial u}{\partial y} \tag{4}$$

The governing equations under Boussinesq approximation in terms of the dimensionless variables may then be written as:

$$\omega = -\nabla^2 \psi \tag{5}$$

$$\begin{aligned} u\frac{\partial \omega}{\partial x} + v\frac{\partial \omega}{\partial y} &= \frac{1}{Re}\left(\frac{\partial^2 \omega}{\partial x^2} + \frac{\partial^2 \omega}{\partial y^2}\right) + \frac{Gr}{Re^2}\left(\cos\phi\frac{\partial \theta}{\partial x} - \sin\phi\frac{\partial \theta}{\partial y}\right) \\ &+ \frac{Ha^2}{Re}\left[\cos\varphi\left(\sin\varphi\frac{\partial u}{\partial x} - \cos\varphi\frac{\partial v}{\partial x}\right) + \sin\varphi\left(\sin\varphi\frac{\partial u}{\partial y} - \cos\varphi\frac{\partial v}{\partial y}\right)\right] \end{aligned} \tag{6}$$

$$u\frac{\partial \theta}{\partial x} + v\frac{\partial \theta}{\partial y} = \frac{1}{Re Pr}\left(\frac{\partial^2 \theta}{\partial x^2} + \frac{\partial^2 \theta}{\partial y^2}\right) \tag{7}$$

Here the Reynolds, Prandtl, Grashof and Hartman numbers are defined as follows

$$Re = \frac{UL\rho_0}{\mu} \quad , \quad Pr = \frac{\mu}{\rho_0 \alpha} \quad , \quad Gr = \frac{\rho_0^2 g \beta L^3 (T_1 - T_0)}{\mu^2} \quad , \quad Ha = LB\sqrt{\frac{\sigma}{\mu}} \tag{8}$$

where μ is the viscosity, β is the coefficient of thermal expansion, α is the thermal diffusivity and σ is the electrical conductivity of the fluid respectively. Boundary conditions are

$$u(x,0) = 0 \ , \ v(x,0) = 0 \ , \ u(x,1) = 0 \ , \ v(x,1) = 0 \tag{9}$$

$$u(0,y) = 0 \ , \ v(0,y) = 1 \ , \ u(1,y) = 0 \ , \ v(1,y) = 1 \tag{10}$$

$$\theta(0,y) = 1 \ , \ \theta(1,y) = 0 \ , \ \left.\frac{\partial \theta}{\partial y}\right|_{y=0} = 0 \ , \ \left.\frac{\partial \theta}{\partial y}\right|_{y=1} = 0 \tag{11}$$

The boundary condition given by Eqs. (9) and (10) may also be expressed in terms of the stream function.

$$\psi(x,0)=0\ ,\ \psi(x,1)=0\ ,\ \psi(0,y)=0\ ,\ \psi(1,y)=0 \tag{12}$$

$$\left.\frac{\partial\psi}{\partial y}\right|_{y=0}=0\ ,\ \left.\frac{\partial\psi}{\partial y}\right|_{y=1}=0\ ,\ \left.\frac{\partial\psi}{\partial x}\right|_{x=0}=1\ ,\ \left.\frac{\partial\psi}{\partial x}\right|_{x=1}=1\ . \tag{13}$$

NUMERICAL TECHNIQUE

The dimensionless governing equations were solved for stream function, vorticity and temperature using the differential quadrature method originally proposed by Bellman et al. [18]. The basis of the method is to assume that the unknown functions can be locally approximated as polynomials. The first and second order partial derivatives of a two-variable function F(x, y) at a point (x_i, y_j) can be approximated as

$$F_x(x_i,y_j)=\sum_{k=1}^{N}a_{ik}F(x_k,y_j)\ ,\ F_y(x_i,y_j)=\sum_{k=1}^{N}\bar{a}_{jk}F(x_i,y_k) \tag{14}$$

$$F_{xx}(x_i,y_j)=\sum_{k=1}^{N}b_{ik}F(x_k,y_j)\ ,\ F_{yy}(x_i,y_j)=\sum_{k=1}^{N}\bar{b}_{jk}F(x_i,y_k) \tag{15}$$

Here (N-1) is the degree of the polynomial assumed and the weight functions are given by Shu [19].

The stream function, vorticity and temperature ratio were assumed to behave as polynomials of 4th degree in both x and y directions and N=5 was used. Numerical mesh size h=0.01 was used in both directions to insure a good accuracy. The velocity components at each mesh point were determined by differentiating the stream function using the expressions for the first order derivatives given by Eqs. (14) at each stage of the iterations. Vorticity on the walls of the enclosure was determined according to:

$$\left.\omega\right|_{\eta=0}=\mp\left.\frac{\partial^2\psi}{\partial\eta^2}\right|_{\eta=0} \tag{16}$$

where η is the normal coordinate measured away from the wall and the differentiation was carried out using the expressions for the second order derivatives given by Eqs. (15). Iterations were stopped when the absolute values of the difference between the successive solutions for stream function, vorticity and temperature ratio at each mesh point are less than 10^{-6}.

ENTROPY GENERATION

Local entropy generation is given by [14-16]:

$$S'''_{gen}=\frac{k}{T_0^2}(\nabla T)^2+\frac{\mu}{T_0}\phi \tag{17}$$

The volumetric entropy generation rate for a two-dimensional flow in Cartesian coordinates can be defined as [11]:

$$S'_{gen}=\frac{k}{T_0^2}\left[\left(\frac{\partial\theta}{\partial x}\right)^2+\left(\frac{\partial\theta}{\partial y}\right)^2\right]+\frac{\mu}{T_0}\left(u^2+v^2\right)+\frac{\sigma_e B_0^2}{T_0}u^2 \tag{18}$$

Dimensionless entropy generation is given as

$$N_s = \left[\left(\frac{\partial\theta}{\partial x}\right)^2 + \left(\frac{\partial\theta}{\partial y}\right)^2\right] + \frac{\text{Ecx Pr}}{\Omega}\left[\left(\frac{\partial\psi}{\partial x}\right)^2 + \left(\frac{\partial\psi}{\partial y}\right)^2\right] + \left[\text{Ha}^2\left(\frac{\partial\psi}{\partial y}\right)^2\right] \tag{19}$$

Bejan number (Be), which is the ratio of heat transfer irreversibility to the total entropy generation (N_s) (total irreversibility due to heat transfer and fluid friction), can be expressed as

$$\text{Be} = \frac{\frac{k}{T_0^2}\left[\left(\frac{\partial\theta}{\partial x}\right)^2 + \left(\frac{\partial\theta}{\partial y}\right)^2\right]}{N_s} \tag{20}$$

VALIDATION OF THE CODE AND GRİD DEPENDENCY STUDY

The results of present study were compared with the Davis [20] and Shu et al. [21] results as given in Table 1. As can be shown from the table, the present code shows good agreement with the literature. A grid dependency study was also conducted and a grid of 101x101 was selected [7].

Table 1. Comparison of the results for Re = 0, Ra = 10^4, Pr = 0.71, Ha = 0 and $\phi = 0°$

	$\lvert\psi_{max}\rvert$	$\lvert\psi^*_{max}\rvert$ (m^2/s)
Present study	5.074	1.07 10^{-4}
Davis [20]	5.071	1.07 10^{-4}
Shu et al.[21]	5.084	1.07 10^{-4}

RESULTS AND DISCUSSION

Flow fields, temperature distributions and heat transfer rates and results for entropy generations are examined for different Richardson numbers, inclination angle of the enclosure, Hartmann and Prandtl number. The results were presented as streamlines, isotherms, Nusselt numbers, contours of entropy generation due to heat transfer irreversibility, contours of entropy generation due to fluid friction irreversibility and Bejan number. The Grashof number was kept at a constant value of 10^4 for whole study.

Local entropy generation due to fluid friction and heat transfer irreversibilities is calculated once the values of velocities and temperature values are available from the converged solution of governing equations of continuity, momentum and energy. Figure 2 (a) to (d) show contours for entropy generation due to fluid friction irreversibility (on the left) and heat transfer irreversibility (on the right) at Ha = 0 and $\phi = 0°$ to see the effects of Richardson number. Entropy contours due to fluid flow irreversibility (FFI) are very dense in the corners for Ri = 0.1 and 1 as shown from Figure 2 (a) and (b) (on the left), respectively. Values of contours are not given on the figure to have a track of details but the magnitude of entropy generation due to FFI is considerably smaller than its counterpart corresponding to the heat transfer. The structure of contours does not change considerably for Ri < 1 (forced convection dominant flow) and Ri = 1 (mixed convection). It is shown that as the Richardson number increases, the local entropy generation is increasingly confined to the neighborhood of the active walls of the lid-driven enclosures. With an increase in Richardson number, entropy generation is seen almost on the whole domain. However, this domain is decreased with decreasing of Richardson number. The results also show that Hartmann number affects the flow field and temperature distribution and it retards the convection. It is also directly affects the entropy generation. To see this effect on local entropy generation Figure 3 is plotted for different Hartmann number and Ri = 1 and $\phi = 0°$. As shown from the figure, entropy generation becomes weak with increasing of Hartmann number due to weakness of the convection heat transfer. Entropy contours were formed almost at the middle of the enclosure and they distributed from top to bottom due to interaction of hot and cold fluid.

Table 2. Variation of the average Nusselt number and Bejan number

Pr	Ri	Ha	φ	ϕ	$\overline{Nu}$	Be
0.71	0.1	0	-	0°	4.069	0.780
	1.0	0	-	0°	2.192	0.671
	10	0	-	0°	1.990	0.581
	100	0	-	0°	2.007	0.284
0.71	1	10	0°	0°	1.882	0.635
	1	50	0°	0°	1.104	0.403
	1	100	0°	0°	1.033	0.325
0.71	0.1	10	0°	30°	3.372	0.744
	0.1	10	0°	60°	3.384	0.746
	0.1	10	0°	90°	3.392	0.748
	0.1	10	0°	120°	3.387	0.746
	0.1	10	0°	180°	3.353	0.741
0.1	0.1	50	0°	0°	1.024	0.390
	1.0	50	0°	0°	1.008	0.382
	10	50	0°	0°	1.002	0.380
	100	50	0°	0°	1.001	0.369
0.71	10	10	0°	30°	1.837	0.577
	10	10	0°	60°	1.743	0.582
	10	10	0°	90°	1.118	0.537
	10	10	0°	120°	1.743	0.582
	10	10	0°	180°	1.693	0.570
1	0.1	50	0°	0°	1.880	0.521
	1.0	50	0°	0°	1.187	0.420
	10	50	0°	0°	1.055	0.392
	100	50	0°	0°	1.039	0.378

Effects of inclination angle on entropy generation were analyzed for the case of Ri = 10, Pr = 0.71 and Ha = 10 as shown in Figure 4. Calculations were performed to see the effect of inclination angle for Ri<1. It was obtained that there is no discernible effect of inclination angle on both heat transfer and entropy generation. The results were not given here to save space in the study. But Table 2 includes the values of Mean Nusselt and Bejan numbers which summarizes the effects of all chosen parameters. As given in the table values are almost same for Ri < 1. On the contrary, entropy generation is affected with changing of inclination angle for values of Ri > 1. Entropy generation occupies the domain for $\phi = 90°$. For other angles, entropy generation is confined to the corners of the active sides.

As indicated above that Bejan number (Be) defines the ratio of heat transfer irreversibility to the total entropy generation (N_s) (total irreversibility due to heat transfer and fluid friction). Table 2 gives Bejan number for all governing parameters. As given in the table that Bejan number decreases with increasing of Richardson number. It means that flow friction becomes more effective with increasing of buoyancy effects. Similarly, it decreases with increasing of Hartmann number due to decreasing of convection heat transfer and flow strength. It becomes almost same with changing of inclination angle for Ri < 1, inclination angle of enclosure affects the Bejan number. Effects of fluid flow irreversibility on entropy generation become strong for $\phi = 90°$. Fluid flow irreversibility is more effective for lower Prandtl number when other parameters become constant.

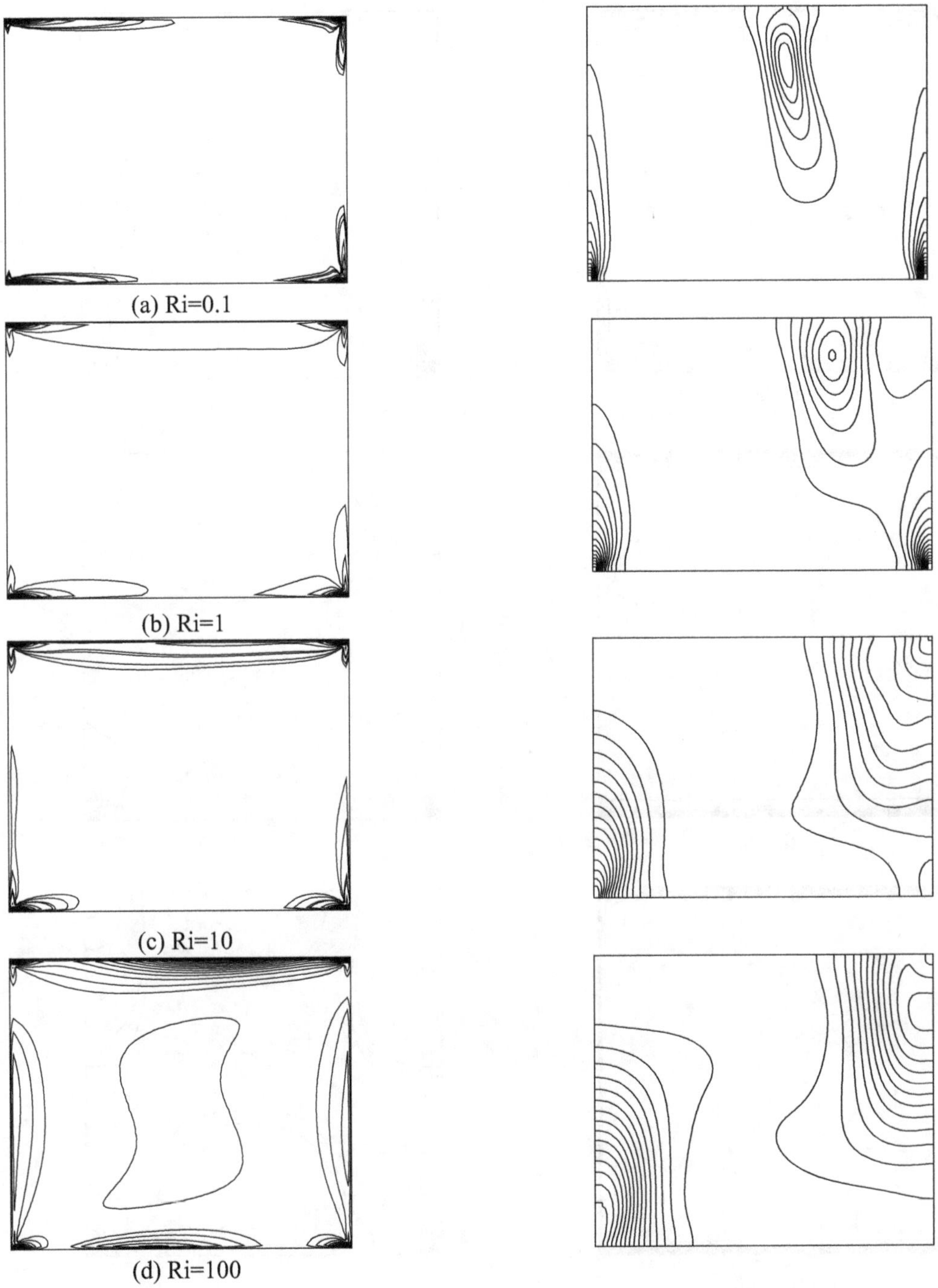

Figure 2. Entropy production due to fluid friction (on the left) and heat transfer (on the right) at Ha = 0, $\phi = 0^{o}$, (a) Ri = 0.1, (b) Ri = 1, (c) Ri = 10, (d) Ri = 100

(a) Ha=10

(b) Ha=50

(c) Ha=100

Figure 3. Entropy production due to heat transfer (on the left) and fluid friction (on the right) at Ri = 1 and $\phi = 0^{\circ}$, (a) Ha = 10, (b) Ha = 50, (c) Ha = 100

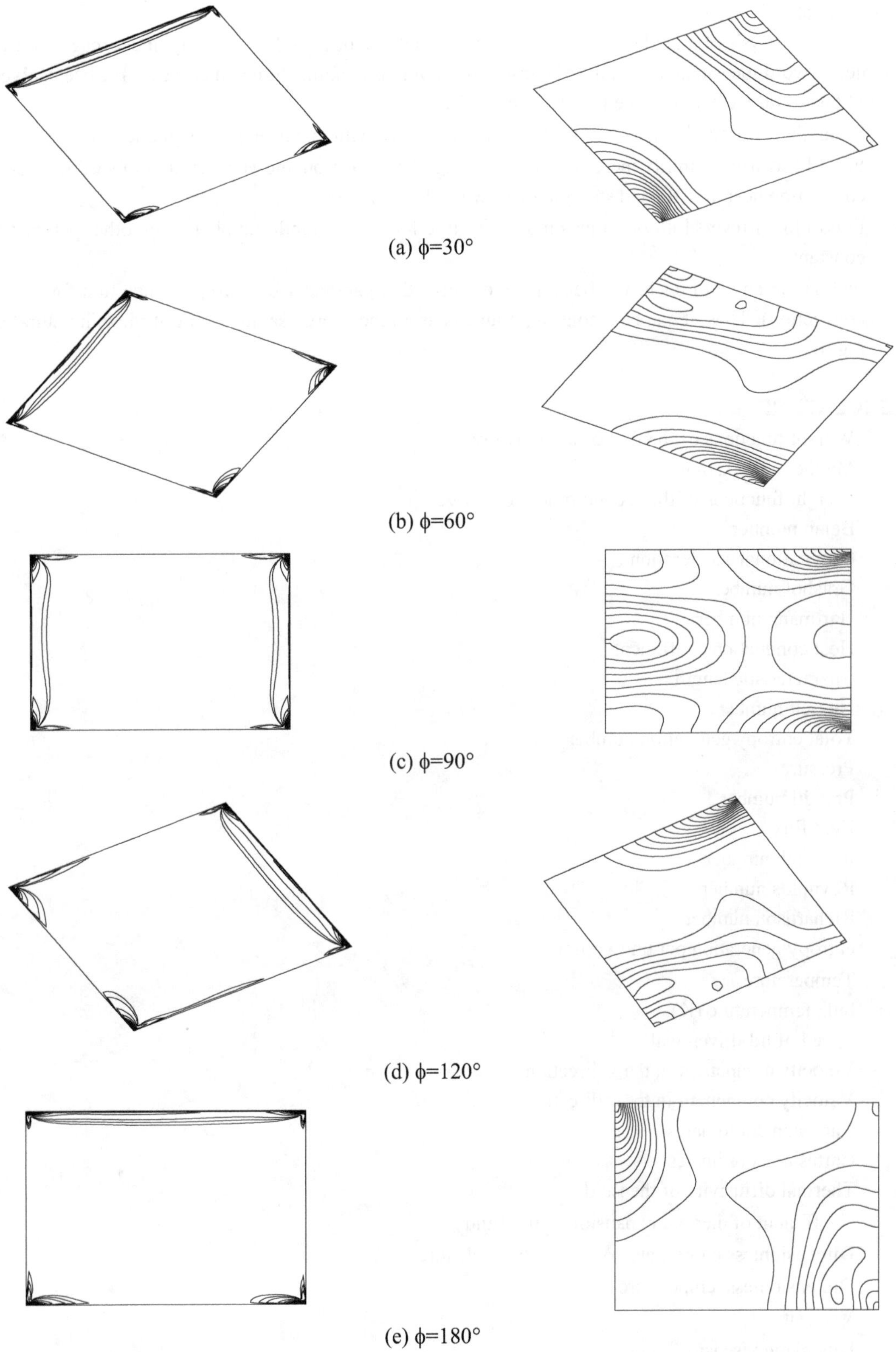

(a) ϕ=30°

(b) ϕ=60°

(c) ϕ=90°

(d) ϕ=120°

(e) ϕ=180°

Figure 4. Entropy production due to heat transfer (on the left) and fluid friction (on the right) at Ri = 10, Pr = 0.71 and Ha = 10, (a) ϕ = 30°, (b) ϕ = 60°, (c) ϕ = 90°, (d) ϕ = 120°, (e) ϕ = 180°

CONCLUSIONS

Second law analysis of mixed convection in inclined double-sided lid-driven cavity under magnetic field has been studied. Two-dimensional equations of conservation of mass, momentum and energy have been solved using differential quadrature technique. The results showed that

- Strength of flow and convection heat transfer decreases with increasing of magnetic field.
- Fluid Flow irreversibility increases of increasing of Richardson and Hartmann number due to decreasing of convection heat transfer and strength of flow field.
- Fluid Flow irreversibility becomes more effective for lower Prandtl number when other parameters become constant.
- Inclination angle of enclosure becomes more effective parameter on entropy generation for $Ri > 1$ (natural convection). Thus, it can be a control parameter from the energy saving point of view in natural convection regime.

NOMENCLATURE

a	Weight functions of the first order derivative
B	Magnetic field strength
b	Weight functions of the second order derivative
Be	Bejan number
g	Gravitational acceleration
Gr	Grashof number
Ha	Hartmann number
k	Heat conduction coefficient
L	Characteristic length
Nu	Nusselt number
N_s	Total entropy generation number
p	Pressure
Pr	Prandtl number
q	Heat flux
Ra	Rayleigh number
Re	Reynolds number
Ri	Richardson number
S'''_{gen}	entropy generation per unit volume
T	Temperature
T_0	bulk temperature (K)
U	Speed of lid-driven wall
u	Velocity component in the x direction
v	Velocity component in the y direction
x	Cartesian coordinates
y	Cartesian coordinates
α	Thermal diffusivity of the fluid
β	Coefficient of thermal expansion of the fluid
η	Dimensionless coordinate normal to the enclosure wall
θ	Dimensionless temperature
μ	Viscosity
ν	Kinematic viscosity
ρ	Density of the fluid
σ	Electrical conductivity of the fluid
ϕ	Angle of inclination of the enclosure, irreversibility.

φ Angle of direction of the magnetic field with respect to the coordinate system

ψ Stream function

ω Vorticity

REFERENCES

[1] Shankar, P. N., Deshpande, M. D. (2000). Fluid mechanics in the driven cavity. Annual review of fluid mechanics, 32(1), 93-136.

[2] Ghia, U. K. N. G., Ghia, K. N., Shin, C. T. (1982). High-Re solutions for incompressible flow using the Navier-Stokes equations and a multigrid method. Journal of computational physics, 48(3), 387-411.

[3] Sharif, M. A. R. (2007). Laminar mixed convection in shallow inclined driven cavities with hot moving lid on top and cooled from bottom. Applied thermal engineering, 27(5-6), 1036-1042.

[4] Torrance, K., Davis, R., Eike, K., Gill, P., Gutman, D., Hsui, A., Zien, H. (1972). Cavity flows driven by buoyancy and shear. Journal of Fluid Mechanics, 51(2), 221-231.

[5] Iwatsu, R., Hyun, J. M., Kuwahara, K. (1993). Mixed convection in a driven cavity with a stable vertical temperature gradient. International Journal of Heat and Mass Transfer, 36(6), 1601-1608.

[6] Mohamad, A., Viskanta, R. (1993). Flow and thermal structures in a lid-driven cavity heated from below. Fluid Dynamics Research, 12(3), 173.

[7] Ogut, E. B. (2017). Effects of Prandtl number and Magnetic field on Mixed Convection Heat Transfer in an Inclined Lid-driven Enclosure. Karaelmas Science and Engineering Journal, 7(1), 145-153.

[8] Ece, M. C., Büyük, E. (2006). Natural-convection flow under a magnetic field in an inclined rectangular enclosure heated and cooled on adjacent walls. Fluid Dynamics Research, 38(8), 564.

[9] Chamkha, A. J. (2002). Hydromagnetic combined convection flow in a vertical lid-driven cavity with internal heat generation or absorption. Numerical Heat Transfer: Part A: Applications, 41(5), 529-546.

[10] Chamkha, A. J. (1998). Mixed convection flow along a vertical permeable plate embedded in a porous medium in the presence of a transverse magnetic field. Numerical Heat Transfer, Part A Applications, 34(1), 93-103.

[11] Mahmud, S., Fraser, R. A. (2004). Magnetohydrodynamic free convection and entropy generation in a square porous cavity. International Journal of Heat and Mass Transfer, 47(14-16), 3245-3256.

[12] Yilbas, B. S., Shuja, S. Z., Gbadebo, S. A., Al-Hamayel, H. A., Boran, K. (1998). Natural convection and entropy generation in a square cavity. International journal of energy research, 22(14), 1275-1290.

[13] Magherbi, M., Abbassi, H., Brahim, A. B. (2003). Entropy generation at the onset of natural convection. International Journal of Heat and Mass Transfer, 46(18), 3441-3450.

[14] Narusawa, U. (2001). The second-law analysis of mixed convection in rectangular ducts. Heat and Mass Transfer, 37(2-3), 197-203.

[15] Mansour, R. B., Galanis, N., Nguyen, C. T. (2006). Dissipation and entropy generation in fully developed forced and mixed laminar convection. International journal of thermal sciences, 45(10), 998-1007.

[16] Tasnim, S. H., Mahmud, S. (2002). Mixed convection and entropy generation in a vertical annular space. Exergy, An International Journal, 2(4), 373-379.

[17] Mahmud, S., Fraser, R. A. (2002). Analysis of mixed convection—Radiation interaction in a vertical channel: Entropy generation. Exergy, an International journal, 2(4), 330-339.

[18] Bellman, R., Kashef, B. G., Casti, J. (1972). Differential quadrature: a technique for the rapid solution of nonlinear partial differential equations. Journal of computational physics, 10(1), 40-52.

[19] Shu, C. (2000). Mathematical Fundamentals of Differential Quadrature Method: Linear Vector Space Analysis and Function Approximation. In Differential Quadrature and Its Application in Engineering (pp. 1-24). Springer, London.

[20] De Vahl Davis, G. (1983). Natural convection of air in a square cavity: a bench mark numerical solution. International Journal for numerical methods in fluids, 3(3), 249-264.

[21] Shu, C., Ding, H., & Yeo, K. S. (2003). Local radial basis function-based differential quadrature method and its application to solve two-dimensional incompressible Navier–Stokes equations. Computer methods in applied mechanics and engineering, 192(7-8), 941-954.

6

EXPERIMENTAL INVESTIGATION AND PERFORMANCE OPTIMIZATION OF A CROSS FLOW HEAT EXCHANGER BY ENTROPY GENERATION MINIMIZATION APPROACH

T. Mohapatra[1], S. K. Rout[1,*]

ABSTRACT

The exergy loss and performance optimization of a cross flow heat exchanger (CFHE) with air and hot water as working fluid have been experimentally investigated. Experiments are performed on various mass flow rates of hot water and air over a ranges of 0.015 $kg.s^{-1}$– 0.04 $kg.s^{-1}$ and 0.117 $kg.s^{-1}$ - 0.763 $kg.s^{-1}$ respectively. To validate the results of the present study, they are compared with available data in literature through which a reasonably a good agreement is obtained between them. This study demonstrates the successful application of Taguchi approach for optimal design of the (CFHE). The effects of design parameters and two different operating conditions such as the fin pitch, the inside tube diameter and the mass flow rate of water and air on exergy loss are investigated. In the Taguchi experimental design method, exergy loss is considered as performance parameter and it is revealed that parameter combinations such as$\dot{m}_h$=0.0163 $kg.s^{-1}$, $\dot{m}_a$=0.1175 $kg.s^{-1}$, d_i=0.0199 m and F_p=0.00259 m resulted optimum performance. Contribution ratios of each parameter on exergy loss are assessed. Mass flow rate of air is determined to be the most effective parameter on exergy loss with a contribution ratio of 51.26 % followed by mass flow rate of hot water, fin pitch, and diameter contribution ratios of 35.91 %, 7.02 % , 5.79 % respectively.

Keywords: *Experimental Investigation, Cross Flow Heat Exchanger, Performance Optimization, Exergy Analysis, Taguchi Method*

INTRODUCTION

In various industries, power plants, chemical process plants, petroleum refineries and air conditioning sectors etc. the use of cross flow heat exchangers is vast. The efficiency of the heat exchangers affects the performance of the systems significantly. Hence optimization of heat exchanger parameters and low cost have been important research topics since last two decades or more. Various approaches adopted by researchers are shape of heat exchanger, types of extended surfaces, mal distribution, Nano fluid, disturbance elements etc. to enhance the heat transfer using various methods like entropy generation principle[1], genetic algorithm[2], Taguchi method[3], exergy loss method [4] and so on. Sahin et al. [5] used the Taguchi method to conduct an experimental investigation for identifying the effects of the longitudinal fin pairs and lateral separations of consecutively on the thermos hydraulic performance. Yakut et al. [6] studied the performance of the hexagonal fins, span-wise and stream-wise distances between fins and flow velocity on heat transfer and pressure drop characteristics using Taguchi method. Kotcioglu et al. [7] experimentally analyzed the heat transfer and pressure drop in a cross-flow heat exchanger with hexagonal, square and circular pin-fin arrays. Gunes et al. [8] determined of the optimum values of the design parameters in a tube with equilateral triangular cross sectioned coiled wire inserts. Bahiraeiet al. [9] evaluated the entropy generation of a biological nanofluid containing silver nanoparticles in a liquid block heat sink for cooling of an electronic processor. They found from the analysis that irreversibility in the whole liquid block decreases with increasing either concentration or Reynolds number, which is a positive result based on second law of thermodynamics. Using CFD approach Bahiraeiet al. [10] studied the irreversibilities caused by heat transfer and friction for a power-law Nano fluid in a minichannel having chaotic perturbations were examined through calculation of entropy generation rates. Heshmatian et al [11] reported the irreversibilities caused by heat transfer and friction for the water-TiO_2 nanofluid flow in a circular microchannel by evaluating entropy generation rates. Heat transfer characteristics and second law of thermodynamics are evaluated by Bahiraei et al. [12] for the water–Al_2O_3 nanofluid flow in a triangular minichannel under constant wall heat flux. The convective heat transfers and analysis of the second law of thermodynamics for the non-Newtonian nanofluid flow

This paper was recommended for publication in revised form by Regional Editor Sandip Kale
[1]Department of Mechanical Engineering,C. V. Raman College of Engineering, Bhubaneswar, Odisha, India.
**Email: sachindra106@gmail.com*

containing TiO_2 nanoparticles within a narrow annulus was investigated by Bahiraei et al. [13].

Based upon the studies mentioned above, all of the process parameters affecting the exergy loss have not been investigated in detail. However, it has been reported that more investigations are required for cross flow of heat exchangers to optimize the design parameter based upon exergy loss. In the present investigation simultaneously both designing parameter and operating conditions are considered as process parameters for analysis using Taguchi approach where exergy loss is considered as the response parameter for designing and predicting of the CFHE.

OBJECTIVE

The objectives of the present work are

a) Validation and verification of the results of experimental approach by comparing the results of tube side convective heat transfer coefficients with the data available in the literature.

b) Performance optimization of the CFHE using entropy minimization approach.

EXPERIMENTS

Experimental setup and it's description

Experimental investigation and performance optimization of (CFHE) were carried out in the experimental set up shown in Fig. 1. The test section of the CFHE (16"x 22") is simply a fin and a tubetype cross flow heat exchanger. It consists 279 aluminum fins with a fin pitch of 2 mm and fin thickness of 0.3 mm.Tubes used in CFHE test sections were manufactured from "Type L" copper with inside tube diameter of 0.008525 m and number of tube rows was 2. Recent study was carried out for two different fluids i.e. hot water and air. Hot water was flowing in a closed loop through the tubes of CFHE test section transferring heat into air, which was forced through the fins of the CFHE test section in an open loop by a fan working as the cooling fluid.

Hot water loop consists of a steel tank, a centrifugal pump, a flow control valve, a rotameter, two digital temperature indicator and two pressure gauges. Hot water was prepared in a steel tank of 45 liter capacity equipped with an immersion heater of 1.5 kW and circulated along the hot water loop by a centrifugal pump of 0.5 HP capacity. For flow regulation and measurement of the hot water one flow control valve and the rotameter were connected after the pump. The operating range of the rotameter is 20 LPH to 200 LPH at a temperature is up to 90°C. Two digital temperature indicators (-50 °C to 200 °C) and two pressure gages are connected along the hot water loop at the inlet and outlet of the CFHE test section to measure the pressures and temperatures.

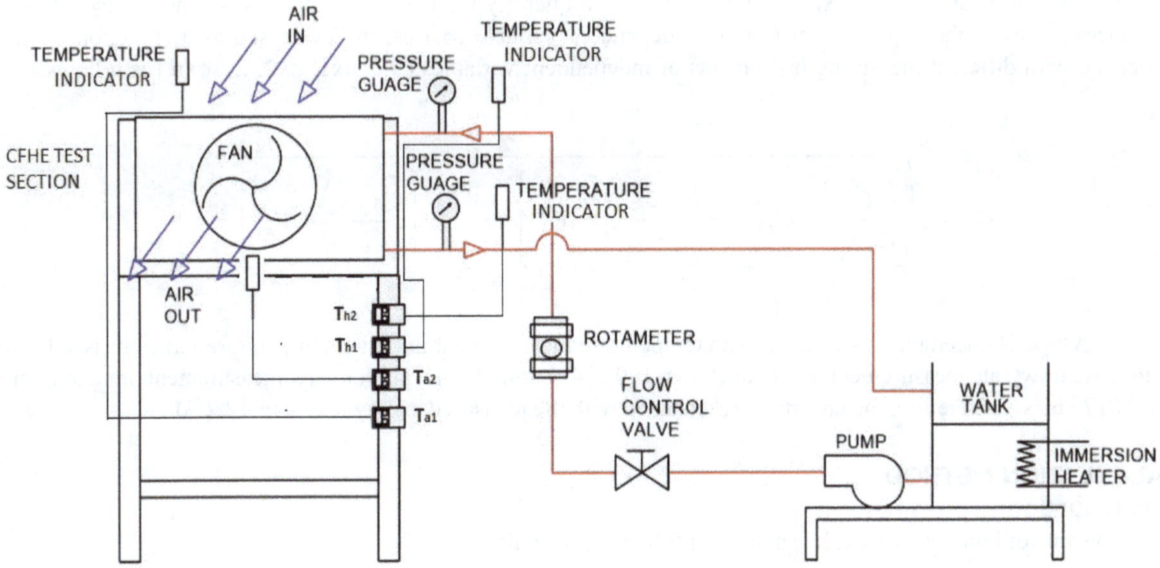

Figure 1.Schematic layout of a cross flow heat exchanger

Air was forced through the CFHE test section by a fan of 14" diameter to take away the heat from the copper tubes through which hot water was flowing as shown in Fig. 2. A speed regulator was provided for the fan motor to supply air at three different speeds. Two digital temperature indicator (-50 oC to 200 oC) were connected along the open air loop to measure the temperatures of air at the inlet and outlet of the CFHE test section. An anemometer was used at the upstream i.e. in between supply fan and cross flow heat exchanger for the measurement of air velocity and at downstream air velocity decreases, when it passes over the fins and copper tubes of the CFHE.

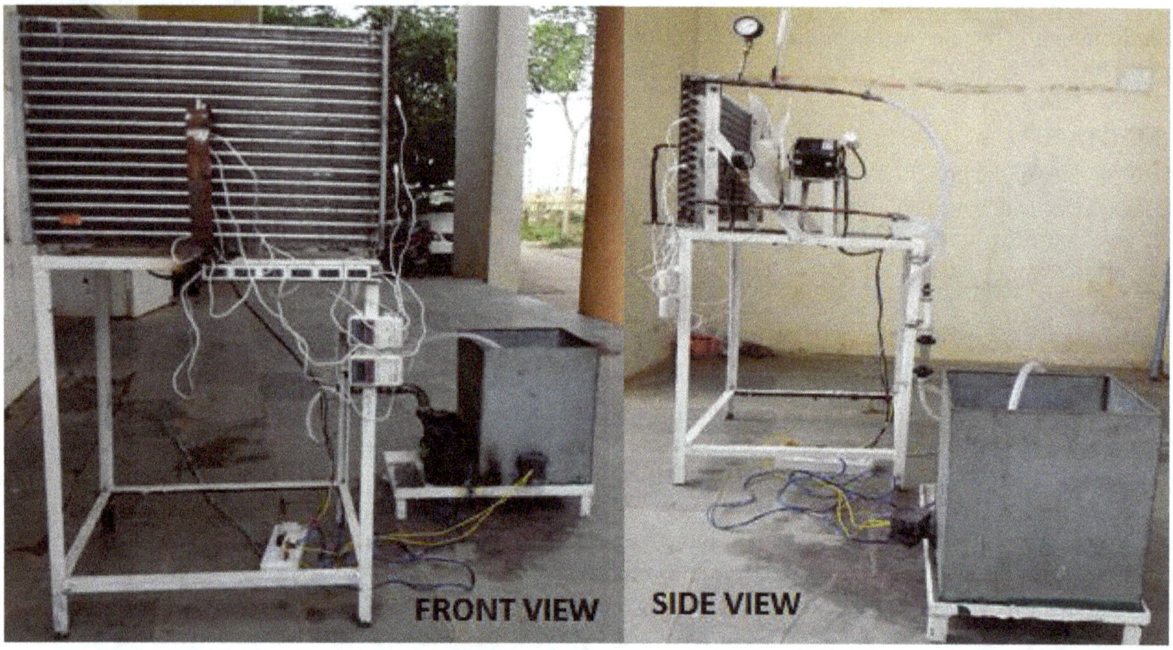

Figure 2.Experimental set up the (CFHE)

UNCERTAINTY OF MEASUREMENT

In experimental studies, the results are typically associated with measurement errors. Uncertainty analysis is generally used to determine the experimental error and to quantify the degree of goodness of the measured values. The uncertainties in the experimental results or dependent variable (σr) can be expressed as a function of errors associated with different measuring instruments or independent variables (i.e. σx1, σx2, ..., σxn) as follows:

$$\sigma_r = \sqrt{\left[\left(\frac{\partial f}{\partial x_1}\right)\times\sigma_{x1}\right]^2 + \left[\left(\frac{\partial f}{\partial x_2}\right)\times\sigma_{x2}\right]^2 + \dots + \left[\left(\frac{\partial f}{\partial x_n}\right)\times\sigma_{xn}\right]^2} \quad (1)$$

A type-B uncertainty associated) with temperature measurement in digital temperature indicator is ± 1.2 oC, volumetric flow rate measurement in rotameter is ± 0.324 L.min-1 and air velocity measurement in anemometer is ± 0.0176 m. s^{-1}. Therefore, the uncertainty associated with the measured exergy loss is (± 12.2%).

CALCULATION METHOD

Data reduction

Mass flow rate of both fluids are calculated using following formulas:
Mass flow rate of hot water:

$$\dot{m}_h = \rho_h . \dot{V}_h \quad (2)$$

Mass flow rate of air:

$$\dot{m}_a = \rho_a . \dot{V}_a \tag{3}$$

Heat transfer rate is calculated by
Hot water heat rejection rate:

$$\dot{Q}_h = \dot{m}_h . c_{p,h} . (T_{h,1} - T_{h,2}) \tag{4}$$

Air heat gain rate:

$$\dot{Q}_a = \dot{m}_a . c_{p,a} . (T_{a,2} - T_{a,1}) \tag{5}$$

Reynolds number are obtained from:
Hot water Reynolds number:

$$\mathrm{Re}_h = \frac{\rho_h . V_h . d_i}{\mu_h} \tag{6}$$

Air Reynolds number:

$$\mathrm{Re}_a = \frac{\rho_a . V_a . D}{\mu_a} \tag{7}$$

As it is observed hot water flow inside the tube is turbulent, the heat transfer coefficient, hh is calculated from Nusselt number correlation was provided by Petukhov-Popov.

$$Nu_h = \frac{(f/8)\,\mathrm{Re}_h\,\mathrm{Pr}_h}{K_1 + K_2 (f/8)^{1/2} (\mathrm{Pr}_h{}^{2/3} - 1)} \quad 10^4 < \mathrm{Re}_h < 5\times10^6 \; 0.5 < \mathrm{Pr} < 2000 \tag{8}$$

Now heat transfer co-efficient inside the condenser tube:

$$h_h = \frac{Nu_h . K_h}{d_i} \tag{9}$$

Similarly the heat transfer coefficient ha can be calculated from Nusselt number correlation for flow over cylindrical tube.

$$Nu_a = 0.027\,\mathrm{Re}_a{}^{0.805}\,\mathrm{Pr}_a{}^{0.334} \tag{10}$$

Now heat transfer co-efficient outside the condenser tube:

$$h_a = \frac{Nu_a . K_a}{d_o} \tag{11}$$

Overall heat transfer coefficient outside of the helical coil when heat transfer takes from hot water to air:

$$\frac{1}{U_o}=\frac{d_o}{h_h.d_i}+\frac{A_o.\ln\left(\frac{d_o}{d_i}\right)}{2\pi k_{cu}L}+\frac{1}{h_a} \tag{12}$$

ENERGY BALANCE

For the present study, we can write a steady state energy balance as follows.
From conservation of energy, it can be represented that
Heat rejected from hot water per unit time = Heat gain by air per unit time + Heat loss/time

$$\dot{Q}_w=\dot{Q}_a+\dot{Q}_{loss} \tag{13}$$

Where,

$$\dot{Q}_w=\dot{m}_w.c_{p,w}.(T_{w,1}-T_{w,2}) \tag{14}$$

$$\dot{Q}_a=\dot{m}_a.c_{p,a}.(T_{a,2}-T_{a,1}) \tag{15}$$

$\dot{Q}_{loss}$ = Heat loss/time (Which is may be radiation or any other unobserved heat loss/time)

It is observed during study that negligible heat loss/time takes place from CFHE to surrounding corresponding to increment in mass flow rates of hot water; therefore, the heat exchange process between hot water and air can be assumed to be adiabatic.

EXERGY BALANCE

Maximum useful work that is obtained reversibly in a specified environment is known as exergy. In the current study, assuming heat transfer process between hot water and air in CFHE is steady and adiabatic; overall exergy loss can be expressed as the summation of individual exergy loss related to the hot water and the air.
Considering $s=s(p,T)$, the change in specific entropy can be written as:

$$ds=\left(\frac{\partial s}{\partial T}\right)_p dT+\left(\frac{\partial s}{\partial p}\right)_T dp \tag{16}$$

Using thermodynamic relations, the change in specific entropy can also be written as:

$$ds=\frac{c_p}{T}dT-\left(\frac{\partial v}{\partial T}\right)_p dp \tag{17}$$

For a differential element the entropy variation of fluid can be given as:

$$dS=\dot{m}\left[\frac{c_p}{T}dT-\left(\frac{\partial v}{\partial T}\right)_p dp\right] \tag{18}$$

The total entropy generation in a heat exchanger is considered to be equal to the total entropy variation of both fluids. For a cross flow heat exchanger, integrating the above equation, i.e., Eq. (15), from the inlet to the outlet of air and hot water, the entropy generation rate can be written as follows:

$$S_g = \left[\int_1^2 \dot{m} c_p \frac{dT}{T}\right]_h + \left[\int_1^2 \dot{m} c_p \frac{dT}{T}\right]_a + \left[-\dot{m}\int_1^2 \left(\frac{\partial v}{\partial T}\right)_p dp\right]_h + \left[-\dot{m}\int_1^2 \left(\frac{\partial v}{\partial T}\right)_p dp\right]_a \quad (19)$$

Rearranging Eq. (16) yields:

$$S_g = S_{g,\Delta T} + S_{g,\Delta p} \quad (20)$$

Where, $S_{g,\Delta T}$ is the entropy generation contributed by heat transfer and $S_{g,\Delta p}$ is the entropy generation contributed by fluid friction.

As we know, the irreversibility of fluid flow due to viscous friction is negligible as compared to that of the irreversibility of heat transfer due to finite temperature difference. Therefore the irreversibility of fluid flow is neglected in this analysis and exergy [$T_o\left(S_g\right)$] balance performed.

$$E = E_h + E_a \quad (21)$$

Exergy losses related to the hot water flow inside the copper tubes can be written as:

$$E_h = T_o\left\{\dot{m}_h \left(\Delta S\right)_h\right\} \quad (22)$$

Where, $\left(\Delta S\right)_h = S_{h,2} - S_{h,1} = C_{p,h} \ln\left(\frac{T_{h,2}}{T_{h,1}}\right)$ (23)

Similarly, exergy loss related to air flow through the fins can be written as:

$$E_a = T_o\left\{\dot{m}_a \left(\Delta S\right)_a\right\} \quad (24)$$

Where, $\left(\Delta S\right)_a = S_{a,2} - S_{a,1} = C_{p,a} \ln\left(\frac{T_{a,2}}{T_{a,1}}\right)$ (25)

By substituting Eq. (20) in to Eq. (19) and Eq. (22) into Eq. (21), overall exergy loss can be expressed as $E = T_o\left\{C_{p,h} \ln\left(\frac{T_{h,2}}{T_{h,1}}\right) + C_{p,a} \ln\left(\frac{T_{a,2}}{T_{a,1}}\right)\right\}$ (26)

Dimensionless exergy loss can be determined:

$$e = \frac{E}{T_o C_{\min}} \quad (27)$$

Where, $C_{\min} = \text{Min}\left\{\dot{m}_h c_{p,h} \text{and}\, \dot{m}_a c_{p,a}\right\}$ (28)

TAGUCHI METHOD

Thermodynamic performance of CFHE can be enhanced considerably by using a reasonable optimization approach [13, 15]. Numbers of factors affect the thermal performance of the CFHE and out of these factors; the principal factors are determined by Taguchi method for optimum design. As reported by Gunes et al. [13]. Taguchi method is a standardized approach for determining the best combination of input parameters to get optimum result. This is accomplished through the design of experiments (DOE). It provides a method for quantitatively identifying just the right parameters that go together for optimal performance or service. Taguchi method is associated with three design stages i.e system design, parameter design and tolerance design. In this paper, the optimum values of system parameters or thermal performance parameters i.e. E is determined from parameter design stage. Signal to noise ratio (SNR or S/N ratio) measures the variations in performance parameters. S is the signal stands for mean and N is the noise stands for standard deviation in Taguchi approach. Three types of quality characteristics "larger is the better"; "nominal is the best" and "smaller is the better" are used to analyze S/N ratio. For the current study following steps of Taguchi technique are used.

Step 1 Identification of the objectives: Minimum exergy loss for the flow in the CFHE is identified as the objective of this work.

Table 1. Design factors and their levels

Control Factors	$\dot{m}_h$, kg.s^{-1}	$\dot{m}_n$, kg. s^{-1}	d_i, m	F_p, m
Level 1	0.0163	0.1175	0.0085	0.0016
Level 2	0.0273	0.2051	0.0138	0.002
Level 3	0.0382	0.0138	0.0199	0.0025

Step 2 Selection of characteristics: "Smaller is the better" quality characteristic is preferred for exergy loss, E. As available energy is higher at minimum exergy loss, hence "Smaller is the better" quality characteristics is preferred.Step 3 Selection of various factors and their levels: In this study, mass flow rate of hot water , mass flow rate of air , tube diameter (d_i) and fin pitch (F_p) are selected as the control factors for influencing exergy loss in the CFHE. The levels of control factors are represented in Table 1.

Table 2. L9 orthogonal array for control factors

No.	$\dot{m}_h$, kg.s^{-1}	$\dot{m}_h$, kg.s^{-1}	d_{i}, m	F_{p}, m
1	0.01638	0.1175	0.008525	1.64
2	0.01638	0.2051	0.0138	2
3	0.01638	0.7801	0.0199	2.59
4	0.0273	0.1175	0.0138	2.59
5	0.0273	0.2051	0.0199	1.64
6	0.0273	0.7801	0.008525	2
7	0.03822	0.1175	0.0199	2
8	0.03822	0.2051	0.008525	2.59
9	0.03822	0.7801	0.0138	1.64

Step 4 Selection of an orthogonal array: Taguchi method is well known for design of experiments by employing a standard table known as orthogonal array for study of various design parameters relative to single quality characteristics. A suitable orthogonal array is selected depending on the number of factors with their levels, degrees of freedom, objective of the study, available resources. Minitab 16.0 is used for the analysis. For four control factors with three levels, L9 orthogonal array is generated and shown in Table 2.

Step 5 Conduction of experiments and result analysis: Orthogonal array design provides the idea about the number of experiments to be conducted in Taguchi method analysis. In this study total 9 numbers of experiments were conducted and results are presented in Table 3 for analysis. The effect of control factors on exergy loss of the CFHE are measured by SNR or S/N (signal to noise) ratio.

For exergy loss, E following formula is used during the analysis of SNR or S/N ratio for "smaller is the better" quality characteristics.

$$SNR_S = -10\log_{10}\left(E^2\right) \tag{29}$$

The optimum levels of various design parameters are obtained according to the greatest value of S/N ratio and contribution ratios of these factors on exergy loss of the CFHE are calculated as shown in Table 4.

RESULTS AND DISCUSSIONS

Validation of experimental approach

The results of the experimental approach i.e. convective heat transfer coefficient of hot water, h_h were compared with literature for cross flow configuration and the results are shown in Fig. 3. For verification and validation of the results of this study, the calculated convective heat transfer of hot water, h_h, were compared with the results of correlation values presented by Gnielinski [14]. There is a good agreement observed between the results of the experimental approach with the literature and thus validated.

Table 3. Mean and S/N ratio of exergy loss, E

RUN	$\dot{m}_h$, kg.s^{-1}	$\dot{m}_n$, kg. s^{-1}	d_i,m	F_{p}, m	Exergy loss, E	Responses	
						Mean	S/N Ratio
1	0.01638	0.1175	0.008525	1.64	3.451105949	3.451105949	-10.75916585
2	0.01638	0.2051	0.0138	2	4.328802392	4.328802392	-12.72735522
3	0.01638	0.763	0.0199	2.59	5.867249014	5.867249014	-15.36869041
4	0.0273	0.1175	0.0138	2.59	4.29422205	4.29422205	-12.65768995
5	0.0273	0.2051	0.0199	1.64	5.900487087	5.900487087	-15.41775728
6	0.0273	0.763	0.008525	2	9.006400795	9.006400795	-19.09102539
7	0.03822	0.1175	0.0199	2	4.928097633	4.928097633	-13.85358607
8	0.03822	0.2051	0.008525	2.59	6.832268443	6.832268443	-16.69129843
9	0.03822	0.763	0.0138	1.64	11.73446402	11.73446402	-21.38926515

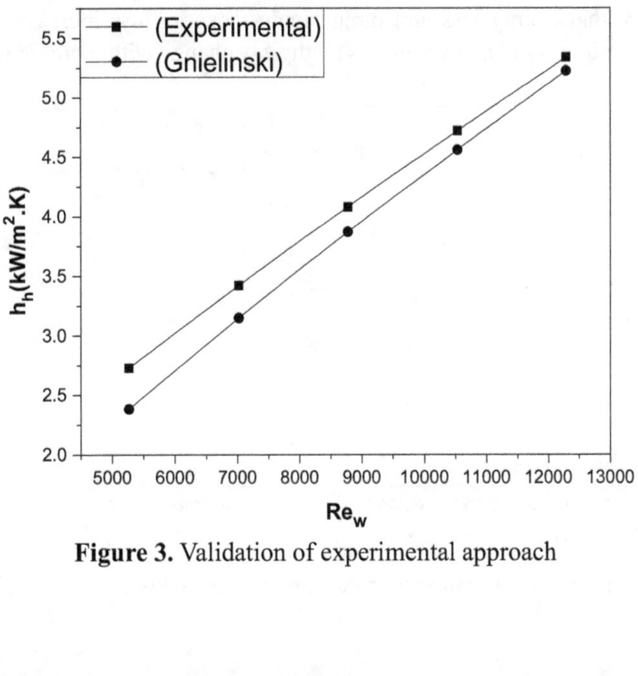

Figure 3. Validation of experimental approach

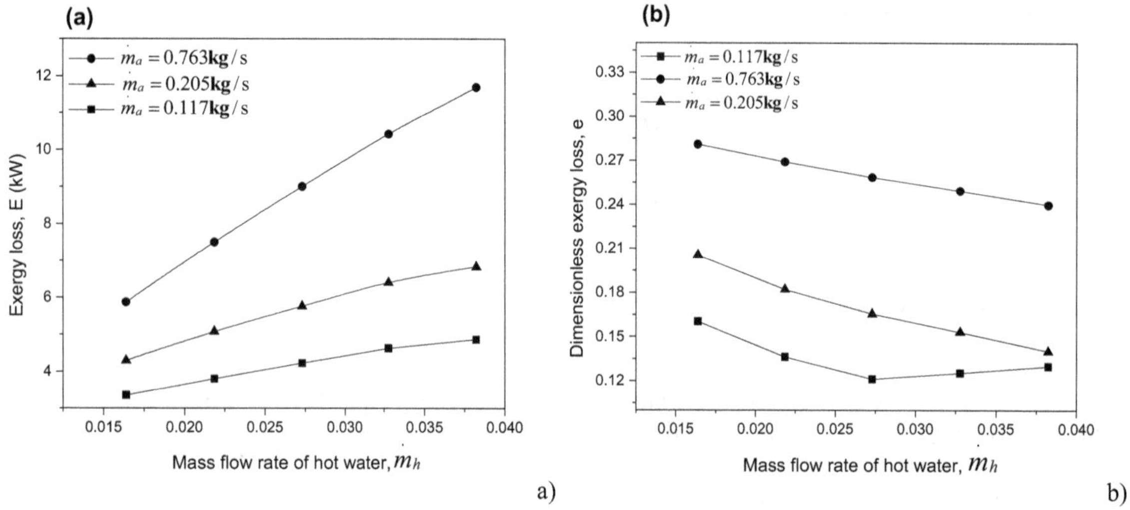

Figure 4. Effect of mass flow rate hot water and air on (a) exergy loss, (b) dimensionless exergy loss

Fig. 4 illustrates the effect of mass flow rate of hot water on exergy loss of the CFHE at different mass flow rates of air. The investigation is conducted at fix dimensions of heat exchanger for all cases. It is observed from Fig. 4 (a) that exergy loss increases with increment in mass flow rates of hot water and air, because more amount of hot water or air flow rate can cause more amount of Reynolds number through the cross flow heat exchanger. It is obvious that higher Reynolds number causes higher heat transfer rate. On the other hand, more heat transfer rate implies more finite temperature difference which is the reason of exergy loss.

In Fig. 4 (b), it is observed that dimensionless exergy loss decreases with the increase of hot water flow rate and increases with the increase of air flow rate. As mentioned in Eq. (23), Dimensionless exergy loss is dependent upon exergy loss and product of ambient temperature, minimum heat capacity. Therefore increment or decrement in dimensionless exergy loss is dependent upon the value of numerator and denominator of Eq. (23). It is observed during analysis that increment of denominator is dominant over increment of numerator for enhancement of hot water flow rate; whereas for enhancement of air flow rate it is reversed. Hence, enhancement of air flow rate increases the dimensionless exergy loss and the increment of hot water flow rate decreases the dimension less exergy loss.

THE EFFECT OF FIN PITCH AND TUBE DIAMETER ON EXERGY LOSS

Fig. 5 demonstrates the effect of fin pitch and tube diameter on exergy loss of the cross flow heat exchanger. The investigation is made at fix of other dimension of heat exchanger and air flow rate for all cases. It is observed from this figure that exergy loss and dimensionless exergy loss increases with increase in inside tube diameter, but exergy loss and dimensionless exergy loss negligibly change with increase in fin pitch.

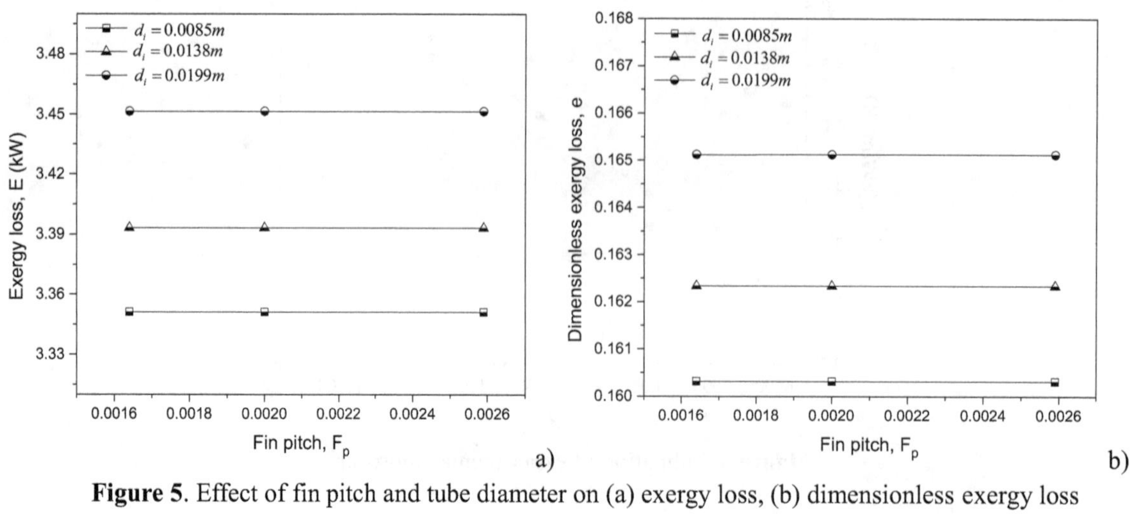

Figure 5. Effect of fin pitch and tube diameter on (a) exergy loss, (b) dimensionless exergy loss

OPTIMUM HEAT TRANSFER ENHANCEMENT CONDITION

The calculated SNR or S/N ratios for 9 numbers of experiments are shown in Table 3. The contribution of each factor on exergy loss of the CFHE is presented in Table 4, where R is the difference between the maximum and minimum S/N ratio for each factor, contribution ratio is the ratio of R value of any factor to the total R values of all factors and rank represents the orderly assignment of factor with respect to minimum exergy loss. The optimum condition is calculated for a combination of levels of control factors having the largest S/N ratio.

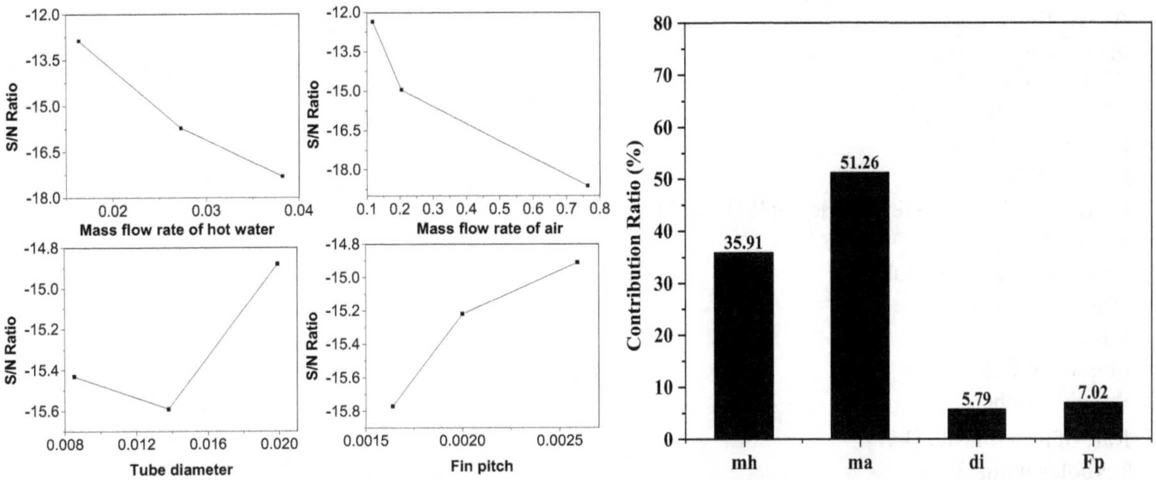

Figure 6. Effects and contribution ratio of each design parameter on exergy loss, E

The effect of each design parameters on exergy loss, E, is shown in Fig. 6. The optimum values of these design parameters for minimum exergy loss, E, is obtained for, $\dot{m}_h$=0.0163 kg.s^{-1}, $\dot{m}_a$=0.1175kg.s^{-1}, d_i=0.0199 m and F_p=0.00259 m according to "smaller is the better" quality characteristics. Using these optimum values another experiment is conducted and exergy loss is calculated about 3.35110719 kW. The contribution ratio of each parameter on exergy loss is shown in Fig. 6 and Table 4. It is clear from figure that the mass flow rate of hot water is the most effective parameter on exergy loss with a contribution ratio of 51.26 % followed by mass flow rate of air , fin pitch, Fp and tube diameter, di with the contribution ratio of 35.91 %, 7.02 % and 5.79 % respectively. It is observed from the results that tube diameter and fin pitches have the least effect on exergy loss.

Table 4. Contribution of design parameters of the CFHE on exrgy loss, E

Table 4. Contribution of design parameters of the CFHE on exrgy loss, E

	Level	$\dot{m}_h$	$\dot{m}_a$	d_i	F_p
S/N Ratio	1	-12.87	-12.34	-15.43	-15.77
-	2	-15.72	-14.95	-15.59	-15.22
-	3	-17.31	-18.62	-14.88	-14.91
R	-	4.4	6.28	0.71	0.86
Rank	-	2	1	4	3
Contribution Ratio %		0.3591	0.5126	0.0579	0.0702

CONCLUSIONS

In this study, experimental investigation and performance optimization of the CFHE was performed. Taguchi method was used for determination of optimum design parameters for minimum exergy loss in CFHE. The results of this work can be summarized as follows.

1. For minimum exergy loss in CFHE, the optimum value of different design parameters were obtained as mass flow rate of hot water is 0.0163 kg/s, mass flow rate of air is 0,117 kg/s, tube diameter is 0.0199 m and fin pitch is 0.00259 m.
2. Contribution ratios of each design parameter on exergy loss were determined. Mass flow rate of hot water is found out the most effective parameter on exergy loss with a contribution ratio of 51.26 % followed by mass flow rate of air, fin pitch, Fp and tube diameter, di with the contribution ratio of 35.91 %, 7.02 % and 5.79 % respectively.

NOMENCLATURE

A	Area [m^2]
c_p	Specific heat [kJ/kgK]
C	Heat capacity [kW/ K]
d	Diameter of the tube [m]
E	Exergy loss [kW]
F_p	Fins pitch [m]
h	Convective heat transfer coefficient [kW/m^2K]
K	Thermal conductivity [kW/mK]
L	Length of the copper tube [m]
$\dot{m}$	Mass flow rate [kg/s]
Nu	Nusselt number
p	Pressure [kPa]
Pr	Prandtl number
Q	Rate of heat transfer [kW]
Re	Reynolds number
S	Entropy [kJ/K]
s	Specific entropy [kJ/kgK]
SNR	Signal to noise ratio
T	Temperature, K
U	Overall heat transfer coefficient [Kw/m^2K]
V	Velocity [m/s]
V	Volumetric flow rate [m^3/sec]
ρ	Density [kg/m^3]
υ	Kinematic viscosity [m^2/s]
μ	Dynamic Viscosity [Ns/m^2]
1	Inlet
2	Outlet
a	air
Cu	Copper
h	Hot water
I	Inner
Min	Minimum
o	Outer
S	Smaller
w	Water

REFERENCES

[1] Ko, T.H. (2006). Thermodynamic analysis of optimal curvature ratio for fully developed laminar forced convection in a helical coiled tube with uniform heat flux. International Journal of Thermal Sciences, 45(7), 729- 737.
[2] Imran, M., Pambudi, N.A., Farooq, M. (2017).Thermal and hydraulic optimization of plate heat exchanger using multi objective genetic algorithm. Case Studies in Thermal Engineering, 10, 570-578.
[3] Kotcioglu, I., Cansiz, A., Nasiri Khalaji, M. (2013). Experimental investigation for optimization of design parameters in a rectangular duct with plate-fins heat exchanger by Taguchi method. Applied Thermal Engineering, 50, 604-613.
[4] Ipek, O., Kılıc, B., Gurel, B. (2017). Experimental investigation of exergy loss analysis in newly designed compact heat exchangers. Energy, 124, 330-335.

[5] Sahin, B. (2005). Optimum design parameters of a heat exchanger. Applied Energy, 82(1), 90-106.

[6] Yakut, K. (2006). Experimental investigation of thermal resistance of a heat sink with hexagonal fins. Applied Thermal Engineering, 26(17), 2262-2271.

[7] Kotcioglu, I., Caliskan, S., Baskaya S. (2011). Experimental study on the heat transfer and pressure drop of a cross-flow heat exchanger with different pin–fin arrays. Heat and mass transfer, 47(9), 1133.

[8] Bahiraei, M., Heshmatian, S. (2017). Application of a novel biological nanofluid in a liquid block heat sink for cooling of an electronic processor: Thermal performance and irreversibility considerations. Energy Conversion and Management, 149, 155–167.

[9] Bahiraei, M., Gharagozloo, K., Alighardashi, M, Mazaheri, N. (2017). CFD simulation of irreversibilities for laminar flow of a power-law nanofluid within a minichannel with chaotic perturbations: An innovative energy- efficient approach. Energy Conversion and Management, 144, 374-387.

[10] Heshmatian, S., Bahiraei, M. (2017). Numerical investigation of entropy generation to predict irreversibilities in nanofluid flow within a microchannel: Effects of Brownian diffusion, shear rate and viscosity gradient. Chemical Engineering Science, 172, 52-65.

[11] Bahiraei, M., Mohammadi, S. (2016). Prediction of entropy generation for nanofluid flow through a triangular minichannel using neural network. Advanced Powder Technology, 27, 673-683.

[12] Bahiraei, M., Alighardashi, M. (2016). Investigating non-Newtonian nanofluid flow in a narrow annulus based on second law of thermodynamics. Journal of Molecular Liquids, 219, 117-127.

[13] Gunes, S., Manay, E., Senyigit, E., Ozceyhan, V. (2011). A Taguchi approach for optimization of design parameters in a tube with coiled wire inserts. Applied Thermal Engineering, 31, 2568-2577.

[14] Gnielinski, V. (1976). New equation for heat and mass transfer in turbulent pipe and channel flow, International Chemical Engineering, 16, 359-368.

[15]Rout S.K., Choudhury B. K, Sahoo R. K., Sarangi S. K. (2014). Multi-objective parametric optimization of Inertance type pulse tube refrigerator using response surface methodology and non-dominated sorting genetic algorithm.Cryogenics 62, 71-83.

7

OPERATION OF HVAC SYSTEM FOR ENERGY SAVINGS AND ECONOMIC ANALYSIS

M. Kassas[1], W. M. Hamanah[1], O. Al-Tamimi[1], A. Sahin[2], B. S. Yilbas[2,*], and C. B. Ahmed[1]

ABSTRACT

In recent years, energy savings in air-conditioning (A/C) systems has become one of the hot topics of applied energy towards innovative management and efficient utilization of the operating systems. Achieving thermal comfort with minimum energy consumption is the main concern for innovative designing of an air conditioning system. The A/C system is one of the chief contributors to energy consumption in warm and hot environments. An innovative design of an A/C operating system is essential to satisfy a high thermal performance and maintain the desired thermal comfort level. The aim of this work is to introduce innovative design of an operating system to simulate and experiment the thermal performance of A/C units for two identical houses located in Dhahran area of Saudi Arabia. In this case, the thermal model for both houses has been developed incorporating two different air-conditioning operating systems. In the analysis, several physical properties and parameters, such as climate conditions and heat gain/loss, have been taken into account inside the house. Matlab/Simulink software is used to simulate the ON/OFF and the VFD air conditioning controller systems. LabView platform with the data acquisition is utilized for the experimental work to monitor the real time climate and electrical power data.

Keywords: *Air-Conditioning, Energy Savings, Variable Frequency Drives VFD, ON/OFF Cycle, Modeling*

INTRODUCTION

Energy savings and efficient utilization of the cooling systems require innovative design of the thermal units that meets the demand with reasonable comfort levels. The performance of variable refrigerant flow systems and best practices for waste energy minimization are the recent interests [1-2] and several approaches were proposed to achieve energy efficient systems [3-6]. On the other hand, many residential areas around the world are characterized by hot and humid atmosphere, such as in the Arabian Gulf, Africa, S America, S. Asia, and part of USA. For example, the primary energy consumption in Saudi Arabia is the HVAC system which accounts for about 60% of the energy used in the houses [7]. Space-conditioning has gone on the rise throughout the Kingdom of Saudi Arabia (KSA) and also consumption of more energy has increased due to the severe hot temperature, high humidity, and dust storms. The significant focus in designing HVAC systems is ensuring the thermal comfort and reducing the energy consumption. To evaluate the energy consumptions used by the space-conditioning, thermal modeling and simulation of the house must be developed and two types of HVAC system operations will be presented. One HVAC system operates by an ON/OFF cycle and the other HVAC system operates by using a variable speed driver (VFD).

The concept of using simulation as a tool for performance validation and energy analysis of HVAC systems is studied by Salsbury and Diamond [8]. They described one way of making use of new technology by applying simulations configured to represent optimum operation to monitor data. The idea is to use simulation predictions as performance targets with which to compare monitored system outputs for performance validation and energy analysis. They presented results from applying the concepts to a large dual-duct air-handling unit installed in an office building in San Francisco. Muratori et al conducted an estimation of required energy to get a coziness level which was developed by using Matlab/Simulink depending on the predicted temperature variation [9]. They compared their results with the IEA Building Energy Simulation Test (IEA BESTTEST) results. In addition, the results are validated with actual metered residential load data. A simple-physically-based model is used to simulate heating and electric energy consumption throughout the year for domestic HVAC systems. They depend on

This paper was recommended for publication in revised form by Regional Editor Ahmet Selim Dalkilic
[1] Department of Electrical Engineering, KFUPM, Dhahran, Saudi Arabia
[2] Department of Mechanical Engineering, KFUPM, Dhahran, Saudi Arabia
**E-mail address: bsyilbas@kfupm.edu.sa*

thermodynamic and heat transfer to regulate air volume in the household. Matlab easily simulates diver scenarios to implement the model. A validation against actual metered residential load data provided by American Electric Power (AEP) is reported by Karmacharya et al [10]. The thermal model of the house is developed by Wen and Burke [11]. They used the autoregressive model with external inputs or with exogenous inputs (ARX model) to obtain the prediction state of the HVAC system. The proposed method is validated by experimentation in a particular home using the GE Nucleus energy management system for data aggregation and algorithm implementation. An ARX model and thermostat regulator is used to simulate and validate their data.

Energy consumption in residential areas is always the main concern and retains a lot of research attention. Zhu has conducted a case study by using a computer simulation code to assess different energy saving approaches. The researcher found that the simulation can offer a reusable tool for energy efficiency researches [12]. Another case study was investigated by Zhou and Park [13] to reduce energy consumption of the building by using the control optimization strategy. Yu et al demonstrated a study for a low energy envelope design of a residential building in the hot summer and cold winter zone in China. They reported that the best energy saving strategy for the HVAC system could be developed through energy modeling to evaluate the effects of envelope design on energy consumption of air conditioners [14]. Nasution et al conducted an experiment for the energy efficiency by using a variable speed driver for a centralized air conditioning system. They controlled the HVAC system using a PID controller as a variable speed compressor and a conventional ON/OFF controller. The energy consumption for both controllers was investigated, and the result shows that an excellent energy saving and temperature control can be achieved by using a variable speed compressor [15].

Thermal analysis of the HVAC system incorporating the ON/OFF controller and the VFD controller are carried out, both experimentally and numerically, to assess the energy consumption of the HVAC system. The experimental set-up with Lab-View based data acquisition monitoring and measurement system is developed while the thermal model and simulation of the house is carried out incorporating different types of A/C system controller (ON/OFF, VFD) units. Predictions of power requirements of the HVAC system are validated through the measurements. Analysis related to the assessment of energy consumption, energy saving, and the cost with the payback period is provided.

ANALYTICAL MOLDING

Thermal Model of the House

In developing strategies to minimize energy consumption in houses, it is crucial to understand the dynamics of working, heat energy generation and losses [16]. This section aims to investigate some of the contributions of heat generation and losses through the developed empirical models. From these models, a thermal model is derived to allow the researcher to build the relationship of heat flows with the variations in temperature [10]. The developed models have adjustable parameters corresponding to different contributions of the heat budget by understanding the form of variation temperatures. Therefore, we aim to determine the most important factors in the energy consumption and production within the house.

The thermal model of a house is presented in Figure 1 [17, 18]. The aim of this section is to show the parameters extracted in real-time were a reasonable representation of the house, in that they could be used to control the cooling plant of a real house. This means that, at any instant, the model would have to represent the thermal aspects of a house in its present condition, thus allowing predictions to be made.

The proposed thermal model of the house includes three layers for the wall, roof and windows. The temperature can be found at each desired point on the surface or center of the plaster, concrete, and outside wall of the plaster or stucco. Moreover, using three layers can enhance the thermal house parameters such as the type of building materials and thickness depending on the desired temperature in the house.

Thermal energy absorbed by air in the house, heat transfer via floor, wall and windows are represented in following equations.

$$Q_{\text{Absorbed in the House Air}}(t) = \frac{\mathrm{d}T_H}{\mathrm{d}t} \times C_{\text{Air}} \tag{1}$$

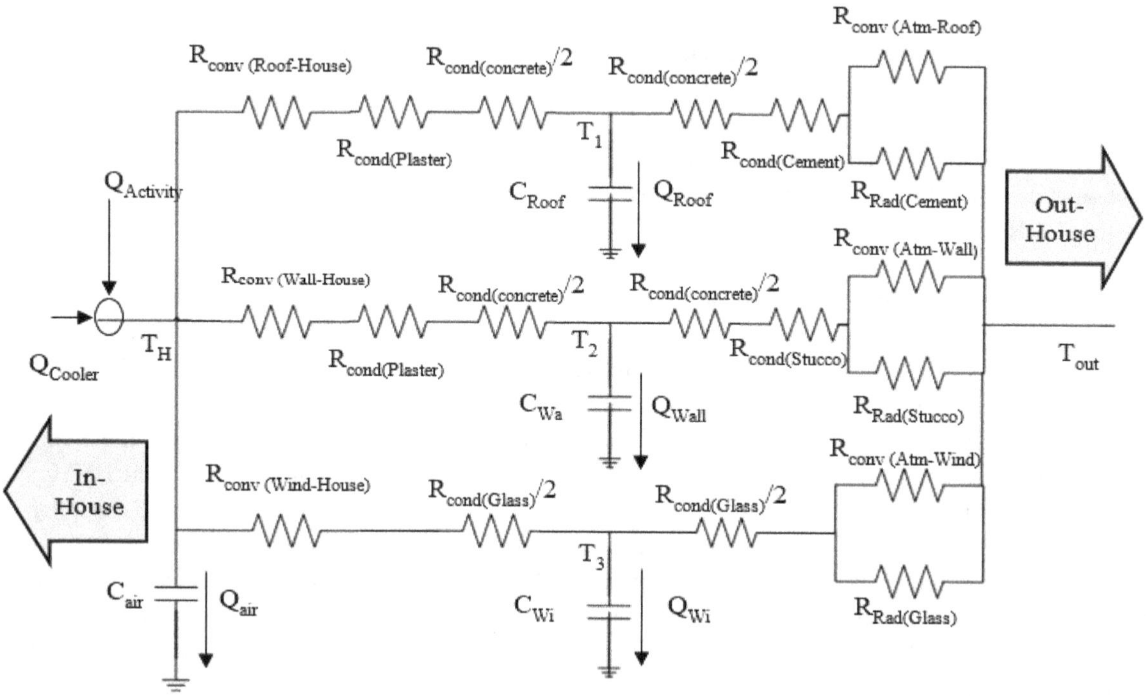

Figure 1. Thermal circuit model of house

$$Q_{\text{Absorbed in the Roof}}(t) = \frac{dT_1}{dt} \times C_{\text{Roof}} \tag{2}$$

$$Q_{\text{Absorbed in the Walls}}(t) = \frac{dT_2}{dt} \times C_{\text{Wall}} \tag{3}$$

$$Q_{\text{Absorbed in the Windows}}(t) = \frac{dT_3}{dt} \times C_{\text{Window}} \tag{4}$$

$$Q_{\text{Absorbed in the House Air}}(t) = (Q_{\text{Heat Passes to the House}}(t) + Q_{\text{Heat Sources in the House}}(t) - Q_{\text{Cooler}}) - (Q_{\text{Absorbed in the Roof}}(t) + Q_{\text{Absorbed in the Walls}}(t) + Q_{\text{Absorbed in the Windows}}(t)) \tag{5}$$

The heat transfer through the roof, walls and windows and the absorbed heat in roof, wall and windows are expressed in the following equations:

$$Q_{\substack{\text{Heat Passes} \\ \text{to the House}}}(t) = Q_{\substack{\text{Heat Passes} \\ \text{though Roof}}}(t) + Q_{\substack{\text{Heat Passes} \\ \text{through Walls}}}(t) + Q_{\substack{\text{Heat Passes} \\ \text{through Windows}}}(t) \tag{6}$$

$$Q_{\substack{\text{Heat Passes} \\ \text{though Roof}}} = Q_{\text{Absorbed in the Roof}}(t) + U_{11}(T_1(t) - T_H(t)) \tag{7}$$

$$Q_{\substack{\text{Heat Passes} \\ \text{though Walls}}} = Q_{\text{Absorbed in the Walls}}(t) + U_{21}(T_2(t) - T_H(t)) \tag{8}$$

$$Q_{\substack{\text{Heat Passes} \\ \text{though Windows}}} = Q_{\text{Absorbed in the Windows}}(t) + U_{31}(T_3(t) - T_H(t)) \tag{9}$$

Finally, the following equations present the thermal model incorporating the parameters related to the house:

$$Q_{\substack{\text{Absorbed in}\\ \text{the House Air}}}(t) = U_{11}(T_1(t) - T_H(t)) + U_{21}(T_2(t) - T_H(t)) + U_{31}(T_3(t) - T_H(t)) + Q_{\text{Heat Sources in the House}}(t) - Q_{\text{Cooler}} \quad (10)$$

$$\dot{T}_H(t) = \frac{1}{C_{Air}}\begin{pmatrix} (U_{11} \times T_1(t)) + (U_{21} \times T_2(t)) + (U_{31} \times T_3(t)) - T_H(t)(U_{11} + U_{21} + U_{31}) \\ + Q_{\text{Heat Sources in the House}}(t) - Q_{\text{Cooler}} \end{pmatrix} \quad (11)$$

$$\dot{T}_1(t) = \frac{1}{C_{Roof}}(U_{12}(T_{out}(t) - T_1(t)) - U_{11}(T_1(t) - T_H(t))) \quad (12)$$

$$\dot{T}_2(t) = \frac{1}{C_{Wall}}(U_{22}(T_{out}(t) - T_2(t)) - U_{11}(T_1(t) - T_H(t))) \quad (13)$$

$$\dot{T}_3(t) = \frac{1}{C_{Window}}(U_{32}(T_{out}(t) - T_3(t)) - U_{31}(T_3(t) - T_H(t))) \quad (14)$$

Factors Affecting Cooling Load

The amount of cooling required depends on a large number of factors. These include the dimensions of the house, outdoor temperature, lights, the outdoor humidity, wind speed, the level of insulation in the house, the amount of air leakage in the house, the amount of southern, eastern, and western facing glass in the house, whether this glass is single, double, or triple glazed, and whether window treatments (curtains or blinds) are kept closed or open. Other factors include the amount of shading from trees, roof overhang, awnings, or buildings and how much heat is generated in the house by people during cooking, sleeping and some electric equipments inside the house that produce heat [19].

The operation electrical load by the compressor in one hot day in Dhahran city (KFUPM Campus) has been studied by taking into account the cooling effective factors. Figure 2 shows the outdoor temperatures of the same hot day which is measured for 24 hours. It can be observed that the temperature starts around 35 °C, and it goes up dramatically to 51 °C at midday, and then it drops slowly until it reaches to 33 °C at midnight. Figure 2 also presents the daily internal heat load for one hot day in the house as there are only two people living in it.

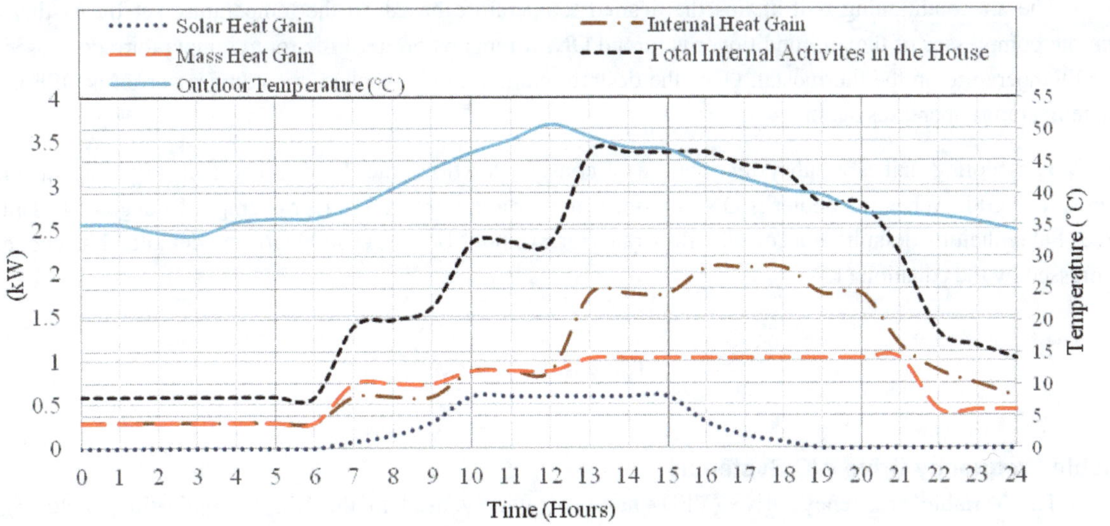

Figure 2. Outdoor temperature with daily internal heat load

House Thermal Model Using Simscape Physical System

The thermal model of both identical houses is built by applied Simscape Physical components in Simulink/Matlab as shown in Figure 3. The model is developed with four thermally distinguishable parts: inside air, house walls, windows, and roof. The house exchanges heat with the environment through its walls, windows, and roof. Each path is simulated as a combination of a thermal convection, thermal conduction, and thermal radiation. This model has been integrated with two different units of an air conditioning system; the 1st house has been equipped with VFD HVAC unit and the 2nd house has been equipped with an ON/OFF HVAC unit.

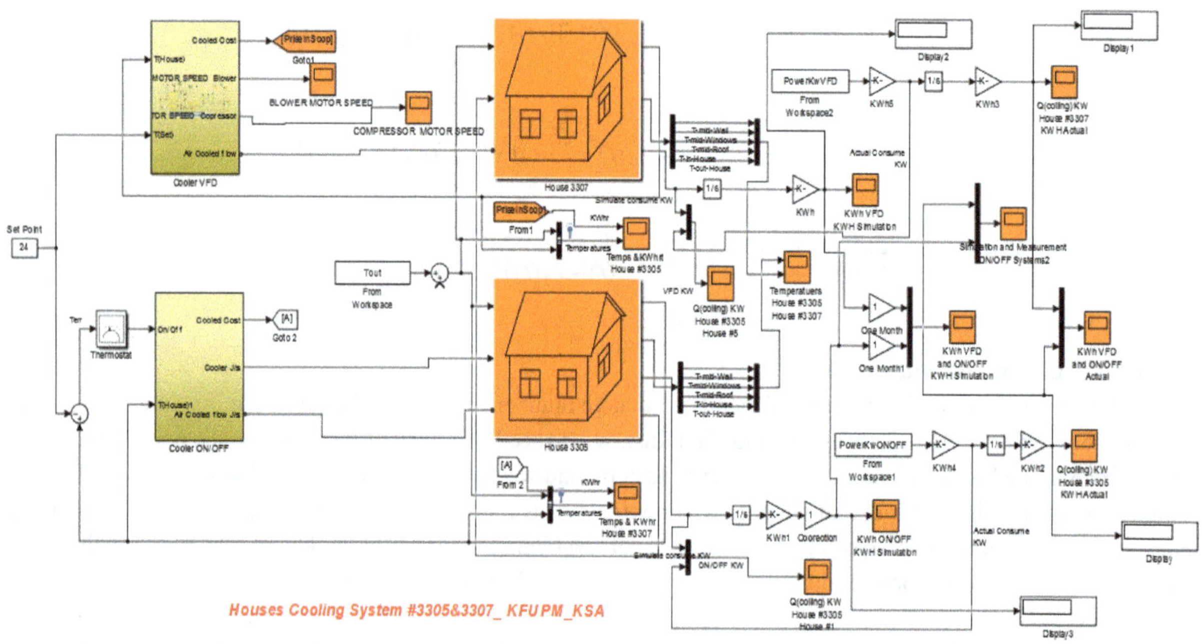

Figure 3. House thermal model integrated with the VFD & ON/OFF HVAC system using simscape physical components

Conventional A/C System (On/Off Cycle)

The air conditioning unit attains the desired temperature based on the temperature set by its thermostat. When the compressor of the air conditioner is turned ON, it remains on until the room temperature decreases to the desired temperature on the thermostat. Once the desired temperature is reached, the compressor turns OFF until the room temperature increases again.

The cooling unit is a subsystem that has a constant air flow rate, " $\dot{M}$ ". The thermostat signal turns the cooler ON or OFF. When the cooler is ON, it absorbs hot air at a temperature (for example 10 degrees Celsius = 50 degrees Fahrenheit by default) at a constant flow rate (for example 1kg/sec = 3600kg/hr by default). The cool airflow is expressed by the equation (15).

$$\frac{dQ_{Cooler}}{dt} = \left((T_{HVAC} - T_{House}) \times \dot{M}_{HVAC} \times C_P \right) \tag{15}$$

Variable Frequency Drive A/C Systems

The Variable Frequency Drive (VFD) has been widely used in the HVAC application, including fans, pumps, compressors, etc. A better understanding of the VFD will lead to improved application and selection of both equipment and HVAC systems. A Variable Frequency Drive (VFD) is a type of adjustable speed drive used to control the speed of the motor based on the system load requirements and operation schedule, resulting in a dramatic cut in energy consumption [20]. VFD technology allows the air conditioner to automatically vary its power output to specifically maintain room temperature at a desired or comfortable level. A non-inverter appliance maintains the temperature by repeatedly switching the power ON and OFF, which consumes much more electrical energy upon starting.

The VFD cooler has two options; constant value and variable air flow rate, " $\dot{M}_{HVAC}(t)$ ". Based on equation 16, the variable airflow rate is delivered from the cooler system using a PI system. The air specific heat is multiplied with airflow rate to produce cooler gain by equation 17. The HVAC coil temperature (T_{HVAC}) uses variable values that are controlled by the PI system. Equation 18 is used to calculate the absorbed heat by the air in the house. In the second subsystem equation 19, T_{House} and $T_{Setting}$ are the inputs signal for the HVAC coil temperature subsystem and the cooling air flow subsystem. The cooling airflows (Q_{Cooler}) into the house are produced from the difference between indoor temperature (T_{House}) and the HVAC coil temperature (T_{HVAC}) and then multiplied by cooling gain.

$$\dot{M}_{HVAC} = K_0 \frac{d(T_{setting} - T_{House}(t))}{dt} + K_1 T_{House}(t) \tag{16}$$

$$\text{Cooler Gain}(t) = \dot{M}_{HVAC}(t) \times C_P \tag{17}$$

$$T_{HVAC} = K_2 \frac{d(T_{setting} - T_{House}(t))}{dt} + K_3 T_{House}(t) \tag{18}$$

$$\frac{dQ_{Cooler}}{dt} = (T_{House} - T_{HVAC}) \times \text{Cooler Gain}(t) \tag{19}$$

EXPERIMENTAL ANALYSIS

Experimental Setup

The major role of the experimental work in this study is to confirm the validly of the simulation work, and to evaluate the performance and energy consumption of the A/C systems. The monitoring and measurement hardware system is comprised of four major blocks: two 5-ton Al-Zamil rooftop units, a Data Acquisition (DAQ) chassis with National Instrument modules (Lab-View), sensors, and a host computer. The National Instrument DAQ-chassis monitoring system has several modules such as voltage measurements, current measurements, and thermocouples. There are four thermocouple sensors (three sensors for indoor and one for outdoor temperature), two humidity sensors (indoor and outdoor), an irradiation sensor, barometric pressure sensor, and three air flow sensors. The host computer has the National Instrument (NI) software, which is the main communicator with DAQ-chassis. The host computer will initiate the execution commands, store the data and display them on the monitor.

Experiment Procedures

The experiments were conducted in the Guest Houses in KFUPM campus, Dhahran, Saudi Arabia. The floor plan and the duct plan are shown in Figure 5.a. Each house consists of two rooms (one living room and one bedroom), kitchen and bathroom. Moreover, all details of the houses which are integrated with ON/OFF and VFD HVAC systems are described in this section and followed by a HVAC monitoring and measurement system.

Sensors Location in the Houses

Different types of sensors have been chosen and placed in both houses to read and measure the required data for the experiment, Figure 4.b, and provide the location of each sensor in the house. Three thermocouple sensors are fixed in the bedroom, corridor and living room. The temperatures are measured from three positions and then we take the average temperature for a house. Airflow sensors are located near inlet air ducts to measure A/C unit flow in kilogram per second. A pressure sensor is also placed inside the house to measure in-house pressure. A thermocouple sensor and an irradiation sensor are placed on the roof of the house to measure the ambient temperature in degree and daily irradiation in watt per meter square.

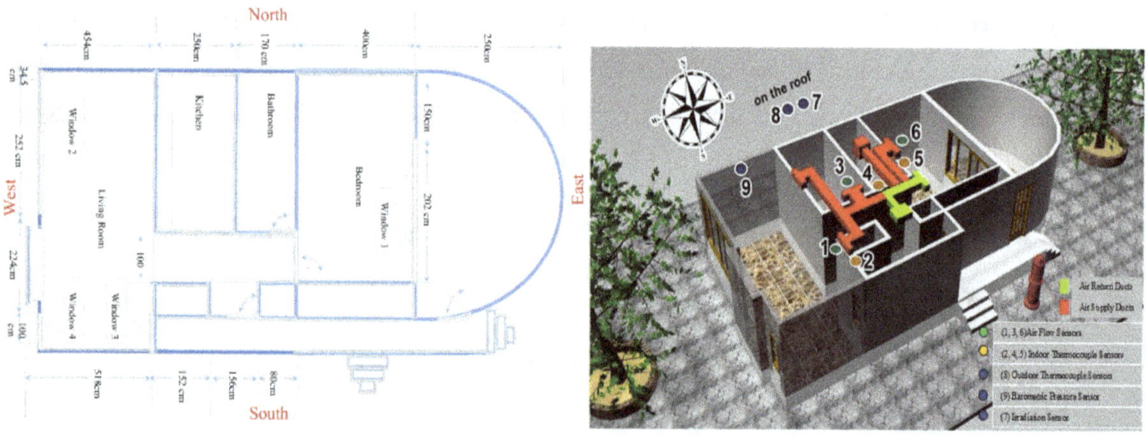

(a) (b)

Figure 4. (a) Floor plan for both houses, (b) Ducts plan and sensor's locations in the houses

Monitoring and Measurement Systems

The monitoring and measurement systems for both units have been developed to achieve different tasks such as displaying several physical and electrical characteristics, as well as environmental conditions. Furthermore, the data acquisition systems incorporate signals, sensors, actuators, signal conditioning, data acquisition devices, and application software. The purpose of data acquisition system is to measure the electrical or physical parameters such as voltage, current, temperature irradiation, humidity, pressure, and wind speed. The PC-based data acquisition system uses a combination of modular hardware, application software (Lab-View), and a computer to take measurements [21]. Figure 5 shows the drawing map of the monitoring and measurement system.

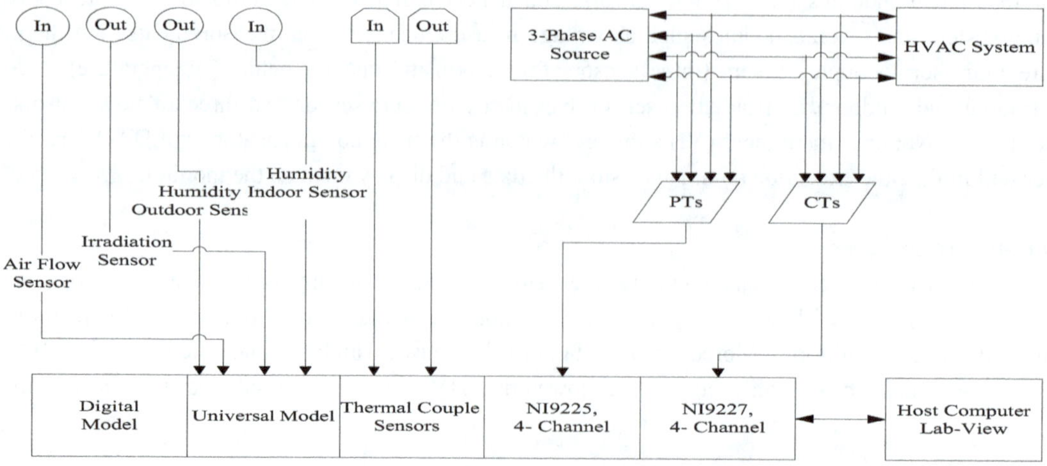

Figure 5. HVAC monitoring and measurement system

The superimposed graph displayed in Figure 6.a shows the instantaneous values of four temperatures and pressure with time and Figure 6.b displays the instantaneous values (peak to peak) of line voltage with corresponding RMS values.

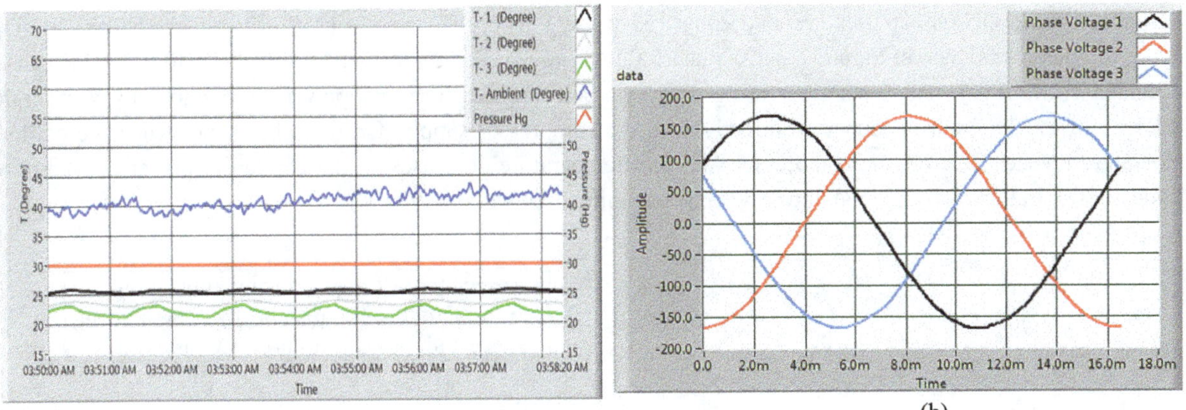

(a) (b)

Figure 6. (a), (b), Monitoring and measurement system (Lab-View Platform)

RESULTS AND DISCUSSION

In this section, the results for the simulation model with validation against the measuring data are provided and the monthly results for variations of the energy consumption, energy savings, and COP variation have been expressed. In addition, the cost analysis of both the ON/OFF and the VFD HVAC system with payback period for each unit have been calculated.

Validation of the Simulation Results with Measurement Data for Both HVAC Systems

Simulation results for in-house air temperature and power consumption for one typical day, with normal activities, (9th of April 2016) have been presented using Simulink/Matlab. The normal internal activities of the house are shown in the Figure 2. The simulation results are produced for ON/OFF and VFD air conditioning units. Measurement data are collected by Lab-View program for both residential houses. These data are validated with the simulation results for both in-house air temperature and power consumption. Set point temperature is 24 °C with a threshold (+1 °C & -1.5 °C) and the initial house air temperature is 24.1 °C.

Typical Day Activities (April 9, 2016) for the On/Off Cycle A/C Unit

Typical day activities validation for indoor temperature is shown in Figure 7. Simulated results are closely following the measurement results where the number of pulses is also equal to 53. The average cycle period is also identical, and the ON-time operation obtained by simulation and measurement are the same (451.05 min = 7.5175 h). The outdoor air temperature affects directly the ON/OFF A/C unit as well as the ON pulse numbers. The outdoor temperature as shown in Figure 7 is still constant and equal to 23 °C from midnight to 6:00 AM and then it increases dramatically to 42.5 °C at 12:30 PM. The temperature decreases again slowly to reach 27.5 °C at 7:30 PM. The house has permanent activities (fridge, computer, monitor, Wi-Fi, lights, and printer) and variable activities (cooking, washing machine, open/close door and windows, etc.).

The activities of heat losses begin at 1.087 kW and goes up to reach 1.27 kW. Then it goes down to 0.87kW as the outdoor temperature stays at 23 °C. From 2:30 AM to 4:00 AM, a decrease is observed in duty cycle that is affected by outdoor temperature reduction to 21°C as well as a reduction in in-house activities. While the outdoor temperature increases from 05:30 AM to 10:00 AM, the duty cycle is still low, and it is affected directly by the tree's shadow that is close to the east window of the house. From 10:00 AM to 6:00 PM, the ON/OFF frequency increases due to the increase in the outdoor temperature that reaches 42.5 °C at midday. The social activities (lunch, dinner, washing machine, open/close windows and door, etc.), increase the duty cycle. The ON/OFF frequency again decreases slowly as a result of the outdoor temperature reduction to 26 °C and some of the social activities (cooking dinner, boiling tea water, etc.), so the duty cycle decreases slowly again.

The simulation and the measurement of the power consumption for the ON/OFF cycle unit on April 9, 2016 are presented in Figure 8. The simulated results are similar to the measurement results (4.95 kW rated value), where the number of the ON/OFF pulses are also equal (53 pulses). The OFF time is (989.112 min = 16.4852 h). The outdoor temperature shown in Figure 7 went up at 10:00 AM to 32 °C and reached 42.5 °C at 12:30 PM. This causes the condenser of the A/C unit to start increasing the power consumption. The total power reached to 5.4 kW when the condenser of the A/C unit is connected directly with the compressor supply. On the other hand, when the outdoor temperature reaches 27.5 °C, the compressor A/C unit temperature decreases so the condenser power consumption also decreases. In this case, the total power reaches the nominal power of 4.95 kW.

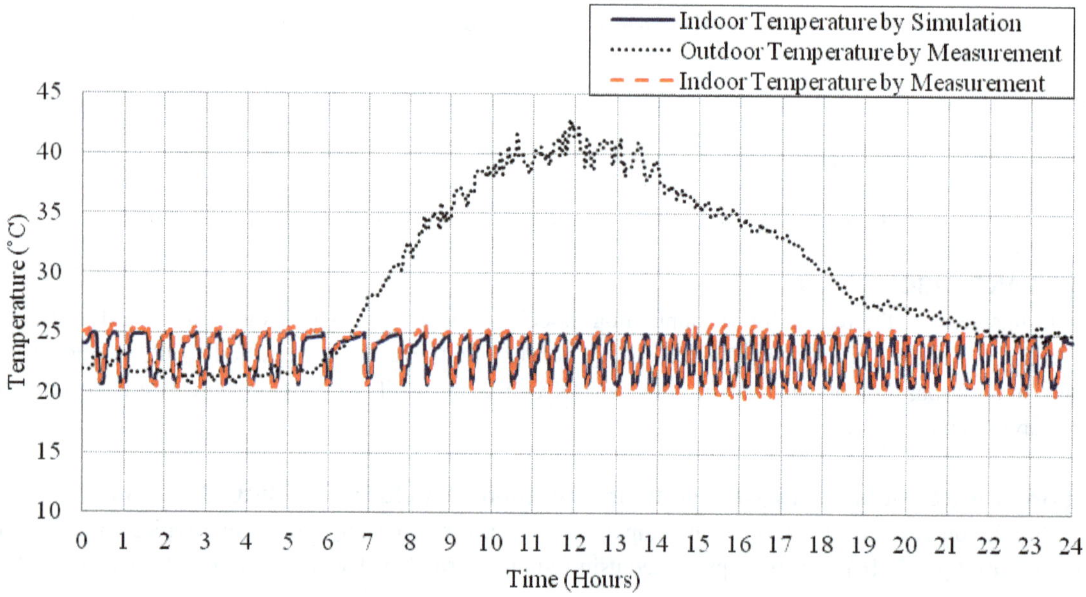

Figure 7. Measured and simulated indoor temperature and the outdoor temperature on 9th of April 2016

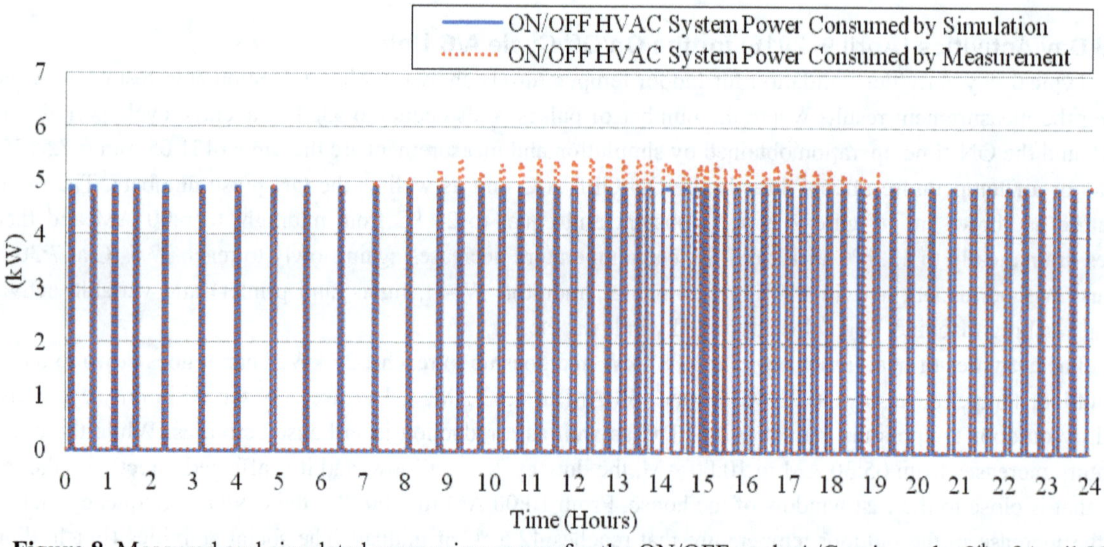

Figure 8. Measured and simulated consuming power for the ON/OFF cycle A/C unit on the 9th of April 2016

Typical Activities (April 9, 2016) for the VFD A/C Unit

Similarly, the typical activities validation for the indoor temperature is displayed in Figure 9. The simulated results are similar to those of the measurement, where the temperature for both values is in the same level (24.5 °C to 23 °C). The outdoor temperature and typical activities (permanent and social heat load) have a direct effect on the VFD A/C unit performance, where the amount of flow rate is increased or decreased to match closely the set point temperature (24 °C).

The simulation and measurement of the power consumption for the VFD A/C unit on the 9th of April, 2016 are shown in Figure 10. The power consumption starts to rise at 05:30 AM and reaches a maximum value (1.7 kW) at 10:00 AM until 12:00 PM because of the in-house activities (cooking breakfast/lunch, washing machine, open/close windows and door, etc.), and rise in the outdoor temperature. The power consumption decreases gradually until it reaches 0.9 kW at 3:00 PM and increases again to 1.3 kW because of the increase in the in-house activities (cooking dinner, washing machine, open/close door, boiling tea water, etc.). It then goes down to 0.4 kW till 09:00 PM and finally goes up to 0.9 kW at midnight.

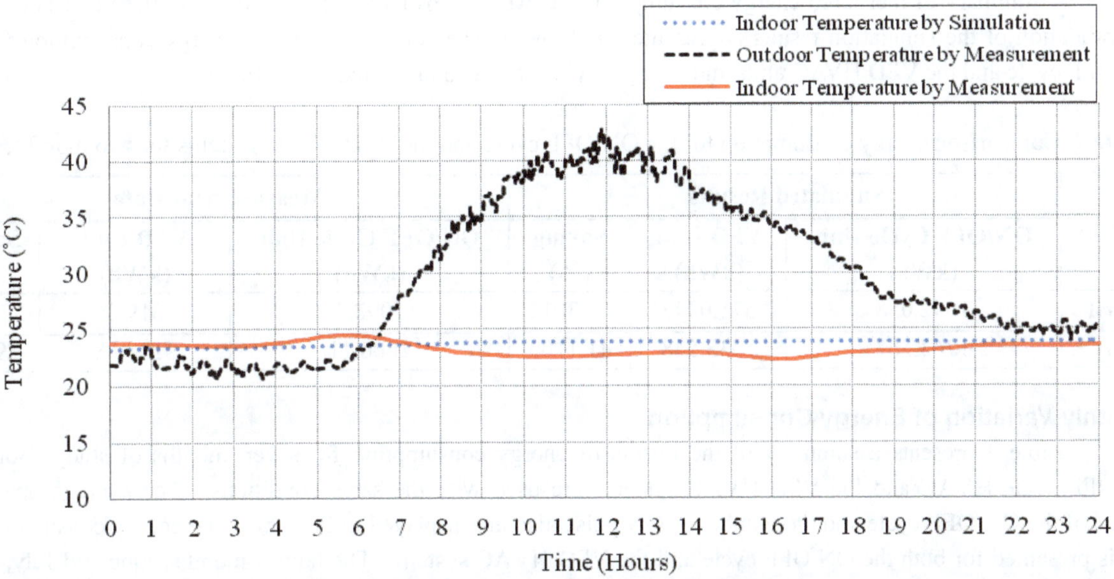

Figure 9. Outdoor measured temperature and indoor measured temperature and indoor simulated temperatures by the VFD HVAC unit on the 9th of April, 2016

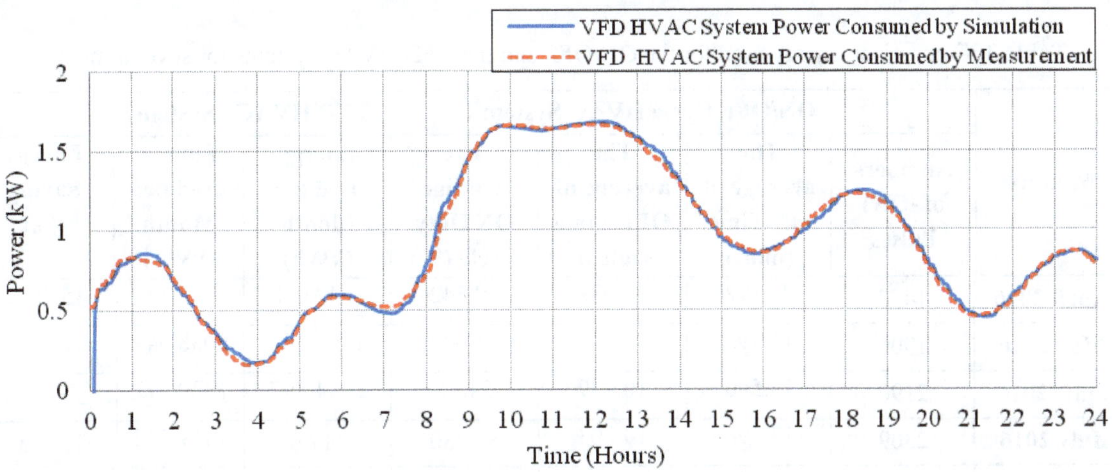

Figure 10. Measured and simulated power consumption for the VFD A/C unit on the 9th of April, 2016

Energy Analysis for Both On/Off Cycle and VFD HVACSystems

The energy consumption for the ON/OFF cycle and the VFD HVAC units is calculated by integrating the area under power consumption curve as shown in Figures 8 & 10 for typical activities (9th of April, 2016) for both simulation results and data measurement. Table 1 shows the energy consumption of both ON/OFF and VFD HAVC for one typical day.

Table 1. Energy analysis for the ON/OFF cycle and the VFD HVAC systems for one day

Day	Simulated Results			Measurement Data		
	ON/OFF Cycle Unit (kWh)	VFD Unit (kWh)	Saving %	ON/OFF Cycle Unit (kWh)	VFD Unit (kWh)	Saving %
Typical activities (April 9, 2016)	37.212	22.4	39.53	38.44	23.71	38.31

A summary of measured energy consumption for two months (April 2016 and May 2016) that is used for the validation of the simulation results is presented in Table 2. The comparison of the energy consumption for the ON/OFF cycle and the VFD HVAC units and energy savings are also presented in the Table.

Table 2. Comparison energy consumption for the ON/OFF cycle and the VFD HVAC systems for two months, 2016

Months	Simulated Results			Measurement Data		
	ON/OFF Cycle Unit (kWh)	VFD Unit (kWh)	Saving (%)	ON/OFF Cycle Unit (kWh)	VFD Unit (kWh)	Saving %
April	1042.058	571.0345	45.2012	1098.3	519.3	52.7184
May	1763.567	1038.548	41.1109	1606.11	1130.88	29.589

Monthly Variation of Energy Consumption

Table 3 presents a summary of the results of energy consumption for seven months of study. Both the ON/OFF cycle HVAC and the VFD HVAC systems are used with the same conditions. The average operation period of the ON/OFF cycle and the number of ON pulse also are displayed in the table. The energy consumption in kW is presented for both the ON/OFF cycle and the VFD HVAC systems. The hottest months, June and July, show the higher duty cycle per month. The comparison of the energy consumption for both systems is shown in Figure 11 and the energy savings per month is presented in Figure 12. Table 3 shows that a considerable amount of savings of energy occur from April 2016 to Oct. 2016.

Table 3. Comparison energy used for the ON/OFF and the VFD HVAC systems for several months

Months	ON/OFF Cycle HVAC System					VFD HVAC System	
	Numbers of (ON) Pulses	The average of ON Time (mins)	The average of OFF Time (mins)	The average ON Duty Cycle (%)	Energy used per Month (kWh)	Energy used per Month (kWh)	Energy Saving (%)
April 2016	1450	7.0172	16.982	50.342	1042.058	571.0345	45.2012
May 2016	1906	11.492	12.507	53.072	1763.567	1038.548	41.1109
June 2016	2190	13.500	10.499	54.15	2004.897	1329.972	33.6638
July 2016	2309	13.796	10.203	54.380	2117.09	1419.172	32.9659
Aug. 2016	2208	13.046	10.953	53.166	2002.028	1326.967	33.7188

Table 3. (Cont.) Comparison energy used for the ON/OFF and the VFD HVAC systems for several months

Months	ON/OFF Cycle HVAC System				VFD HVAC System		
	Numbers of (ON) Pulses	The average of ON Time (mins)	The average of OFF Time (mins)	The average ON Duty Cycle (%)	Energy used per Month (kWh)	Energy used per Month (kWh)	Energy Saving (%)
Sep. 2016	2150	11.732	12.267	53.333	1742.216	1142.922	34.3983
Oct. 2016	1540	7.9502	16.049	51.633	1219.962	756.8483	37.9613

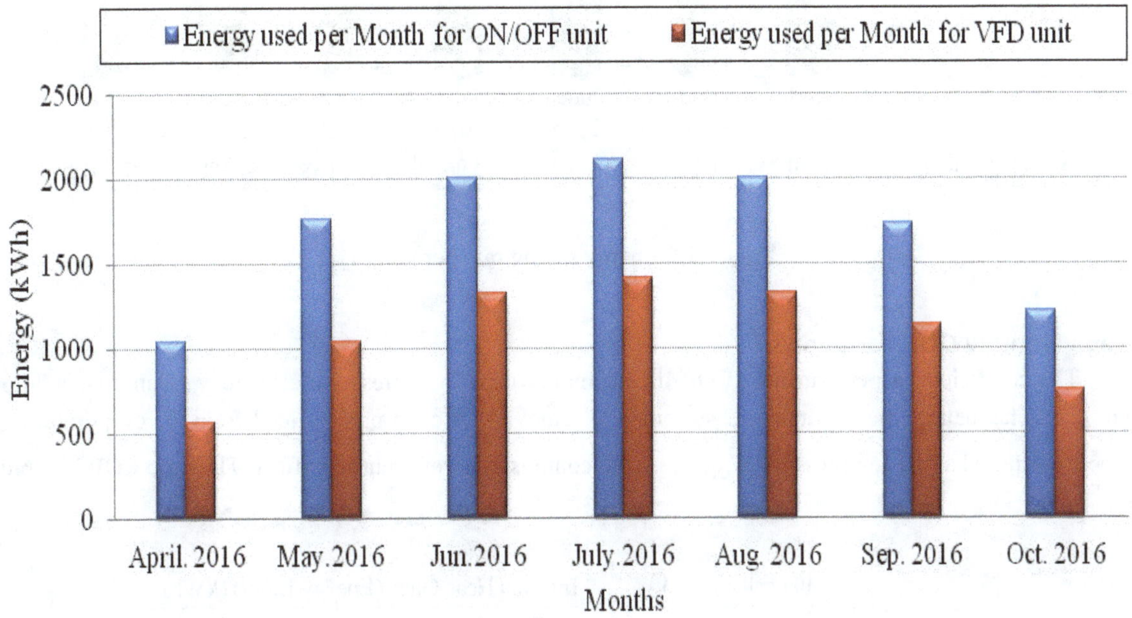

Figure 11. Energy consumption for the ON/OFF cycle and the VFD HVAC system

Monthly Energy Savings

The energy consumption is the power (*P*) multiplied with the time of operation (*t*) of the AC system [22], whereas the percentage of energy savings is calculated based on the difference between energy consumed using the ON/OFF control and energy consumed using the VFD control.

$$\text{Power} = \frac{V \times I \times PF}{1000} (\text{kW}) \tag{20}$$

where I is the current (Ampere), V is voltage (Volts), and PF is the power factor.

$$\text{Energy} = \text{Power} \times t \qquad (\text{kWh}) \tag{21}$$

$$\text{Energy Savings} = \frac{[\text{ON/OFF HVAC Energy}]-[\text{VFD HVAC Energy}]}{[\text{ON/OFF HVAC Energy}]} \times 100 \tag{22}$$

The energy savings per month is shown in Figure 12.

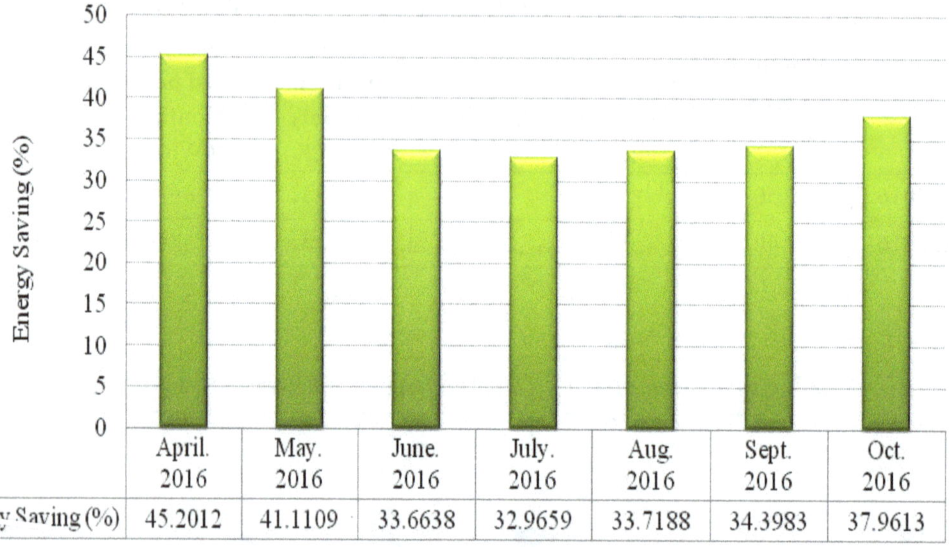

Figure 12. Energy saving rate per month

Average Monthly COP Variations

The coefficient of performance (COP) is commonly used to express the efficiency of an air-conditioning system [22]. The main purpose of the A/C system is to remove heat or to process load from the evaporator (Q_L). The energy required at the compressor (W_{com}) is to accomplish the refrigeration effect. Thus, the COP is expressed as:

$$COP = \frac{(h_1 - h_4)}{(h_2 - h_1)} = \frac{Q_L}{W_{com}} = \frac{\text{Internal Heat Gain (Energy Input) (kWh)}}{\text{Total Energy Consumption (kWh)}} \quad (23)$$

where, h_1 and h_2 (kJ/kg) are the enthalpy at the compressor inlet and that of the compressor outlet, respectively, h_4 (kJ/kg) is the enthalpy at the evaporator inlet, Q_L (kWh) is the refrigerating effect, and W_{com} (kWh) is the compression work.

The maximum theoretical COP for the air conditioning system is expressed by Carnot's theorem given by the following equation:

$$COP_{Maximum} = \frac{(T_C)}{(T_H - T_C)} \quad (24)$$

where T_C is the cold temperature and T_H is the hot temperature. The air conditioning system cools the house to 22.5 °C for (ON/OFF) and 24 °C for (VFD). If the average outdoor temperature is 35.55 °C the theoretical maximum COP is:

$$COP_{Max(ON/OFF)} = \frac{(T_C)}{(T_H - T_C)} = 1.72 \quad (25)$$

$$COP_{Max(VFD)} = \frac{(T_C)}{(T_H - T_C)} = 2.10 \tag{26}$$

The total internal heat gain for the 9th of April 2016 is calculated based on the solar heat gain, mass heat gain, and the internal house activities in both houses. On the other hand, the monthly average COP variation is calculated based in the measured energy and the internal Heat Gain (April 2016).

The daily average of internal Heat Gain for the month of April for both houses = 42.922 kWh

The daily average of energy used for the month of April by the ON/OFF HVAC unit = 36.61 kWh

The daily average of energy used for the month of April by the VFD HVAC unit = 20.82 kWh

$$COP_{(ON/OFF)} = \frac{Q_L}{W_{com}} = \frac{42.922\,kWh}{36.61\,kWh} = 1.172 \tag{27}$$

$$COP_{(VFD)} = \frac{Q_L}{W_{com}} = \frac{42.922\,kWh}{20.82\,kWh} = 2.062 \tag{28}$$

Cost Analysis of Both the On/Off and the VFD HVAC System

The payback period is calculated by counting the number of years it will take to recover the cash invested in installing the VFD HVAC unit instead of the conventional ON/OF HVAC unit. The payback period for setting the temperature to 24 °C can be calculated by the following formula.

$$\text{Payback Period} = \frac{\text{Additional Cost}}{\text{Annual Saving}} \tag{29}$$

The cost difference between the ON/OFF unit and the VFD unit is about 40% (8750SR) in which the VFD unit is higher cost than the conventional ON/OFF unit.

Based on the calculation performed in Table 4, the payback period is six years and 6 months, so if the user installs a VFD HVAC system, he can recover the extra paid money within 6 years and 6 months. During the time after the payback period, the user will start to get the benefit of having the VFD HVAC in the house as the energy consumption is going to be less than that of having an ON/OFF unit.

Table 4. Payback period for a residential HVAC

Electrical Energy Cost [23]		Annual Consumption				
Residential Sector		ON/OFF	VFD	Annual Energy Savings	Annual Cost Saving	Payback Years
kWh	Halalah					
1-2000	5	4	8734.92 kWh	34% 4481.484 kWh	4481.484 kWh* 0.3 Hala=1325.54SR	6 years & 6 months
2001-4000	10					
4001-6000	20					
6001- 8000	30					
8000 <	30					

CONCLUSIONS

The HVAC systems incorporating the ON/OFF controller and the VFD controller are considered to determine the compressor power consumptions for the HVAC unit in residential areas located at the King Fahd University of Petroleum and Minerals campus in the Kingdom of Saudi Arabia. Thermal analysis is carried out in line with the experimental conditions to assess the energy consumption of the HVAC system. The experimental set-up is developed incorporating the LabVIEW based data acquisition monitoring and measurement system. The thermal model considers different types of A/C system controller (ON/OFF, VFD) units. The predictions of the power consumptions of the HVAC system have been compared with those of the measured data. The developed mathematical thermal house model is integrated with the cooling source model of an ON/OFF cycle and VFD air conditioning systems. The developed thermal models are simulated by using Simscape physical components in Matlab/Simulink environment. In addition, the analysis pertinent to the assessment of energy consumptions, energy savings, and the cost of the payback period have been evaluated. In general, the variable frequency drive controller provides higher efficiency and low cost of A/C system. In addition, the specific conclusions derived from the present study are listed as follows:

- The energy consumptions have been examined for several months and it is demonstrated that the energy savings reach between 32% - 45% during the period from April 2016 to October 2016. It has been noticed that the performance system simulation is affected by the outdoor temperature and the indoor activities during the day. Energy savings with VFD reached 45% in a medium outdoor temperature.
- The simulation results matched significantly with the experimental data measurement with an error of about 3%.
- The daily and monthly average COP variations of actual and ideal system are calculated and the findings revealed that VFD unit has higher COP than the ON/OFF cycle.
- The cost analysis and the payback period have been calculated. It is found that the user needs only 6 years and a half to recover the additional cost. In the near future, if there is any further drop in the price of the VFD unit, it will become a catalyst in the promotion of energy conversion through the use of a VFD unit.
- The paper has shown that installing a VFD for a residential HVAC system gives considerable energy savings over the ON/OFF cycle.

ACKNOWLEDGMENT

The authors acknowledge the support by the DSR funded project RG1205-1&2 at King Fahd University of Petroleum and Minerals, Dhahran, Saudi Arabia for this research work. Also, the authors acknowledge the donation of two 5-tons A/C systems from the Al-Zamil Air-Conditioning Company, Industrial City, Dammam, Saudi Arabia.

NOMENCLATURE

H	Heat transfer coefficient, $W/m^2.K$
T_{out}	External house temperature, K
T_H	Internal house temperature, K
T_1	Internal roof temperature, K
T_2	Internal wall temperature, K
T_3	Internal windows temperature, K
C_{Air}	Specific heat of air, J/kg
C_{Roof}	Specific heat of roof, J/kg
C_{Wall}	Specific heat of wall, J/kg
C_{Wind}	Specific heat of window, J/kg
Q_{Air}	The rate of energy stored in the volumetric space of the house, J
Q_{Roof}	The rate of energy stored in the wall, J
Q_{Wall}	The rate of energy stored in the roof, J
Q_{Wind}	The rate of energy store in the windows, J
$\dot{M}_{HVAC}(t)$	Mass flow rate of supply air, kg/s
C_P	Air specific heat, J/kg. K
T_{House}	In-House temperature, K
T_{HVAC}	HVAC coil A/C temperature, K
K_0	The effective cooler gain, kg/K (default $K_0 = 1$ kg/K)
K_1	The effective time cooler gain, kg/s.K (default $K_1 = 1$ kg/s. K)
K_2	The effective HVAC gain, default $K_2 = 0.005$

K_3	The effective time HVAC gain, S^{-1}(default $K_3 = 0.9\ S^{-1}$)
$\frac{dQ_{Cooler}}{dt}$	The absorbed heat flow rate by cooler, J/s
$Q_{\substack{Heat \\ House \\ Source}}$	The heat source, J

REFERENCES

[1] Salmi, W., Vanttola, J., Elg, M., Kuosa, M., Lahdelma, R. (2017). Using waste heat of ship as energy source for an absorption refrigeration system. Applied Thermal Engineering, 115, 501-516.

[2] Aly, W. I., Abdo, M., Bedair, G., Hassaneen, A. E. (2017). Thermal performance of a diffusion absorption refrigeration system driven by waste heat from diesel engine exhaust gases. Applied Thermal Engineering, 114, 621-630.

[3] Wang, J., Wang, B., Wu, W., Li, X., Shi, W. (2016). Performance analysis of an absorption-compression hybrid refrigeration system recovering condensation heat for generation. Applied Thermal Engineering, 108, 54-65.

[4] Xu, Y., Jiang, N., Wang, Q., Chen, G. (2016). Comparative study on the energy performance of two different absorption-compression refrigeration cycles driven by low-grade heat. Applied Thermal Engineering, 106, 33-41.

[5] Sun, L., Han, W., Jin, H. (2015). Energy and exergy investigation of a hybrid refrigeration system activated by mid/low-temperature heat source. Applied Thermal Engineering, 91, 913-923.

[6] Farsi, A., Mohammadi, S. H., Ameri, M. (2017). Thermo-economic comparison of three configurations of combined supercritical CO2 refrigeration and multi-effect desalination systems. Applied Thermal Engineering, 112, 855-870.

[7] Al-Shaalan, A., Ahmed, W., Alohaly, A. (2014). Design guidelines for buildings in Saudi Arabia considering energy conservation requirements. In Applied Mechanics and Materials (Vol. 548, pp. 1601-1606). Trans Tech Publications.

[8] Salsbury, T., Diamond, R. (2000). Performance validation and energy analysis of HVAC systems using simulation. Energy and buildings, 32(1), 5-17.

[9] Muratori, M., Marano, V., Sioshansi, R., Rizzoni, G. (2012, July). Energy consumption of residential HVAC systems: a simple physically-based model. In 2012 IEEE Power and Energy Society General Meeting (pp. 1-8). IEEE.

[10] Karmacharya, S., Putrus, G., Underwood, C., Mahkamov, K. (2012, June). Thermal modelling of the building and its HVAC system using Matlab/Simulink. In 2012 2nd International Symposium On Environment Friendly Energies And Applications (pp. 202-206). IEEE.

[11] Wen, Y., Burke, W. (2013, April). Real-time dynamic house thermal model identification for predicting HVAC energy consumption. In 2013 IEEE Green Technologies Conference (GreenTech) (pp. 367-372). IEEE.

[12] Zhu, Y. (2006). Applying computer-based simulation to energy auditing: A case study. Energy and buildings, 38(5), 421-428.

[13] Zhou, D., Park, S. H. (2012). Simulation-assisted management and control over building energy efficiency–a case study. Energy Procedia, 14, 592-600.

[14] Yu, J., Yang, C., Tian, L. (2008). Low-energy envelope design of residential building in hot summer and cold winter zone in China. Energy and Buildings, 40(8), 1536-1546.

[15] Nasution, H., Dahlan, A. A., Aziz, A. A., Azmi, U., Sumeru, S., Shodiya, S. (2016). Energy Efficiency of A Variable Speed of The Centralized Air Conditioning System Using PID Controller. Jurnal Teknologi, 78(8-4).

[16] Coley, D. A., Penman, J. M. (1996). Simplified thermal response modelling in building energy management. Paper III: Demonstration of a working controller. Building and environment, 31(2), 93-97.

[17] Hamanah,W.M., (2016), "Modeling, Simulation and Energy Performance of VFD and ON/OFF Cycle HVAC Systems," M.S. thesis, KFUPM, Az Zahran, Saudi Arabia.

[18] Hamanah, W. M., Kassas, M., Mokheimer, E. M., Ahmed, C. B., Said, S. A. M. (2019). Comparison of Energy Consumption for Residential Thermal Models With Actual Measurements. Journal of Energy Resources Technology, 141(3), 032002.

[19] Ohyama, K., Kondo, T. (2008). Energy- Saving Technologies for Inverter Air Conditioners. IEEJ Transactions on Electrical and Electronic Engineering, 3(2), 183-189.
[20] Ohyama, K., Kondo, T. (2008). Energy- Saving Technologies for Inverter Air Conditioners. IEEJ Transactions on Electrical and Electronic Engineering, 3(2), 183-189.
[21] Belhadj, C. A., Hamanah, W. M., Kassas, M. (2017, June). LabVIEW based real time Monitoring of HVAC System for Residential Load. In 2017 IEEE International Conference on Computational Intelligence and Virtual Environments for Measurement Systems and Applications (CIVEMSA) (pp. 66-71). IEEE.
[22] Affandi, M. (2004). Energy Saving in an Air-Conditioning System Using an Inverter and a Temperature-Speed Controller (Doctoral dissertation, Universiti Teknologi Malaysia).
[23] Council of Ministers, (2016) Consumption Tariff, Saudi Electricity Company, No.95 https://www.se.com.sa/en us/customers/Pages/TariffRates.aspx/

8

RESPONSE SURFACE BASED OPTIMIZATION OF RIBBED ISOSCELES TRIANGULAR TWISTED TAPE HEAT EXCHANGER USING ENTROPY AUGMENTATION GENERATION NUMBER WITH AL_2O_3 NANO WORKING FLUID

K. Pavan Kumar[1*], R. Siddhardha[2], Ramabhadri Raju[3], K. Sudheer Kumar[1]

ABSTRACT

In this work, a combination of passive techniques like providing ribs on the duct surface, inserting twisted tapes were employed at different configurations and analysed using ANSYS Fluent 17.2. The enhancement is probed by placing ribs on the duct surfaces at various angles. Twisted tape inserts were used in conjunction with ribs on the duct and an output parameter, Entropy Augmentation Generation Number (EAGN) is analysed when having an Al_2O_3 nanofluid (ϕ=3%) as working medium. In furtherance, tapes of isosceles triangular projections with variable configuration, such as rib angles ($30^0<\alpha< 90^0$), Internal angle ($30^0<\beta<90^0$) and projection distance (1mm< x< 5mm) were also inserted in place of plain twisted tape inserts to verify the enhancement promised by this alteration. As expected, rate of heat transfer due to the presence of isosceles triangle projections gave decent augmentation. Later Response Surface based optimization was employed with non-parametric regression and genetic algorithm to make an investigative search of all the modified parameters so as to suggest best blend of inputs for low Entropy Augmentation Generation Number. Optimum performance was obtained at rib angle of 30^0, projection distance of 2.1mm and Internal angle of 44.4^0 with entropy augmentation number value of 0.77. The performance of Genetic Algorithm was compared with Micro Genetic Algorithm; it shows that optimized result is obtained less than half the time using Micro Genetic Algorithm.

Keywords: *Ribs, Isosceles Triangular Projections, Entropy Augmentation Generation Number, Non Parametric Regression, Genetic Algorithm, Micro Genetic Algorithm*

INTRODUCTION

In many process industries viscous fluids are used in heat exchangers. Due to low heat transfer coefficient in laminar flow, active or passive heat transfer augmentation techniques are needed to be implemented. This implementation generally reduces size and cost of heat exchanger. The additional advantage of reducing temperature of the fluid can be obtained by increasing the heat transfer coefficient, and by doing this there is a decrease in entropy generation. The pressure drop identified in a heat exchanger is one of the reasons for enhancement in internal heat generation. The combination of heat transfer and pressure drop is postulated in Entropy Augmentation Generation Number (EAGN) based on second law of thermodynamics.

The Performance of Heat exchanger can be enhanced by using twisted tape passive technique [1-8].Swirl induced heat transfer enhancement will increase the frictional losses and hence increasing the pumping power. Attempts have been made to reduce the frictional losses by several approaches [9] using loose fit twisted tape. Eiamsa-ard et.al [10] has suggested that in twisted tape heat exchanger space ratio (free space length to tube diameter) should be maintained less than unity to reduce frictional losses and get better heat transfer rates.Jaisankaret.al [11] found that spacer at trailing edge of twisted tape reduces the frictional losses. Twisted tapes along with ribs have shown better performance than either of them [12]. Saha SK et.al [13] has found that pinching of tapes in place has shown marginal increase in thermo hydraulic performance than that connecting with rods for twisted tape. Experimental Investigation of heat transfer and pressure drop characteristics are reported for square ducts with ribs for short and full length twisted

This paper was recommended for publication in revised form by Regional Editor Sandip Kale
[1] Department of Mechanical Engineering (MCT), M.G.I.T., Hyderabad,INDIA
[2] Department of Mechanical Engineering, G.V.P.S.O.E., Visakhapatnam, INDIA
[3] Department of Mechanical Engineering, S.R.K.R., Bhimavaram, INDIA
**E-mail address: pavankonchada@gmail.com*

tapes. Regularly spaced twisted-tape elements proved to be better than short-length ones when their performance is evaluated [14]. Pathipakka et.al [15] numerically investigated the heat transfer characteristics for circular tube with nano fluid. Enhancement of 31.29% was found with helical twist ratio 2.93 together with Al_2O_3nanofluid with 1.5% volume concentration at 2039 Reynolds number compared to base fluid.A significant improvement of the average Nusselt number was obtained for angled ribs in anti-periodic configuration[16].This improvement can be attributed to the formation of longitudinal vortices that mix near wall and center flows.

Longitudinal finned tube heat exchanger performance with shell side nano fluid is evaluated using the 2nd law of thermodynamics for performance evaluation and the usage of nanofluid has improved the efficiency [17].Performance evaluation of single and three start spirally corrugated tube twisted tape heat exchanger is performed using EAGN for five different pitches and varying the characteristic parameters like diameter and height of the tapes has shown that the heat transfer rate has increased[18, 19].Natural convection process is analyzed based on entropy augmentation generation number [27, 28].

From the literature review it is obvious that different geometry of twisted tape and transverse ribs can enhance the heat transfer, it is evident from above literature that analysis is carried on circular tube having helical twisted tape inserts with Al_2O_3nano fluid as working fluid [15] and ribs of different configurations [12]. But Geometric variations in helical twisted tape and ribbed ducts with nanofluid as working fluid were not present in the previous works. Therefore it is enthralled to potential in enhancement of heat transfer for newly designed twisted tape and ribs on duct with promising geometry. In the present work twisted tape with Isosceles triangular projections (ITP) inserted in a square ductwith staggered ribs are evaluated for Al_2O_3nano fluid ($\phi = 3\%$) as working fluid for Reynolds number 100-400.It is evident that the presence of Al_2O_3 nano particles in fluid decreases the EAGN [26] which indicates the augmentation of heat exchanger efficiency. The combination of twisted tape heat exchanger and entropy augmentation generation number has been less researched in conjunction with each other which has formed the impetus of the present research. The different configurationsof modified twisted tape and duct is designed to have better swirl motion along the tube which is expected to provide better mixing. The configuration optimization was performed based on entropy augmentation generation number using Neuro-Genetic Algorithm.

NUMERICAL SIMULATION

Problem Formulation

In the current work passive augmented technique employed is the change of cross section of twisted tape. It is done by reducing the thickness of tape from 2mm to 1mm, the decrement in thickness is compensated with triangular projections on tape. Different included angles and distances between projections are considered for simulation by maintaining constant surface area. In addition to modification done to twisted tape, ribs are also placed on the duct [16] for augmented geometry.

A layout of square duct with cross sectionH× S (13×13 mm)and length 62 mm with three numbers of staggered ribs of depth H/10 [16] at different anglesis considered as shown in Figure 1.Cross section of ITP twisted tape having a constant twist ratio(y/w = 2.88) with variable internal angleand projections distance is as shown in Figure 2.The Three independent parameters considered for the present work are

1) Rib angle ($30^0 < \alpha < 90^0$)
2) Internal Angle ($30^0 < \beta < 90^0$)
3) Projection Distance (1mm< x< 5mm)

Mathematical Model

A three dimensional numerical simulation of entropy augmentation generation was conducted by ANSYS FLUENT 15.The Governing differential equation 'Continuity equation' derived from conservation of mass principle, the 'Momentum or Navier-Stokes equation' derived from conservation of momentum principle and the 'Energy equation' derived from conservation of energy principle were the three conservation principles used in building the mathematical model. These fundamental equations along with appropriate initial conditions and boundary conditions were solved to understand various fluid and heat flow characteristics prevailing in the heat exchanger.

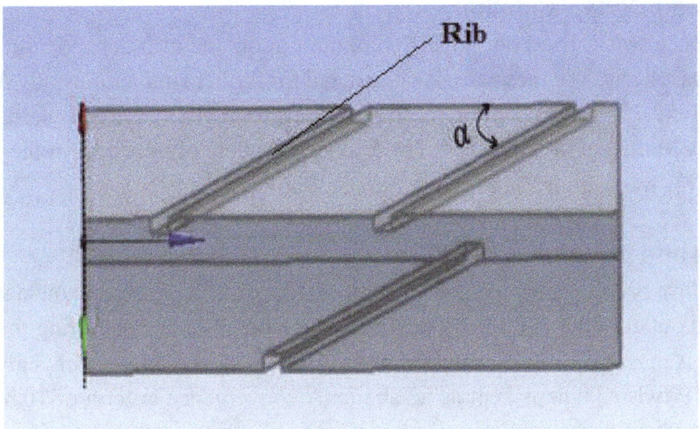

Figure 1. Modelling of duct

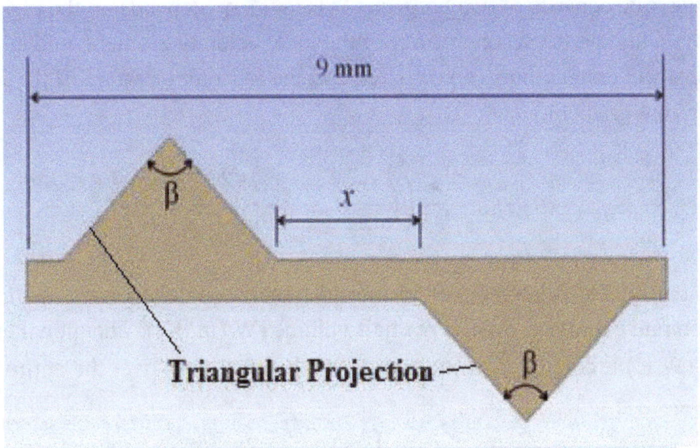

Figure 2. Cross section of twisted tape

The continuity equation given for steady state flow of a compressible fluid is

$$\frac{\partial(\rho U_i)}{\partial x_i} = 0 \tag{1}$$

The momentum equation given for steady state flow of a compressible fluid is

$$U_j \frac{\partial(\rho U_i)}{\partial x_j} = -\frac{\partial P}{\partial x_i} + \frac{\partial}{\partial x_j}\left((\mu).\frac{\partial(\rho U_i)}{\partial x_j}\right) \tag{2}$$

The energy equation given for steady state flow of a compressible fluid is

$$C_p U_j \frac{\partial(\rho T)}{\partial x_j} = \frac{\partial}{\partial x_j}\left((k).\frac{\partial(\rho T)}{\partial x_j}\right) \tag{3}$$

Nanofluid used in the simulation is considered as single phase with equivalent thermo physical properties [17] of two phase fluids. These thermo physical properties were hooked as UDFs to the code.Discretization of convective term is done using the QUICK scheme and pressure velocity coupling by SIMPLE scheme.

Mesh Independence and Code Validation

Mesh independence was carried out for different configurations of independent parameters α, β and xwith average Nusselt number as criteria. The mesh size is considered for further simulation study only if the Nusselt number variation is less than 1% for 9,25,768 tetrahedron mesh size. Code used in this study is validated with theexperimental correlations developed by Mallick and Saha [14]. The Nusselt numbers obtained from the Ansys Fluent 17.2 are of the same order and the error was within 5% compared to correlations.

Boundary Conditions and Solution Procedure

A constant heat flux 10000 w/m^2 was applied on ribbed duct wall and no slip boundary condition was imposed on the wall. The inlet and outlet of the ITP ribbed duct was made periodic by converting them into interface to make the flow fully developed. A mass flow rate corresponding to different Reynolds number is given as input periodic flow condition. The wall of ITP twisted tape is kept as adiabatic. A convergence criteria of 10^{-6} is been set forcontinuity, momentum and energy equation.

Entropy Augmentation Generation Number

Entropy generation rate consists of two parts one "viscous" $\dot{s_v}$,depends on the physical viscosity, on the local temperature of the fluid and on the second power of the local velocity gradient, and another, called "thermal" $\dot{s_T}$, that depends on the physical conductivity, on the square of the local temperature of the fluid and on the second power of the local temperature gradient:

$$\dot{s_{gen}} = \dot{s_T} + \dot{s_v} = \frac{k}{T^2}\left(\left(\frac{\partial T}{\partial x}\right)^2 + \left(\frac{\partial T}{\partial y}\right)^2 + \left(\frac{\partial T}{\partial z}\right)^2\right) + \frac{\emptyset}{T} \tag{4}$$

In our case flow considered is laminar so the entropy generation due to viscous effects is neglected. Note that the entropy generation rate expressed in (4)is per unit volume (W/(m^3 K)). The global entropy generation rate $\dot{S_{gen}}$of the entire domain (W/K) is computed as the integral of the local rates over the entire volume

$$\dot{S_{gen}} = \int_v \dot{s_{gen}} dv \tag{5}$$

Bejan and Pfister [20] proposed that the merit of a given heat transfer augmentation technique may be evaluated by comparing the rate of entropy generation of the heat exchange apparatus before and after the implementation of the augmentation technique. Entropy generation augmentation number is given by [21- 23]:

$$N_{s,a} = \frac{\dot{S_{gen,a}}}{\dot{S_{gen,o}}} \tag{6}$$

For $N_{s,a}<1$ passive techniques employed are thermodynamically advantageous augmentation technique because, in addition to enhancing heat transfer, they reduce the degree of irreversibility of the apparatus.

OUTLINE OF RESPONSE SURFACE BASED OPTIMIZATION

Non-Parametric Regression

Non-parametric regression (NPR) is analgorithm which is initiated by any one of the available Design of Experiments (DOE) and can produce enhanced response quality. For present work, this NPR algorithm of ANSYS Design Xploreris used so as to handle higher non-linear behavior outputs with inputs. This technique uses hyper planes to segregate data groups and thus belongs to Support Vector Method (SVM) type techniques. Although the regression method works almost similarly, it differs in hyper plane which is used to categorize a subset (also called "support vector" set) of the input sample vectors which are adequatelyestimated to represent the output in question. In the current version, the internal parameters of the meta-model are fixed to constant values and are not optimized.

Genetic Algorithm

The technique of Genetic algorithms (GA) is based onevolutionary ideas of natural selection and genetic [24]used for algorithms that take investigative search. The ideology of this algorithm is to rely on chromosome-like data structures for fetching out a possible and appropriate solution to the problem and to apply genetic operators to these structures accordingly [25].The algorithm starts with an initial set of populationencoded inbinary values(0s and 1s) and then decides the worth of a particular member by bringing its objective function value to comparison with that of other members in the population.This comparison is termed as fitness function and later to this; the three basic genetic operators viz., reproduction, crossover and mutation were performed. Reproduction is the selection procedure based on the fitness function of the members to mate them for generation of offsprings. Here members of higher fitness function were more likely to get selected and mated in accordance with Darwinian principal of survival of the fittest.

In crossover, the offsprings from reproduction were allowed to get partial alterations in their strings for generation of new strings under certain crossover probability.In order to bring disparity in the population, and to check the nearby points of the current solution, Mutation of population is exercised. This ensures that the local points surrounding the offsrings were also considered due toinduced mutation probability giving room for variability in population.Population size and the crossover fraction considered in this study is 100 and 0.8 respectively.

Micro Genetic Algorithm

The Micro-Genetic Algorithm (μGA) practices similar approach as that of Genetic Algorithm but for the population size. In μGA, the population size is usually very less than Genetic Algorithm so that the computational time demanded for best fit solution is lessened. The population size considered in this work is 5 for each generation.

RESULTS AND DISCUSSIONS

The local entropy generation as indicated by the contours of Plain twisted tape heat exchanger at its mid spans'transverse section(Figure 3) has higher range of values manifestedto greater extent of cross-section when compared to ribbed ITP twisted tape heat exchanger (Figure 5).

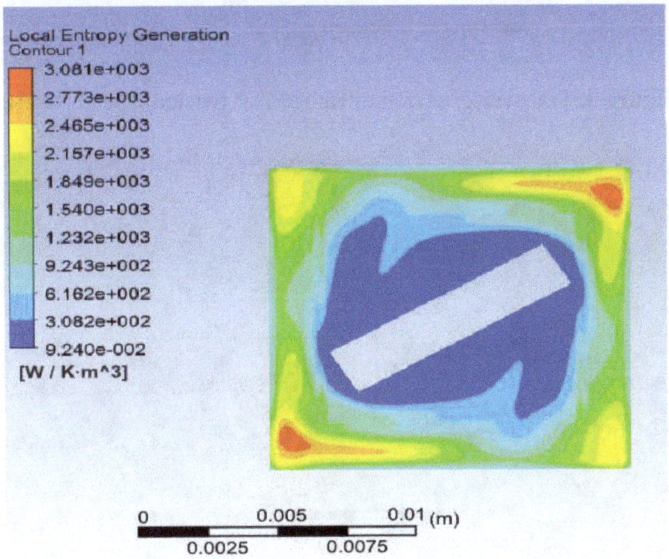

Figure 3. Transverse section of Plain twisted tape at Re=100

The longitudinal section of Plain twisted tape heat exchanger portraits higher range of entropy generation values along the length at various sections (Figure 4) and the influence of higher entropy generation is felt to greater extent when compared to ribbed ITP twisted tape heat exchanger (Figure 6). The presence of ribs and projections of twisted tape ensure quick response in terms of swirl and thereby its influence surpassed the effect of shear and strain.

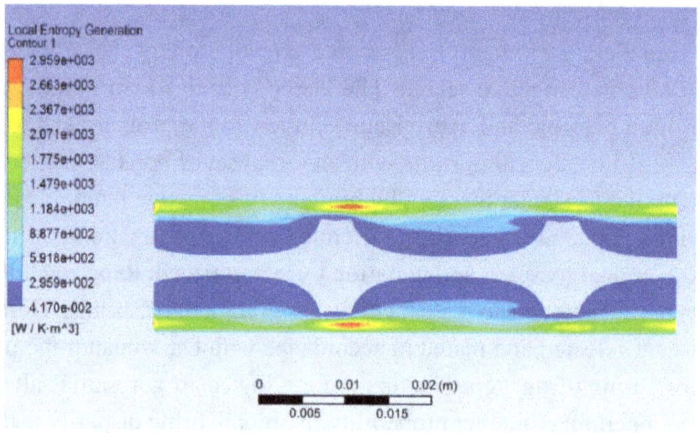

Figure 4. Longitudinal section of Plain twisted tape at Re=100

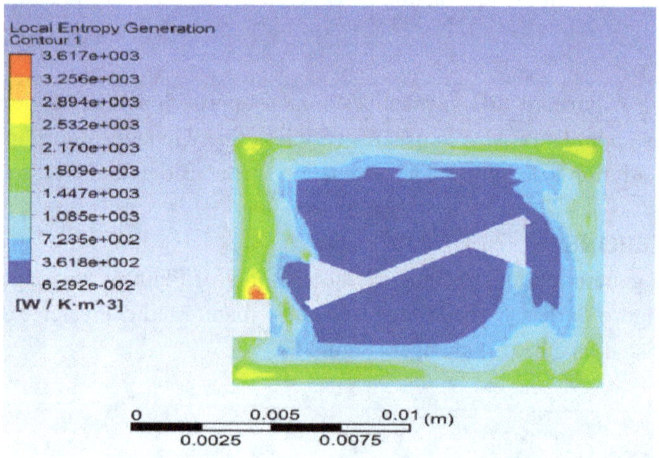

Figure 5. Transverse section of ribbed ITP twisted tape at Re=100

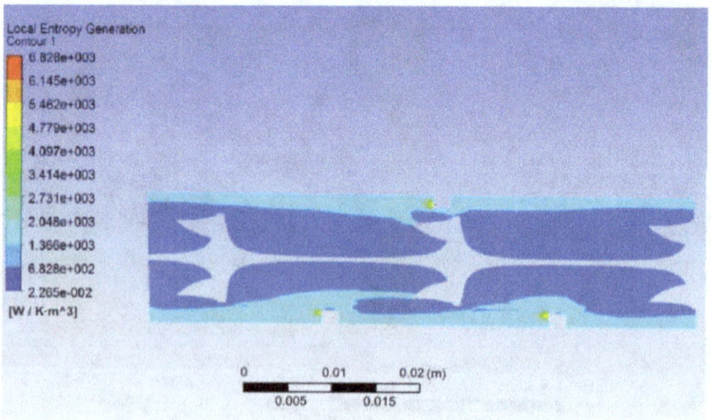

Figure 6. Longitudinal section of ribbed ITP twisted tape at Re=100

Entropy Augmentation Generation number is low for lesser rib angle (Ns,a = 0.8 for $\alpha = 60^0$,Ns,a= 0.815 for $\alpha = 90^0$ at Re =100) by keeping remaining parameters constant due to the intense mixing of fluid as shown in Figure7.As the Rib angle decreases there is enhanced mixing of the fluid and more amount of heat will be extracted and get dispersed uniformly in the working fluid. Thus EAGN value decreases as the temperature gradient decreases.

It can also be observed that as Reynolds number increases decrement in Ns,a is higher which indicates augmentation of efficiency of heat exchanger as Re increases.

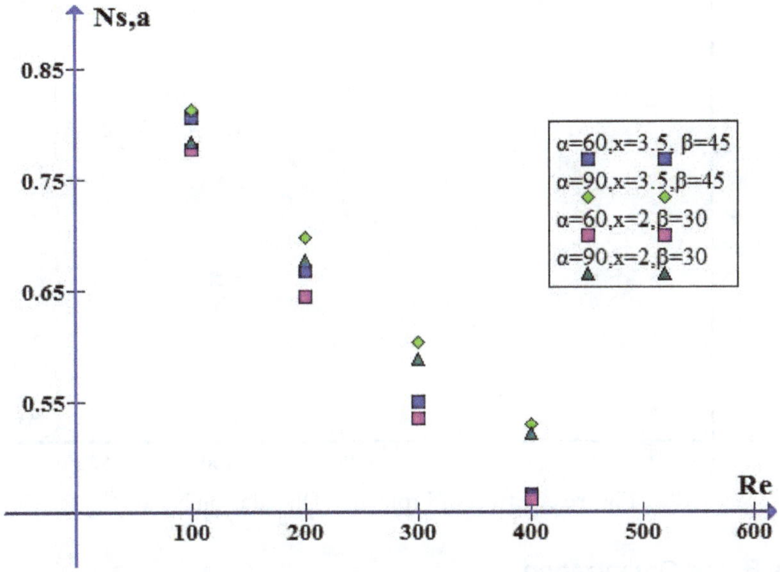

Figure 7. Effect Of Rib Angle on Ns,a

Variation in Internal angle don't have the profound effect on the Ns,a, however for smaller internal angle Ns,a decreases with increase in Re compared to higher angles. The above said is due to slight enhancement in intensity ofmixing (Figure 8) at higher Re values. It is evident from Figure 9 that projection distance have great impact on Ns,a at lower Reynolds number(Ns,a= 0.78 at x= 2 and Ns,a=0.81 at x=3.5 for Re=100) and they are overlapping at Higher Reynolds number. It can also be noted that lower projection distance ensures better performance.

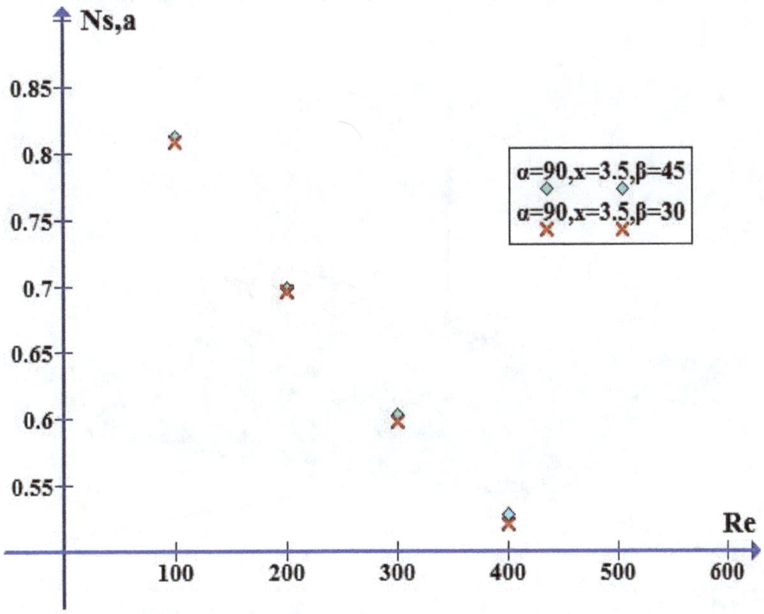

Figure 8. Effect of Internal Angle on Ns,a

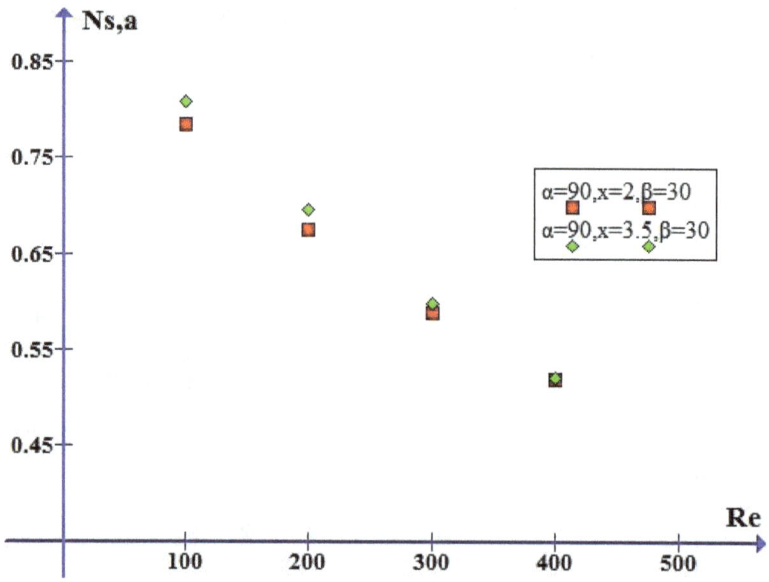

Figure 9. Effect of Projection Distance on Ns,a

Response Surface Based Optimization

A total of 50 simulation results for Reynolds number 100 and volume fraction of nanofluid (ϕ=3%)are fed into NPR by varying Rib angle, internal angle and Projection distance within considered range for response surface generation (Figure 11 and 12) in Ansys Design Xplorer. Figure 11 shows the combined interaction of projection distance on the twisted tape and rib angle of duct on Entropy augmentation generation number where the internal angle is kept at zero level ($\beta=60^0$). It can be observed that EAGN decreases as both projection distance and rib angle decreases. Maximum EAGN is observed at 4 mm projection distance and 90° rib angle and minimum is observed at in between 2 and 2.5 mm projection angle and 30° rib angle.As shown in Figure 13 it can be observed that at projection distance of 2mm and internal angle around 40^0 gives the least EAGN on keeping the rib angle at zero level ($\alpha=60^0$).

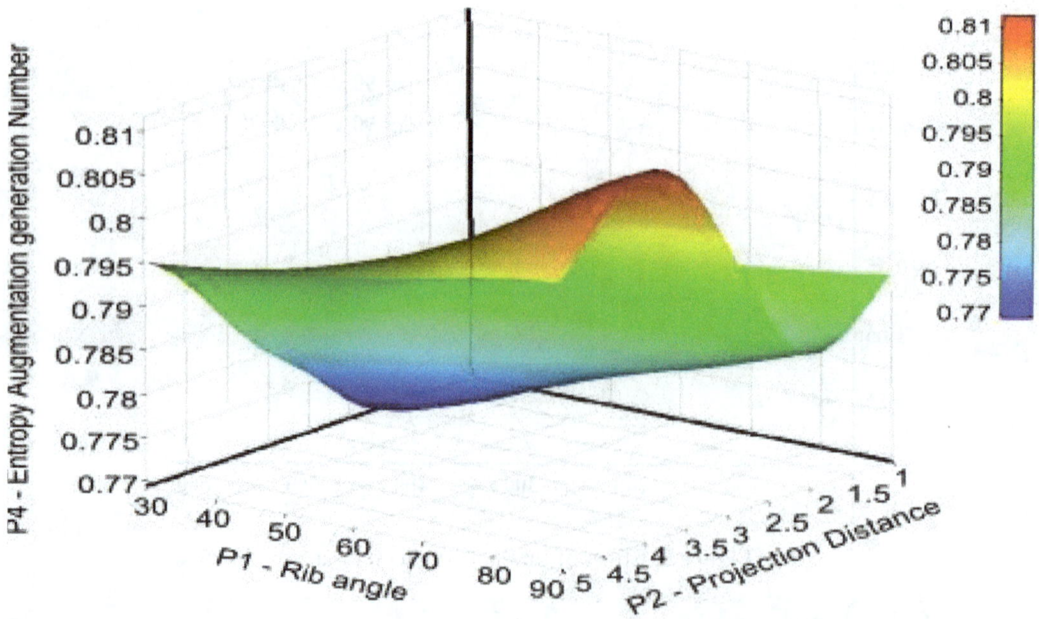

Figure 11. Response Surface For Rib angle and Projection Distance

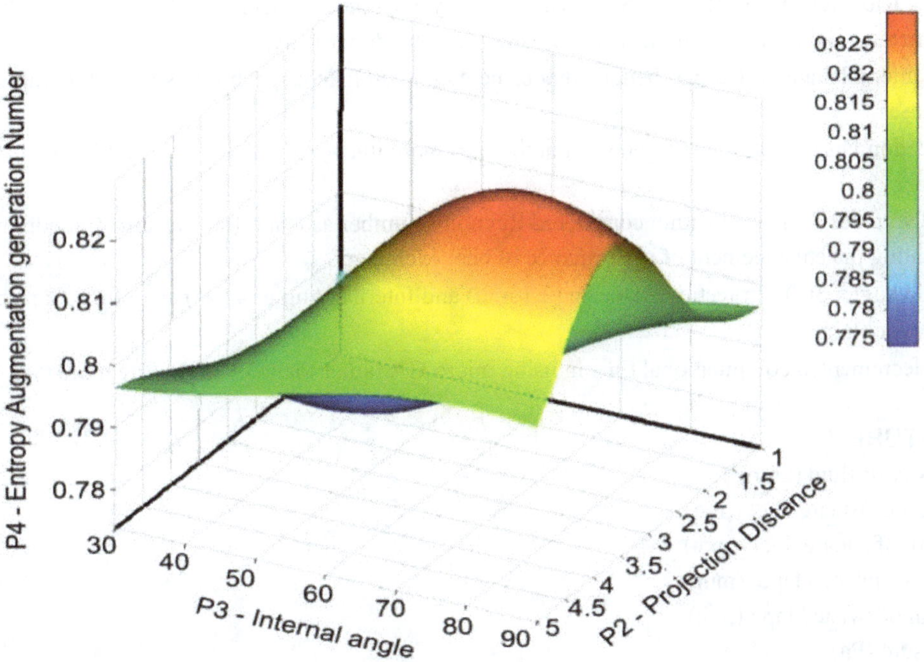

Figure 12. Response Surface For Internal angle and Projection Distance

Results from µGA and GA integrated with Response Surface Generated by NPR

The objective function is Minimization

$$N_{s,a} = f(\alpha, \beta, x) \tag{7}$$

Subjected to constraint

$$\frac{\sqrt{(18 sin\beta)}}{\cos(\beta/2)} + x - 9 \leq 0 \tag{8}$$

From the above NPR response surface 10000 points are fed into Genetic Algorithm of Response surface based optimization of the objective function.

The optimum values of Entropy Augmentation generation number value obtained from GA and Micro GA is 0.766 and 0.771. The computational time has been reduced by more than half in micro GA with less variation in optimum values.

Table 1. The specified value of GA and Micro GA

Type	RibAngle	Projections Distance (mm)	Internal Angle	$N_{s,a}$	Computational Time
GA	30.1^0	2.1097	44.4^0	0.766	18
Micro GA	30.643^0	2.0007	44.371^0	0.771	7

CONCLUSION

The following are the conclusions drawn from the above study:

- Enhancement of performance escalates with increase in Reynolds number

- As the Rib angle decreases EAGN decreases for any Reynolds number for without any variability in other parameters, and EAGN decrements more as Reynolds number increases
- Low internal angle in ITP has better enhancement and EAGN changeability is low as Reynolds number increases
- Projection Distance at lower Reynolds number has more impact on EAGN compared to higher Reynolds number
- EAGN values decreased within considered Reynolds number as Nano fluid volume fraction increased indicating the enhancement of performance of heat exchanger.
- At Rib angle(30^0) ,Projection Distance (2.1 mm) and Internal Angle (44.4^0) an optimized result of $N_{s,a}$ = 0.766 is obtained
- The decremented computational time in using micro GA rather than GA is slightly more than half

NOMENCLATURE

ρ	Density of fluid (kg/m^3)
H	Height of square duct (mm)
S	Width of square duct (mm)
Y	Pitch of twisted tape (mm)
w	Width of twisted tape (mm)
P	Pressure (Pa)
k	Thermal ConductivityW / m °C
U_i	Velocities in different Cartesian direction (m/s)
ϑ	Kinematic viscosity (m^2/s)
α	Thermal Diffusivity (m^2/s)
c_p	Specific Heat at Constant Pressure (kJ/kgK)
$\dot{s}_{gen}$	Local entropy generation number (W/m^{3}K)
$\dot{s}_T$	Entropy Generation Due to heat temperature Gradient(W/m^{3}K)
$\dot{s}_v$	Entropy Generation due to Viscosity(W/m^{3}K)
$\emptyset$	Viscous Dissipation Rate (W/m^3)
$N_{s,a}$	Entropy augmentation generation number
NPR	Non Parametric Regression
GA	Genetic Algorithm
µGA	Micro Genetic Algorithm
Re	Reynolds Number
EAGN	Entropy Augmentation Generation Number
T	Time (s)

REFERENCES

[1] Sarma, P. K., Kishore, P. S., Rao, V. D., Subrahmanyam, T. (2005). A combined approach to predict friction coefficients and convective heat transfer characteristics in A tube with twisted tape inserts for a wide range of Re and Pr. International Journal of Thermal Sciences, 44(4), 393-398.

[2] Eiamsa-ard, S., Thianpong, C., Promvonge, P. (2006). Experimental investigation of heat transfer and flow friction in a circular tube fitted with regularly spaced twisted tape elements. International Communications in Heat and Mass Transfer, 33(10), 1225-1233.

[3] Eiamsa-Ard, S. (2010). Study on thermal and fluid flow characteristics in turbulent channel flows with multiple twisted tape vortex generators. International Communications in Heat and Mass Transfer, 37(6), 644-651.

[4] Eiamsa-Ard, S., Promvonge, P. (2007). Heat transfer characteristics in a tube fitted with helical screw-tape with/without core-rod inserts. International Communications in Heat and Mass Transfer, 34(2), 176-185.

[5] Sundar, L. S., Sharma, K. V., & Ramanathan, S. (2007). Experimental investigation of heat transfer enhancements with Al2O3 nanofluid and twisted tape insert in a circular tube. International Journal of Nanotechnology and Applications, 1(2), 21-28.
[6] Sivashanmugam, P., Suresh, S. (2007). Experimental studies on heat transfer and friction factor characteristics of turbulent flow through a circular tube fitted with regularly spaced helical screw-tape inserts. Applied Thermal Engineering, 27(8-9), 1311-1319.
[7] Eiamsa-ard, S., Pethkool, S., Thianpong, C., & Promvonge, P. (2008). Turbulent flow heat transfer and pressure loss in a double pipe heat exchanger with louvered strip inserts. International Communications in Heat and Mass Transfer, 35(2), 120-129.
[8] Kumar, A., Prasad, B. N. (2009). Enhancement in solar water heater performance using twisted tape inserts. National Journal of the Institution of Engineers, 90, 6-9.
[9] Al-Fahed, S., Chakroun, W. (1996). Effect of tube-tape clearance on heat transfer for fully developed turbulent flow in a horizontal isothermal tube. International journal of heat and fluid flow, 17(2), 173-178.
[10] Eiamsa-ard, S., Thianpong, C., Promvonge, P. (2006). Experimental investigation of heat transfer and flow friction in a circular tube fitted with regularly spaced twisted tape elements. International Communications in Heat and Mass Transfer, 33(10), 1225-1233.
[11] Jaisankar, S., Radhakrishnan, T. K., Sheeba, K. N. (2009). Experimental studies on heat transfer and friction factor characteristics of thermosyphon solar water heater system fitted with spacer at the trailing edge of twisted tapes. Applied Thermal Engineering, 29(5-6), 1224-1231.
[12] Pramanik D., Saha SK (2006).Thermohydraulics of laminar flowthrough rectangular and square ducts with transverse ribs andtwisted tapes. Trans ASME J Heat Transfer, 128, 1070–1080.
[13] Saha, S. K., Dutta, A., Dhal, S. K. (2001). Friction and heat transfer characteristics of laminar swirl flow through a circular tube fitted with regularly spaced twisted-tape elements. International Journal of Heat and Mass Transfer, 44(22), 4211-4223.
[14] Saha, S. K., Mallick, D. N. (2005). Heat transfer and pressure drop characteristics of laminar flow in rectangular and square plain ducts and ducts with twisted-tape inserts. Journal of heat transfer, 127(9), 966-977.
[15] Pathipakka, G., Sivashanmugam, P. (2010). Heat transfer behaviour of nanofluids in a uniformly heated circular tube fitted with helical inserts in laminar flow. Superlattices and Microstructures, 47(2), 349-360.
[16] Nonino, C., Comini, G. (2002). Convective heat transfer in ribbed square channels. International Journal of Numerical Methods for Heat & Fluid Flow, 12(5), 610-628.
[17] Konchada, P., Pv, V. Bhemuni, V. (2016). Statistical analysis of entropy generation in longitudinally finned tube heat exchanger with shell side nanofluid by a single phase approach. Archives of Thermodynamics, 37(2), 3-22).
[18] Zimparov, V. (2001). Enhancement of heat transfer by a combination of three-start spirally corrugated tubes with a twisted tape. International Journal of Heat and Mass Transfer, 44(3), 551-574.
[19] Zimparov, V. (2002). Enhancement of heat transfer by a combination of a single-start spirally corrugated tubes with a twisted tape. Experimental Thermal and Fluid Science, 25(7), 535-546.
[20] Bejan A., Pfister P.A. J r (1980). Evaluation of heat transfer augmentation techniques based on their impact on entropy generation.Lett. Heat Mass Transfer, 7, 97–106.
[21] Bejan A. (1982).Entropy Generation through Heat and Fluid Flow, Wiley, New York.
[22] Hesselgreaves J.E. (2000).Rationalisation of second law analysis of heat exchangers.Internat. J. Heat Mass Transfer, 43, 4189–4204.
[23] Zimparov V. (2001). Extended performance evaluation criteria for enhanced heat transfer surfaces: heat transfer through ducts with constant heat flux. International J. Heat Mass Transfer, 44, 169–180.
[24] Golberg, D. E. (1989). Genetic algorithms in search, optimization, and machine learning. Addion wesley, 1989(102), 36.
[25] D. Whitley (1993). A genetic algorithm tutorial.Colorado State University Technical Report CS-93-103.
[26] Razeghi, A., Mirzaee, I., Abbasalizadeh, M. et al. J Braz. Soc. Mech. Sci. Eng. (2017) 39: 2307

[27] Al-Rashed , A. , Kolsi , L. , Hussein , A. K. , Hassen , W. , Aichouni , M. and Borjini. (2017). M. Numerical study of three-dimensional natural convection and entropy generation in a cubical cavity with partially active vertical walls , Case Studies in Thermal Engineering , 10,100-110.
[28] Kolsi , L. , Hussein, A.K. , Borjini, M. , Mohammed , H. and Ben Aïssia ,H.(2014). Computational analysis of three-dimensional unsteady natural convection and entropy generation in a cubical enclosure filled with water-Al2O3 nanofluid , Arabian Journal for Science and Engineering , 39 ,7483-7493.

9

DETERMINATION OF SOME DOMESTIC RADIATORS' THERMAL CAPACITY NUMERICALLY

A. S. Dalkilic[1,*], O. Acikgoz[1], E. Ekici[1], S. Wongwises[2]

ABSTRACT

Free convection and radiation comprise the heat transfer mechanisms through which a hydronic household radiator conveys heat from its surface to air and surrounding surfaces. It should also be noted that their performance could be enhanced by improving surface geometries as well as increasing temperature levels. In the present study, heat transfer rates and convective heat transfer coefficients occurring through the investigated radiators, were numerically examined. To this end, radiators at two different dimensions having two different geometric shapes were drawn and analyzed in the program Ansys 17. The heat transfer rates obtained from the program were validated via radiator producer catalogues. Furthermore, the influence of parameters, such as water velocity in the radiators and thus mass flow rate, temperature difference between water inlet and outlet and also between radiator surface and surrounding air on convective heat transfer coefficient over radiator, were scrutinized.

***Keywords**: Radiator, Heat Transfer, Natural Convection, Radiation, ANSYS*

INTRODUCTION

In today's world energy economy agenda, high energy consumption in buildings, has accounted for the requirement for emerging low-temperature heating devices both in new and old ones. Also, energy saving can be achieved by diminishing supply temperatures and radiant panel heating systems that encompass floor heating, ceiling heating and wall heating in central heating systems.

Nonetheless, improving the performance of central heating in buildings has a significant effect on energy savings, as well. Furthermore, heating by radiators has still been the most prevalent technique for domestic and industrial applications in the world. In the usual design of radiators, the hot water is circulated in the narrow radiator channel and heat is conveyed to the ambient air, and the circulated water in the radiator leaves at a lower temperature. Convectors support to rise the heat transfer from the radiators. Computational Fluid Dynamics (CFD) is one of the significant technique to design and improve the performance of the radiator numerically.

Improving the heat capacity of hydronic central heating systems in buildings can be the core point in terms of energy saving. Existing radiators of these systems have working conditions at constant flow plan with thermostat control device mostly. This sort of operating mode may not be evaluated efficient with regard to energy consumption and thus other working scenarios are necessary to improve the heat capacity of the radiators. Given the issues relevant to having better thermal efficiency from radiators, the literature have been thoroughly reviewed and presented in the following paragraphs.

By means of a particle image velocimetry along with a computational fluid dynamics approach, Calisir et al. [1] analyzed the flow of the air over a panel radiator. The inlet and outlet temperatures of the radiator were selected as 75°C and 65°C, respectively. Experiments were carried out on a radiator of which dimensions were 600x1000 mm, under a stable laboratory environment with a room temperature of 20°C. In conclusion, at discrete parts of the radiator, velocity distribution was obtained and figured out that a non-symmetrical velocity distribution took place over the radiator.

This paper was recommended for publication in revised form by Regional Editor Bekir Yilbas

[1]Department of Mechanical Engineering, Yıldız Technical University, İstanbul, Turkey
[2]Department of Mechanical Engineering, King Mongkut's University of Technology and Thonburi, Bangkok, Thailand
[*]Corresponding Author: dalkilic@yildiz.edu.tr

Embaye et al. [2] aimed to study the influence of panel radiator under pulsed flow conditions in order to have indoor spatial temperature and velocity distributions. They stated that they have succeed energy saving with thermal comfort given by international standards. They benefitted from CFD technique for constant and pulsed flows and validated their results with data in open sources.

Ludumor et al. [3] reviewed automotive engine cooling systems indicating the importance of engine surface temperature regarding with the engine ideal efficiency. They have given latest information on forced engine cooling systems to progress new approaches to increase its efficiency. Parameters affecting the radiator performance from the experimental and numerical works have been summarized briefly.

Calisir et al. [4] aimed to see the influences of the sizes of convectors on heat transfer mainly. Initially, the current convector sizes used in a produced radiator were applied in the code of simulation work. Then, the influences of various sizes about the heat transfer were studied. The primary aim of their work was to investigate an optimum arrangement to have the maximum heat output. The space between two convectors, dimension of the base of the convectors and their tip width was examined, and the maximum heat transfer was achieved for 6.31 mm, 4 mm and 12 mm, correspondingly. It was detected that decreasing the convector height and length has an increasing influence about heat transfer performance. A slight rise in heat transfer was obtained with rising convector thickness, and space between two opposite fins improved the heat transfer to a specified value.

Kayastha [5] used a 55 HP engine radiator data for the numerical analysis. He prepared a model in Pro-E software and solved it in ANSYS-12. Two different pitches around 15 mm and 20 mm are used for helical tubes in the radiator. The comparison is completed for dissimilar mass flow rates in helical type tubes. According to the analyses, maximum temperature drop and minimum pressure drop happen at the mass flow rate of 0.5 kg/sec. It is determined that there is increase in temperature drop and decrease in pressure drop with decreased mass flow rate.

Embaye et al. [6] focused on the effect of pulsed flow parameter on the energy consumption of panel radiators. They modeled two hydronic panel radiators with constant and pulsating flows by CFD technique using the conjugate heat transfer in COMSOL Multiphysics program. The tested radiators were one with single finned surface and one without fins, their sizes were 500 mm long and 300 mm high. They validated their CFD model of the constant flow ones with experimental data in literature successfully. Then, they stated that their simulation results of pulsed flow conditions can decrease the energy consumption of panel radiators in comparison to constant flow ones.

Johansson and Wollerstrand [7] studied on the increase of convection in order to have larger heat output. Their numerical works include the comparison of standard panel radiators with and without add-on-fan blower. They prepared their model by COMSOL software using a 2D-model and the multi physic mode. They tested their model at different temperature levels in the radiator, and at different fan speeds. They aimed to derive new temperature software for the radiator at various fan speeds.

Shi et al. [8] indicated the significance of the optimum design of plate-type radiators. They tried to change the shape of the radiator. They obtained larger efficiencies with the angles of 13-20° for inlet sides of ducts. Their study includes not only numerical simulations, but also it has experimental parts.

Myhren and Holmberg [9] focused on radiator heat output and comfort temperatures in a small office room. They have made tests using different positions for the ventilation air inlet. Their numerical works were performed by CFD software's simulations including visualization of thermal comfort situations. Their conclusions indicated that more stable thermal climate conditions can be obtained by ventilation-radiators under used parameters than the traditional ones. Moreover, they obtained lower radiator surface temperatures by the use of ventilation-radiators and thus energy and environmental savings were able to be obtained.

Sarbu and Sebarchievici [10] studied the thermal performance of various kinds of low temperature heating systems both experimentally and numerically. They compared the heat capacity and comfort of radiator and radiant floor heating systems joined to a ground-coupled heat pump. They prepared a mathematical model with the validation process for the numerical analyses of radiant floors. Furthermore, their comparison work included the energy, environmental and economic performances of floor, wall, ceiling and floor-ceiling heating. They concluded that floor-ceiling heating systems were better than other low-temperature ones relating with having better thermal comfort, lower energy consumption, lower CO_2 emission and lower operating cost.

Aydar and Ekmekci [11] have examined panel radiators in Turkey through CFD codes in three dimensional spaces. They acquired numerical efficiency data and afterward the data were compared with extant catalogue values of panel radiators. Optimum air-side convective heat transfer value was found and as a result of that, computational calculations were carried out in accordance with the determined value.

Chacko et al. [12] issued the airflow distribution in the radiator cover using CFD techniques. A remarkable number of optimization case studies were conducted to determine the optimum configuration of the radiator cover. CFD studies compared with test data have demonstrated that notable fields of re-circulation flows are present in the radiator cover. The optimization provided by means of the CFD work has ensured those recirculation regions to be averted and have augmented the flow within the radiator around 34%.

Sevilgen and Kilic [13] implemented a numerical investigation within a room heated through two radiators, while a virtual manikin having actual dimensions as well as physical shape similar to a real person was sitting. In conclusion, the authors have determined heat interaction between the surfaces of the manikin and the room environment, the local heat transfer coefficients of the manikin and the surroundings etc.

Shati et al. [14] have examined the influence of emissivity and roughness of the surface behind a radiator over the thermal output of the radiator. The numerically and experimentally obtained data have shown that the presence of large scale surface roughness and a high emissivity value enhance not only the heat output, but also the air velocity behind the radiator if it is compared to a smooth surface. The data also demonstrate that the heat transfer may be augmented approximately 26% using a high emissivity saw tooth surface.

Arslanturk and Ozguc [15] developed an analytical model to assess the optimum dimensions of a central heating radiator. They have calculated the optimum geometry that maximizes the heat transfer rate as well as the geometrical limits relevant with manufacturing techniques. Also, the influences of geometrical and thermal parameters over the radiator's efficiency were given.

Menendez-Diaz et al. [16] investigated stoneware panel coverings which are claimed that they enhance the thermal efficiency of radiators. To implement this aim, they carried out a theoretical and experimental work. The results have indicated that within the cooling period, the stoneware panel temperature is a little higher than the corresponding temperature of the aluminum radiator surface of the radiator. Nonetheless, aftermath of the start of cooling, this difference has a tendency to vanish. After 50 minutes cooling has started, the difference was found lower than 2°C.

Brady et al. [17] studied the effect of magnetic decorative covers on the heat output from a radiator. A significant number of case studies were conducted and the heat output a bare radiator was compared with a radiator applied under a magnetic cover, and within a wooden cover. The obtained results have indicated that magnetically applied radiator has a higher efficiency roughly 13-20%, compared to traditional radiator wooden cover.

Kılıç et al. [18] have utilized a CFD approach in order to do the thermal analysis of a steel panel radiator. The study was done according to TS EN442, and the CFD application was executed using finite volume method. It should also be noted that the case studies were implemented under steady-state conditions. As a result, it has been observed that the acquired numerical data are consistent with the relevant experimental data in the literature.

By means of a 3D finite volume CFD code, Jahanbin and Zanchini [19] examined the performance of a thin plane radiator in a real-size room. The case studies they carried out were verified by comparing the mean Nusselt number on the radiator surface with the one attained through the equations derived by some other researchers' in the literature. Using the code, the temperature and velocity fields within the enclosure, the total output of the radiator and the operative temperature have been obtained. As a conclusion, it was understood that a range of 10 cm between radiator and wall provided a small enhancement in air movements within the room.

Beck et al. [20] worked on the effect of wall emissivity behind panel radiators over the heat output. The study was implemented by means of both experimental and numerical methods. The results indicate that the heat transfer can be augmented roughly 20% with the usage of a black wall, instead of a reflective wall. In addition to this, it was stated that the output of single plate radiator would be enhanced approximately 10% and a double radiator roughly 5%.

Furthermore, in his articles Khalifa [21, 22] have scrutinized and presented the convective heat transfer correlations derived by numerous researchers which were found for vertical and horizontal free plates, alongside for enclosures heated through various methods. It should be noted that in the present study, the correlations cited by Khalifa [21, 22] have been selected to compare the results of this study with data in the literature. The correlations chosen, the researchers by whom derived and the conditions they are accurate are presented in Tables 1 and 2.

In case the relevant literature is rigorously scrutinized, it can easily be understood that there has been a significant uninvestigated field regarding household radiators issue. Thus, the influence of numerous parameters that have effects on heat transfer through the radiators, and the convective heat transfer coefficient arising over the radiators have been scrutinized, in the current work. Parameters such as temperature differences at air and water sides, water velocity and mass flow rate were taken in consideration on the calculation of heat output of the tested radiators, and on the convective heat transfer coefficients occurring over them. Moreover, the main purpose of this computational study is to demonstrate the likelihood of validation process of household radiators by means of experimental data, to provide visual clarifications and to produce various outputs using Ansys software's results. The methods of numerically solving this issue and accomplishing this aim consist of examining the internal and external fluid flow of the radiator and then comparing these outputs to the experimental data of a radiator producer company.

NUMERICAL METHOD

Due to long lasting of solution periods, CFD programs were not highly preferable in recent years. Nevertheless, continuously progressing computer processors, in these days allow researchers and building simulators avail themselves of CFD programs in heat transfer applications for heat transfer applications in simulations, such as the radiator systems which consist of different radiator geometries such as in this study. This work, comprised of a broad range of results obtained through varying a number of parameters contains CFD outcomes. To solve the governing equations (continuity, momentum, energy) the CFD program ANSYS 17, which involves FLUENT 17 within, has been preferred. It should also be noted that the outline of the program's solution path is built over the control volume theory through altering relevant equations into algebraic equations for them to be solved. Techniques of control volume are executed by carrying out the integration of the relevant equations on the individual control volume and generating discretization of the equations [23].

One should first be aware of the fact that, to initiate a numerical process, it is significant to separate the given, known data from the unknown one. All sizes for tested radiators are known from producer company's catalogs. The mass flow rates of water regarding the experimental thermal capacities and temperature levels are known. The inlet temperatures for water are taken from catalogs as 75 °C and 90 °C and ambient temperatures varied from 10 °C to 26 °C. The whole numerical process starts with the draw in SolidWorks. Some preferences and automatically determined parameters within the program can be summarized in the following paragraphs:

Among from 5 different size functions, "curvature" has been chosen. Corresponding to Smoothing, Transition, Growth rate options; Medium, Slow and 1.2 selections were made, respectively. Also, minimum edge length of the meshes was $3x10^{-2}$. In addition to this, Transition ratio and Maximum layer values have been determined as 0.272 and 5, respectively. As statistical values, it should be noted that 10895 nodes and 46193 elements occurred within the body of the radiator. A gravitational force value at –y direction has been input as 9.81 m/s^2.

In ANSYS a meshed images of the radiators examined are shown in Figure 1. The next stage is about the entering boundary conditions as water flow velocity, water inlet temperature, ambient temperature and pressure outlet in order to calculate heat transfer rate, surface temperature of the radiator, water outlet temperature, convective heat transfer coefficient of air side, and temperature distribution in the radiator. From this information, it is clear that there are many unknowns for the system. The experimental heat transfer amount is predicted by means of the numerical model in ANSYS 17 of which solution steps are detailed in this section. Some of the obtained numerical results can be seen in Tables 3 and 4 for 900 mm /160 mm and 500 mm/160 mm radiators' 75 °C / 65 °C and 90 °C / 70 °C operating conditions, respectively.

Table 1. The correlations presented Khalifa [21] derived for vertical free plates

Researchers	Conditions	Equation
Churchill and Chu	Range applicable in buildings	$h = \left[0.134(L)^{-\frac{1}{2}} + 1.11(\Delta T)^{\frac{1}{6}}\right]^2$
Alamdari and Hammond	Rayleigh range: 10^4-10^{12}	$h = \left\{\left[1.5\left(\frac{\Delta T}{L}\right)^{1/4}\right]^6 + \left[1.23(\Delta T)^{\frac{1}{3}}\right]^6\right\}^{\frac{1}{6}}$
Fishenden and Saunders	Laminar flow	$h = 1.368(\Delta T\ /\ L)^{1/4}$
Fishenden and Saunders	Turbulent flow	$h = 1.973(\Delta T)^{1/4}$
Griffiths and Davis	For 1.2 m square plate, up to 100°C temperature difference	$h = 1.776(\Delta T)^{1/4}$
Heilman	Discs up to 0.25 diameter	$h = 1.664(\Delta T)^{0.27}$
King	Correlated data from other researchers	$h = 1.517(\Delta T)^{0.33}$
Hottinger	No note available	$h = 2.5(\Delta T)^{1/4}$
Wilkes and Peterson	Two heated plates 2.4 x 0.8 m^2 with 0.1 m air space	$h = 3.05(\Delta T)^{0.12}$
ASHRAE	For laminar flow	$h = 1.42(\Delta T)^{1/4}$
ASHRAE	For laminar flow	$h = 1.31(\Delta T)^{1/3}$
McAdams	For laminar flow	$h = 1.42(\Delta T)^{1/4}$

Table 2. The correlations presented Khalifa [22] derived for enclosures heated through different surfaces

Researchers	Conditions	Equation
Min et al. (heated floor)	Ra range: 10^9-10^{11}	$h = 2.416\,(\Delta T)^{0.31}/D_e^{\,0.08}$
Min et al. (vertical wall)	Ra range: 10^9-10^{11}	$h = 1.873\,(\Delta T)^{0.32}/H^{0.05}$
Min et al. (heated ceiling)	Ra range: 10^9-10^{11}	$h = 4.622\,(\Delta T)^{0.05}/H^{0.85}$
Li et al.	An occupied office room at working conditions	$h = 2.88(\Delta T)^{0.25}$
Khalifa and Marshall	On vertical wall	$h = 2.30(\Delta T)^{0.23}$

Using the program ANSYS FLUENT 17, a laminar model has been selected owing to entire found Rayleigh number values to be smaller than 10^9. Also, a steady state solution approach has been seen appropriate. In addition, in order to solve the equations with a high precision, the second order upwind scheme has been chosen to discrete them. The under relaxation factors corresponding to pressure, density, body forces, momentum and energy have been selected 0.3, 1, 1, 0.7 and 0.9, respectively. The necessary time for each case study to converge at the solution was approximately 2 hours.

The residuals for momentum and continuity equations were determined as 10^{-3}, while the value was 10^{-6} for energy equation. Furthermore, thermo-physical properties of the room air have been calculated at the mean surface temperature of surfaces in each case study, through the tables given by Incropera and DeWitt [24]. Moreover, between two solvers presented by FLUENT, pressure based coupled solver along with a SIMPLE scheme have been selected.

Figure 1. Cornered end radiators used in analyses (a) 500 mm - 160 mm, (b) 900 mm - 160 mm

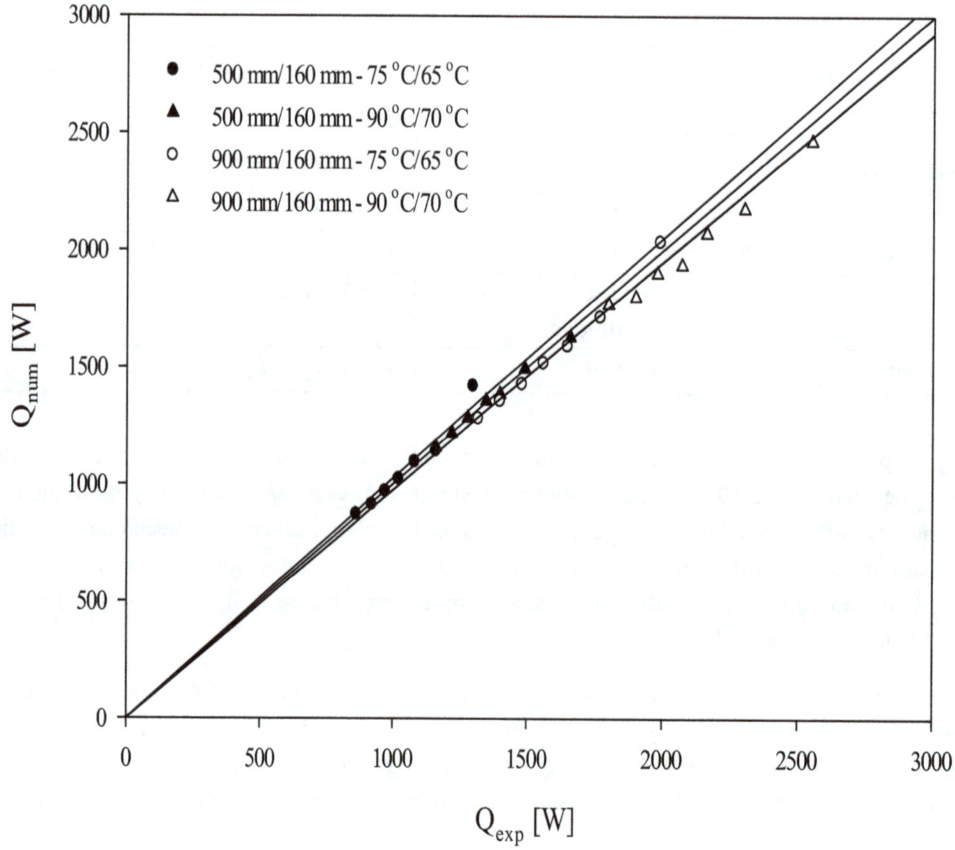

(a)

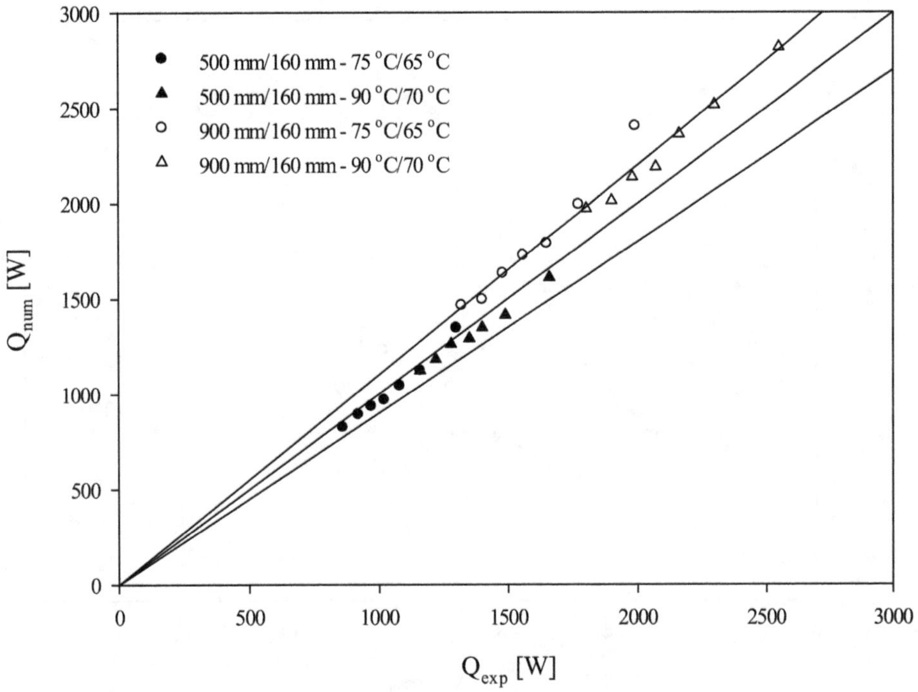

(b)

Figure 2. Comparison between the ANSYS total heat transfer values and experimental data (a) for cornered end, (b) for chamfered end radiators

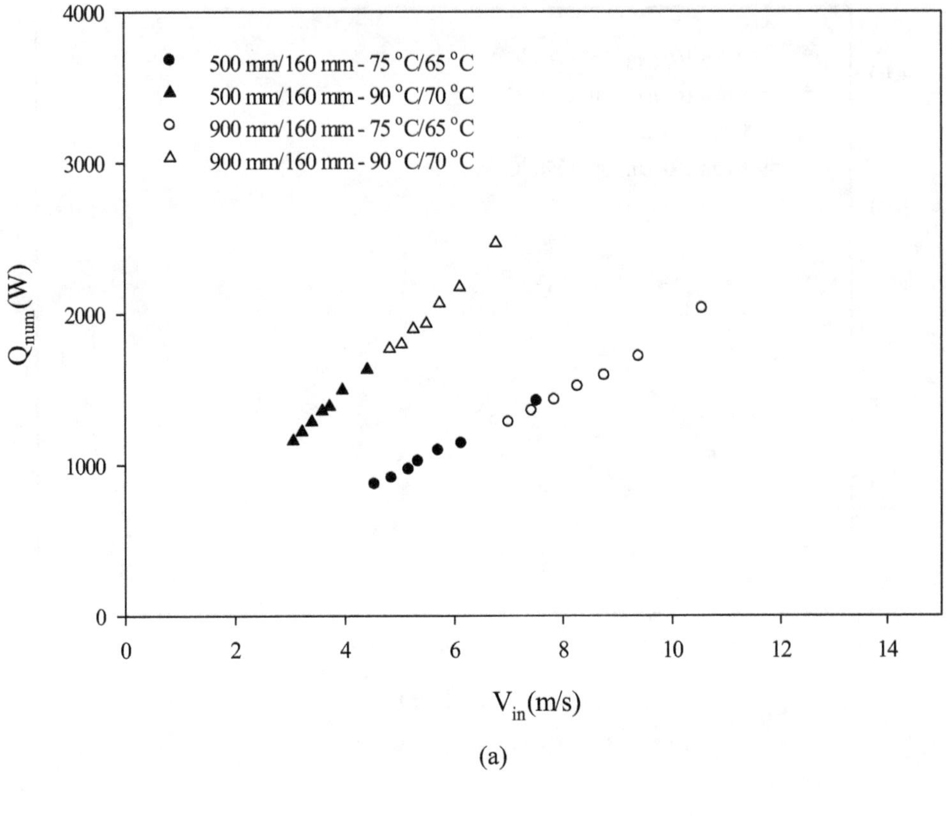

(a)

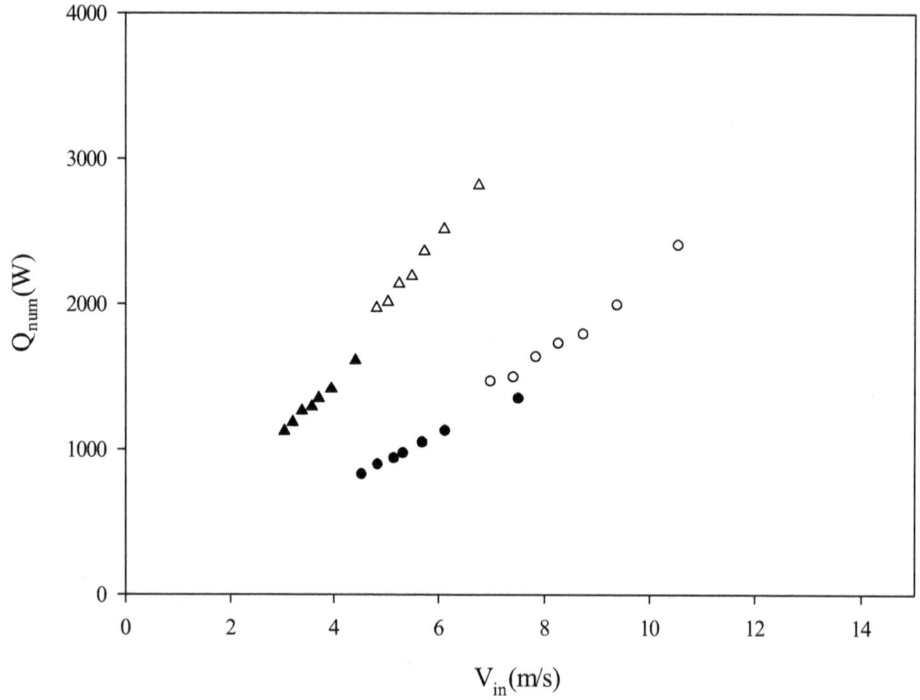

(b)

Figure 3. The variation of total heat transfer rate values with inlet water velocity (a) for cornered end, (b) for chamfered end radiators

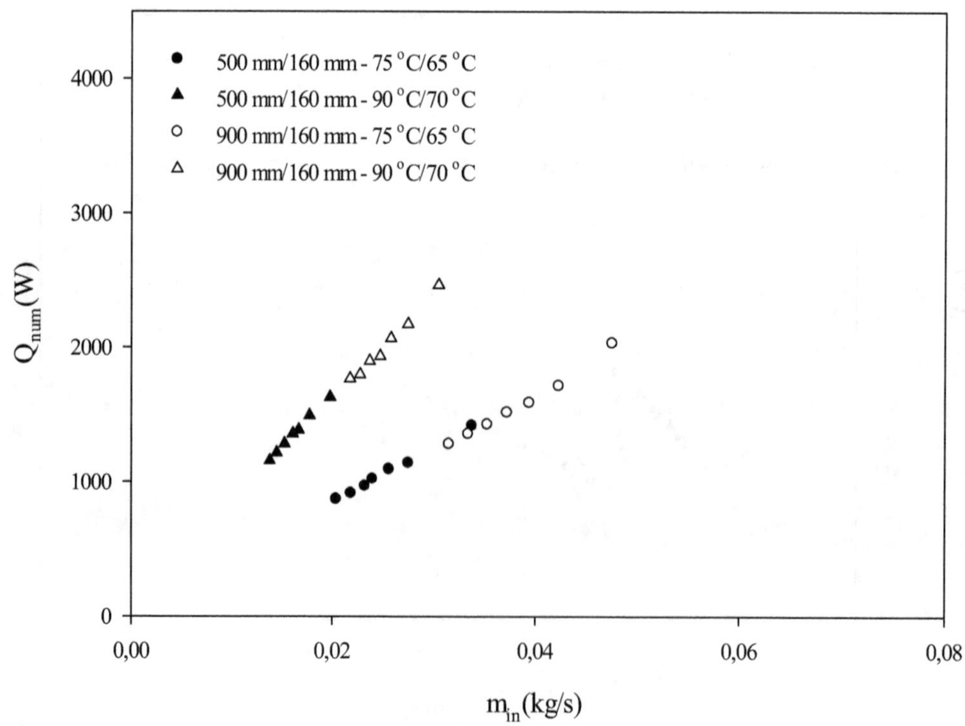

(a)

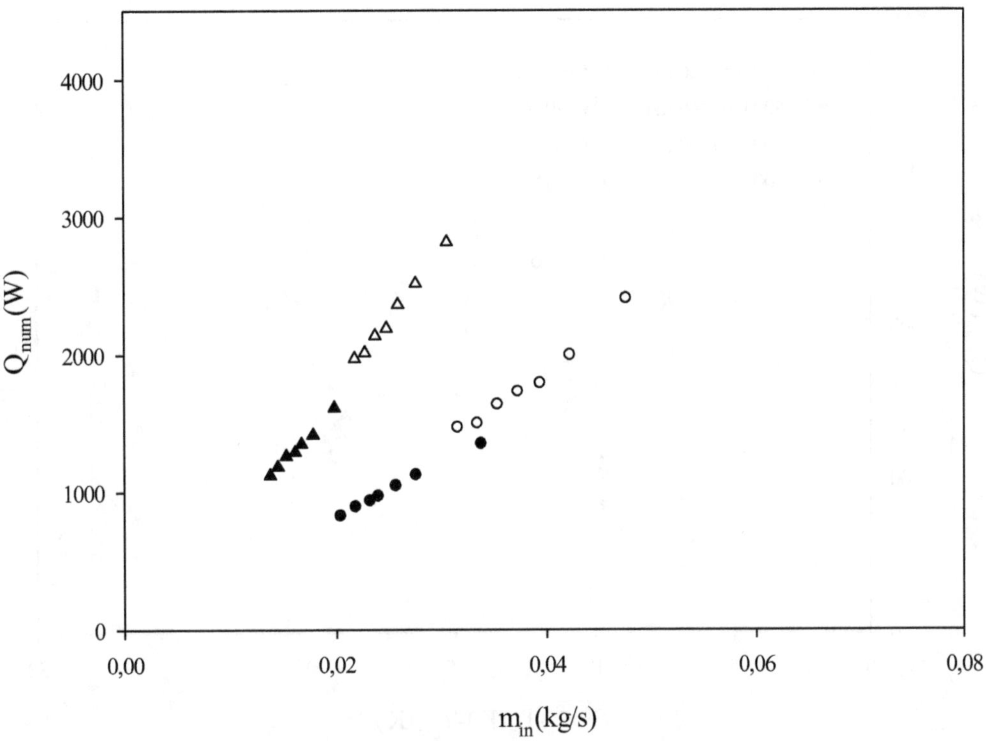

(b)

Figure 4. The change of total heat transfer rate found via ANSYS with mass flow inlet (a) for cornered end, (b) for chamfered end radiators

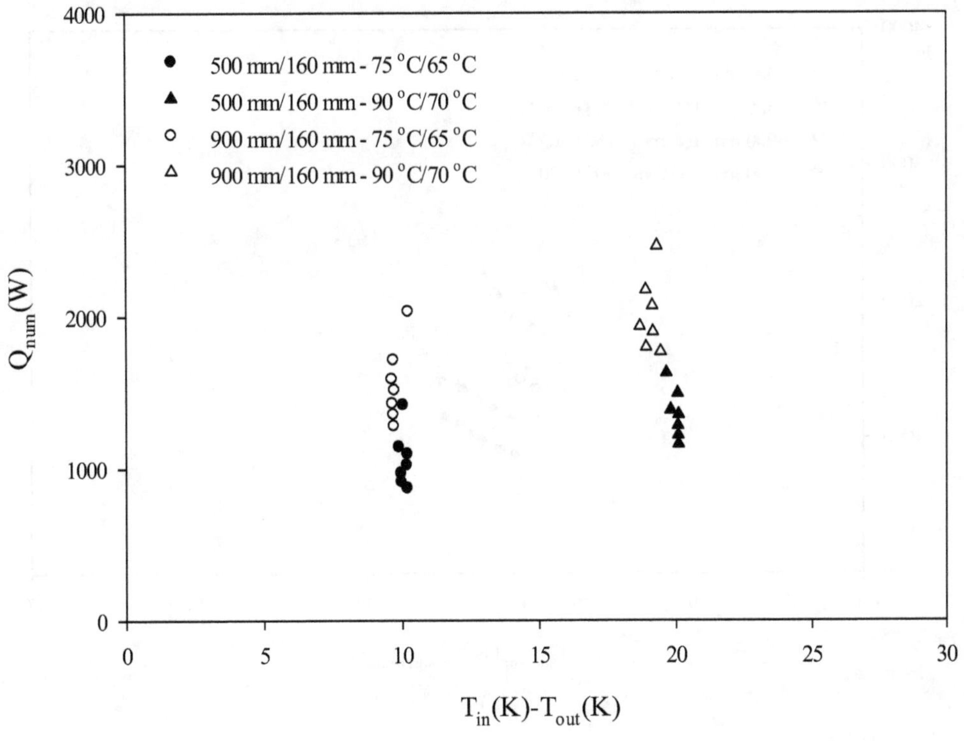

(a)

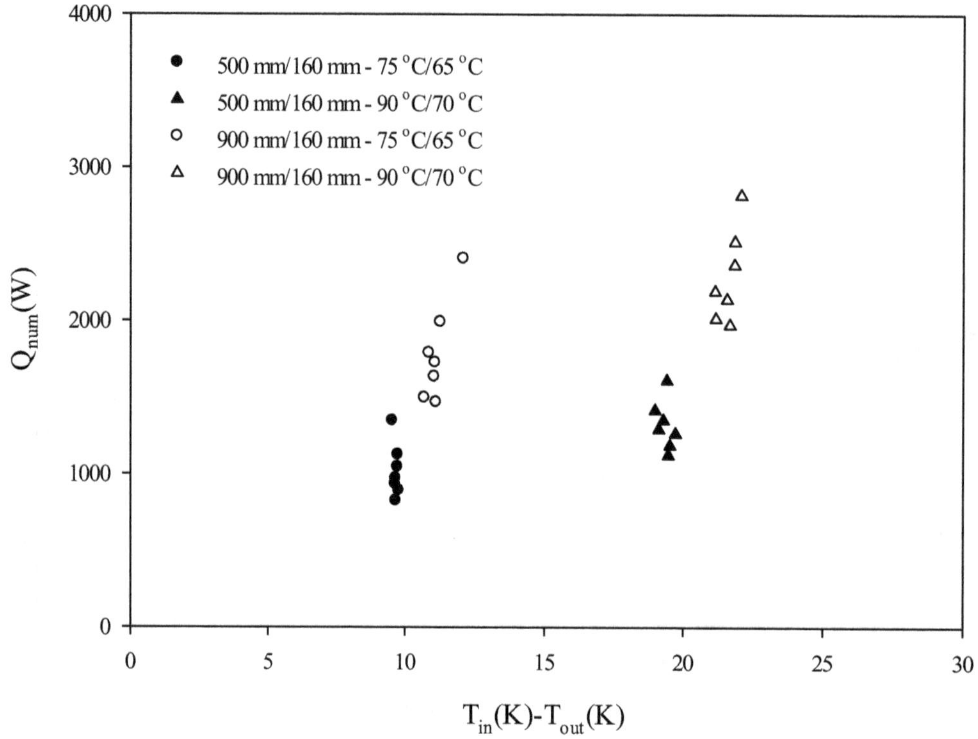

(b)

Figure 5. The alteration of total heat transfer rate obtained via ANSYS with radiator inlet and outlet temperature difference (a) for cornered end, (b) for chamfered end radiators

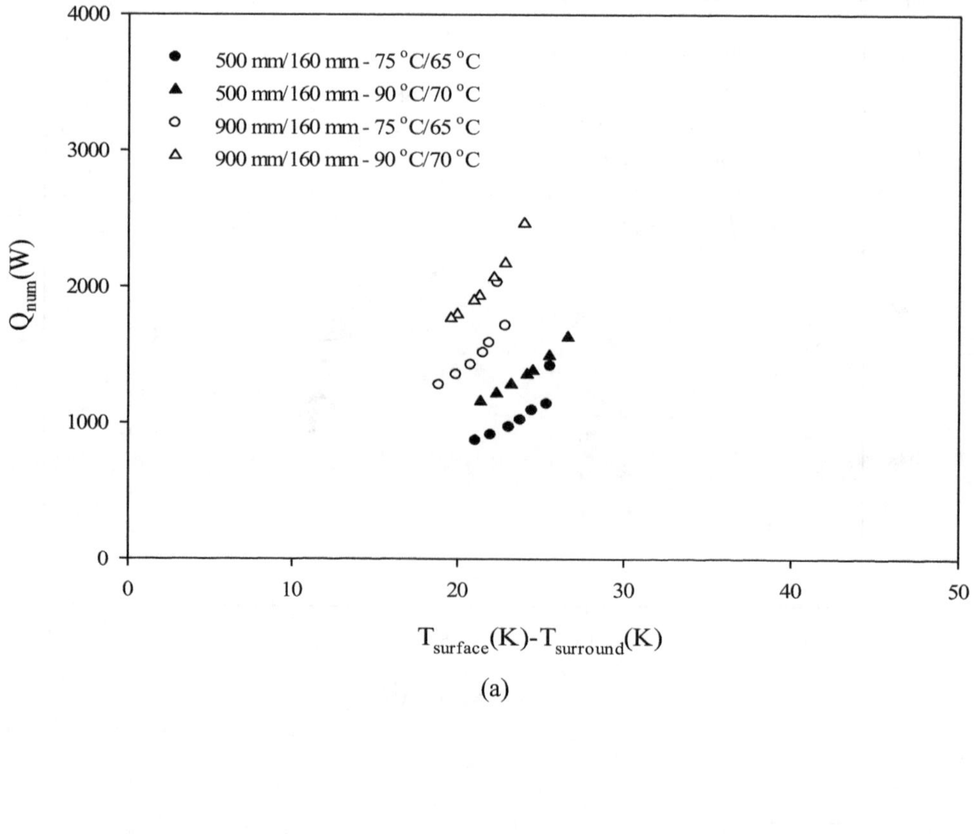

(a)

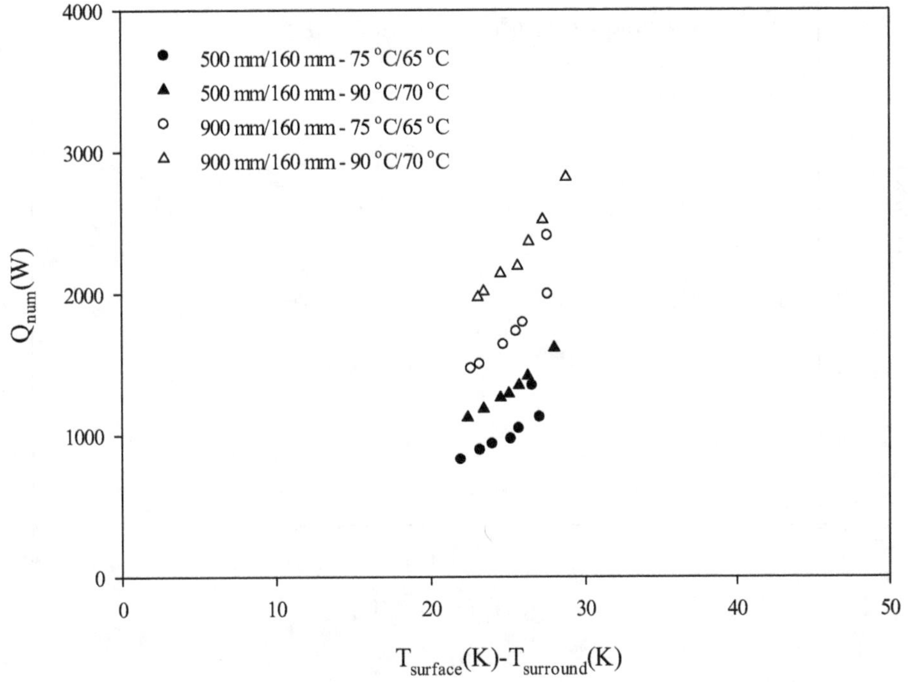

(b)

Figure 6. The variation of total heat transfer rate obtained via ANSYS as the temperature difference between radiator surface and air varies, (a) for cornered end, (b) for chamfered end radiators

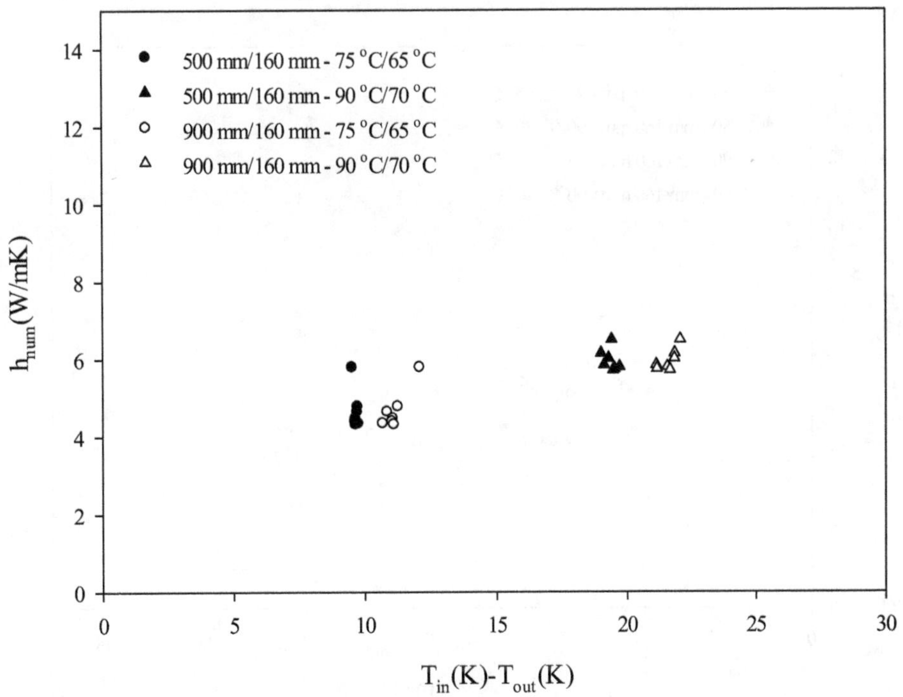

(a)

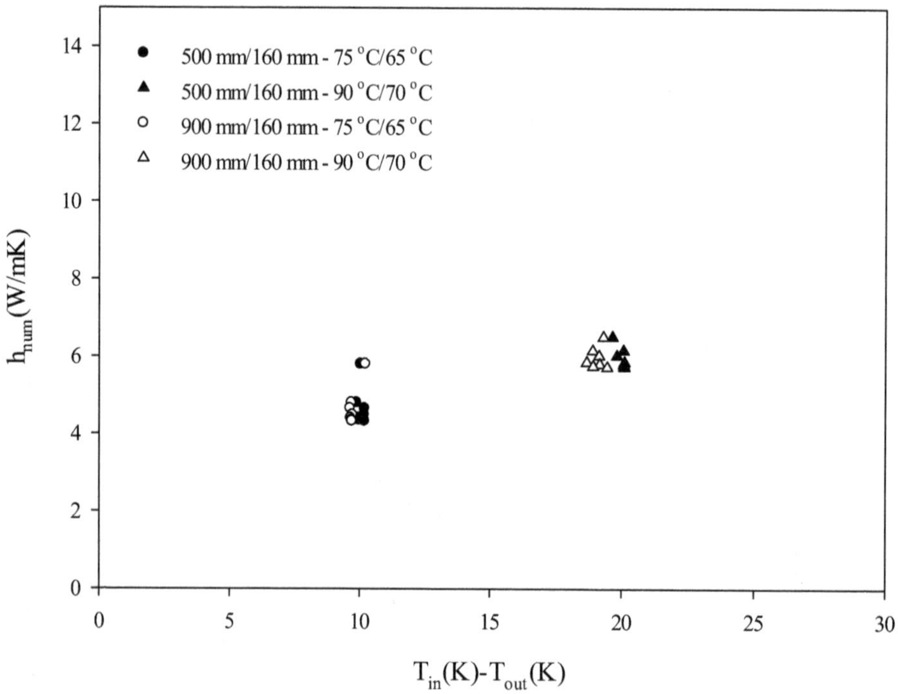

(b)

Figure 7. The change of convective heat transfer coefficient occurring over the radiator, with the variation of the temperature difference between radiator inlet and outlet temperatures, (a) for cornered end, (b) for chamfered end radiators

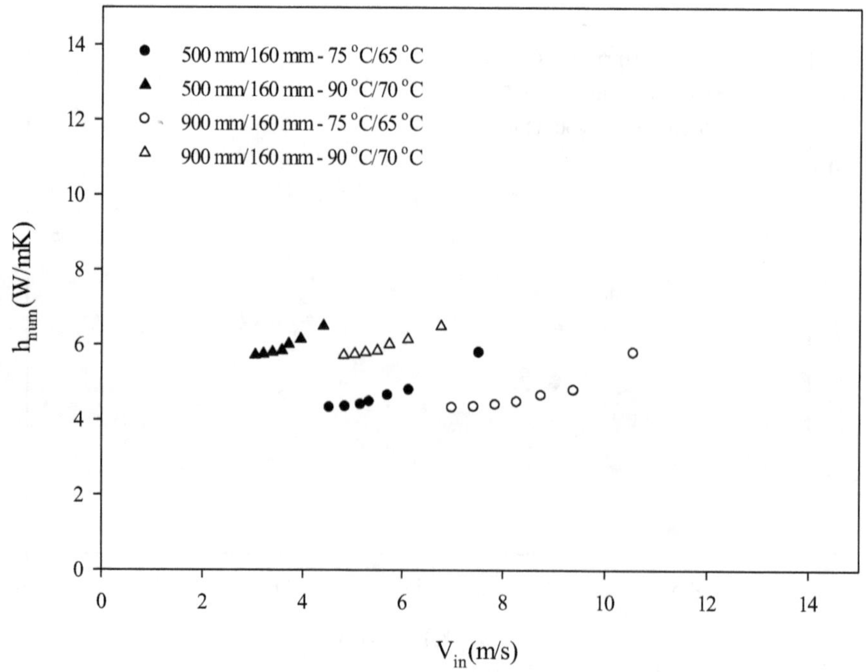

(a)

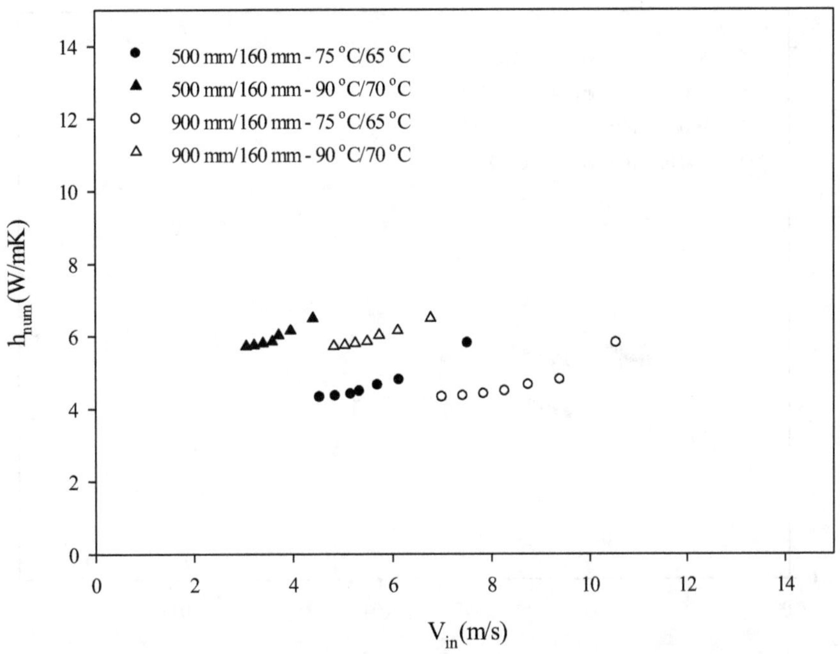

(b)

Figure 8. The variation of convective heat transfer coefficient values with inlet water velocity (a) for cornered end, (b) for chamfered end radiators

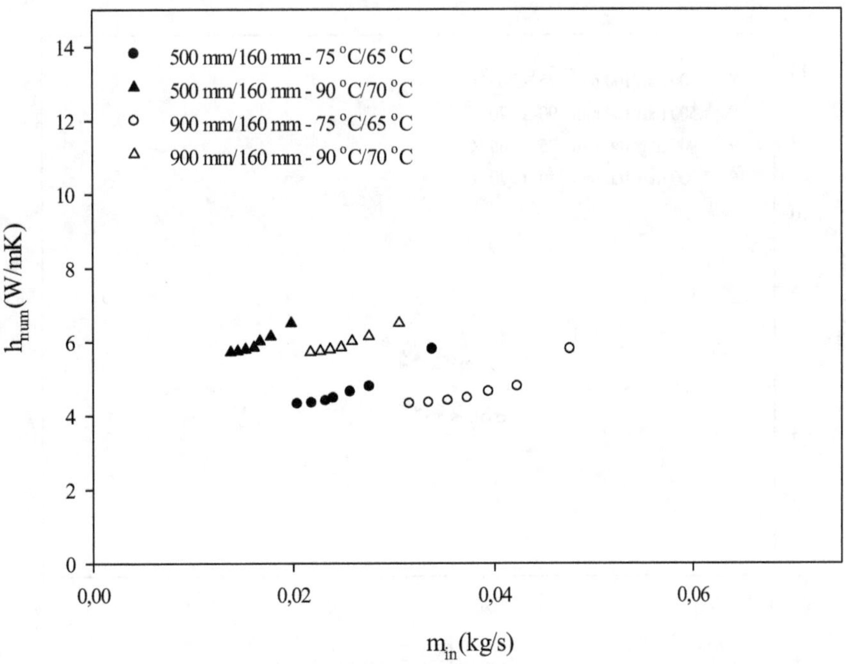

(a)

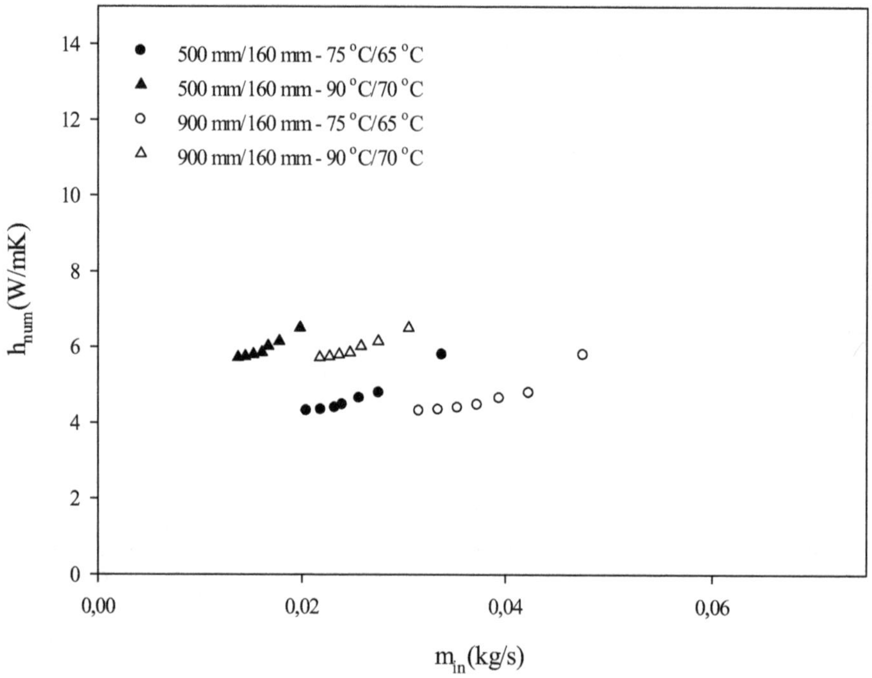

(b)

Figure 9. The change of convective heat transfer coefficient with mass flow inlet (a) for cornered end, (b) for chamfered end radiators

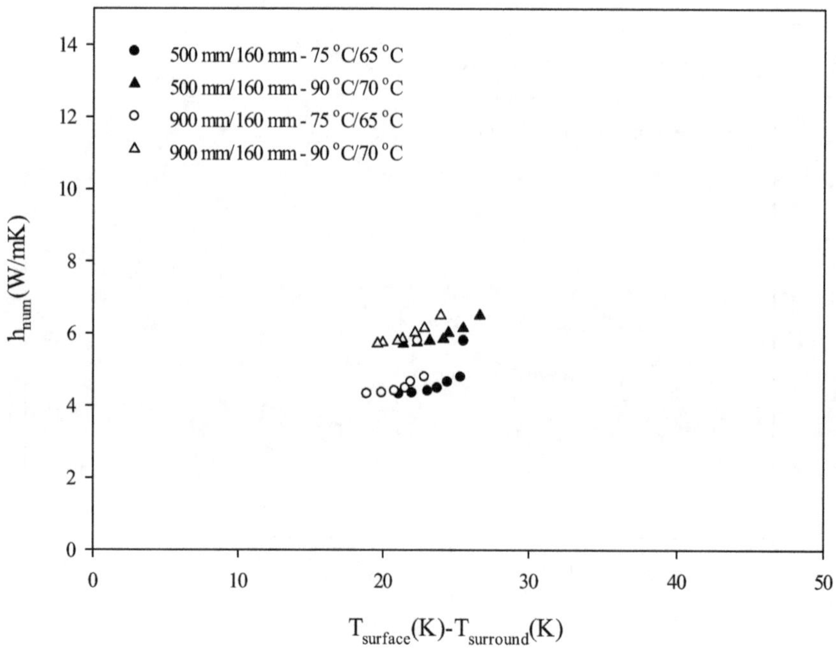

(a)

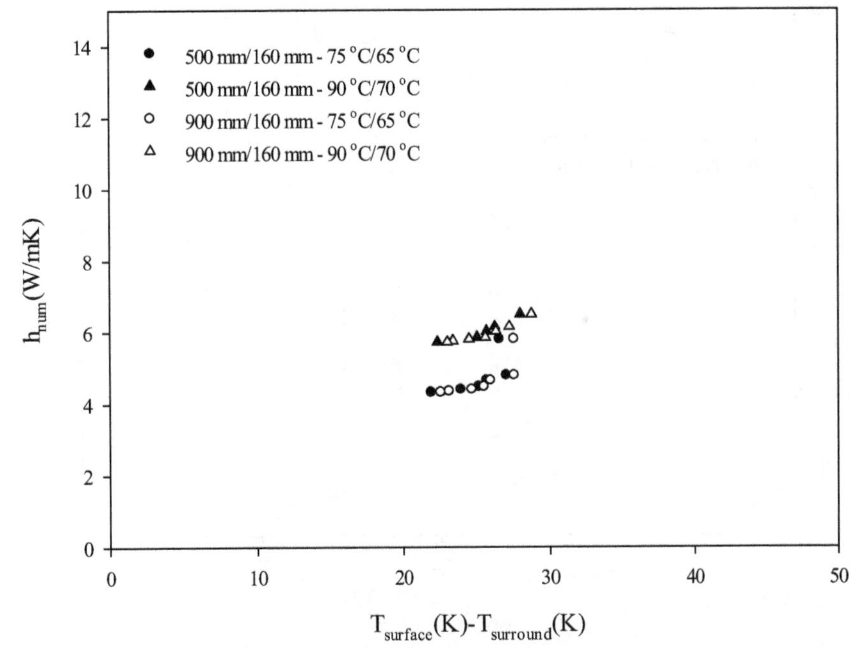

(b)

Figure 10. The variation of convective heat transfer coefficient obtained via ANSYS as the temperature difference between radiator surface and air changes, (a) for cornered end, (b) for chamfered end radiators

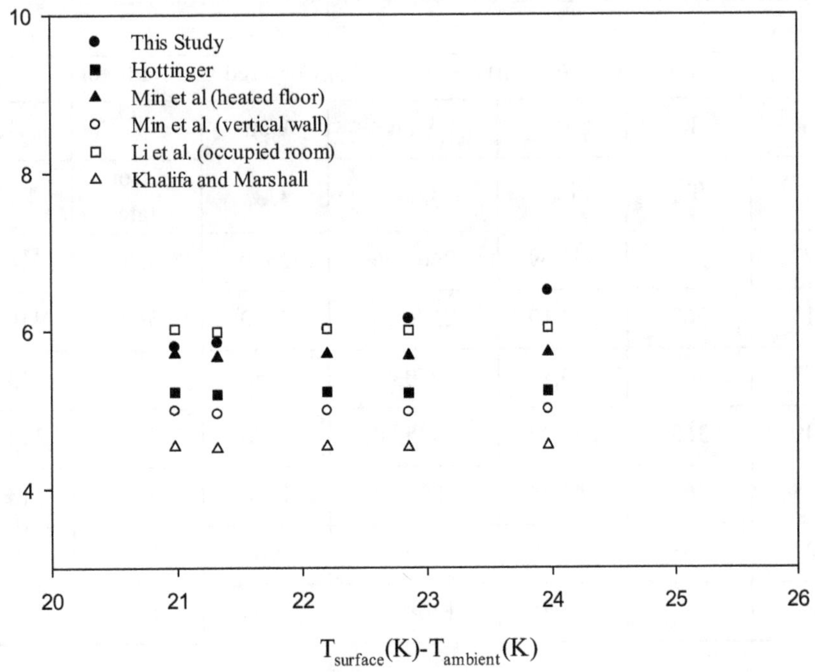

(a)

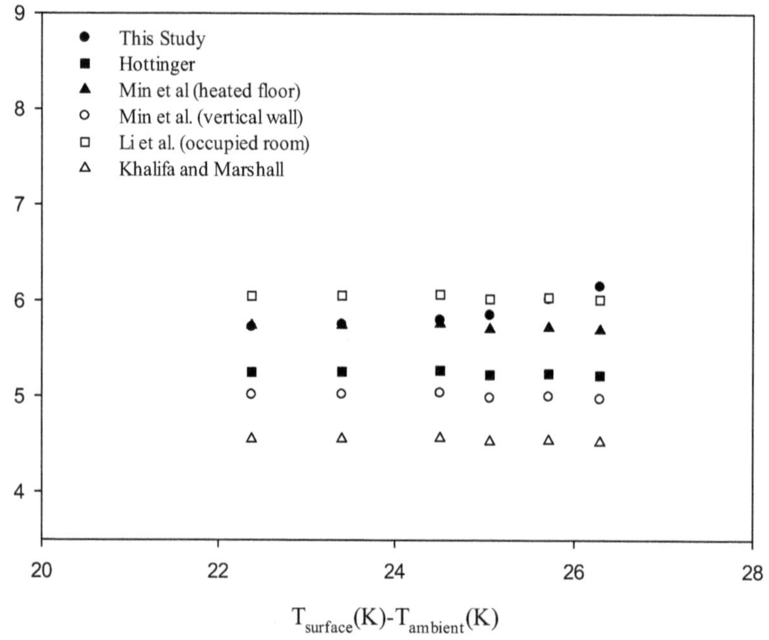

(b)

Figure 11. Comparison between the results of this study and the results obtained using the correlations derived by other researchers, (a) for a cornered end 900 mm/160 mm 90°C/70°C radiator, (b) for a chamfered end 500 mm/160 mm 90°C/70°C

Table 3. Some of the operating conditions and findings of 900 mm -160 mm radiator for 90 °C / 70 °C

(900 mm - 160 mm) (90 °C / 70 °C) Cornered End Radiator								
W/m^2K	m/s	K	K	Watt	Watt	%	K	K
h	v_{in}	T_{in}	T_{out}	Q_{num}	Q_{exp}	Error rate	$T_{surface}$	$T_{ambient}$
6,5	6,77	363	343,69	2465,76	2550	3,30	306.96	283
6,15	6,10	363	344,10	2176,20	2300	5,38	310.86	288
6,02	5,73	363	343,86	2070,41	2160	4,14	313.20	291
5,85	5,49	363	344,32	1935,40	2070	6,50	314.32	293
5,8	5,25	363	343,83	1899,97	1980	4,04	315.98	295
5,75	5,04	363	344,08	1799,20	1900	5,30	316.96	297
5,72	4,82	363	343,55	1768,67	1800	1,74	318.59	299

Table 4. Some of the operating conditions and findings of 500 mm -160 mm radiator for 90 °C / 70 °C

(500 mm - 160 mm) (90 °C / 70 °C) Cornered End Radiator								
W/m²K	m/s	K	K	Watt	Watt	%	K	K
h	V_{in}	T_{in}	T_{out}	Q_{num}	Q_{exp}	Error rate	$T_{surface}$	$T_{ambient}$
6.5	4.41	363	343.35	1629.942	1660	1.81	309.617	283
6.15	3.95	363	342.937	1494.397	1490	-0.29	313.485	288
6.02	3.71	363	343.191	1385.869	1400	1	315.492	291
5.85	3.58	363	342.901	1356.829	1350	-0.5	317.144	293
5.8	3.39	363	342.929	1284.305	1280	-0.33	318.215	295
5.75	3.22	363	342.918	1218.02	1220	0.16	319.357	297
5.72	3.06	363	342.902	1159.013	1160	0.08	320.38	299

RESULTS AND DISCUSSION

The main purposes of this study are to numerically find heat transfer characteristics occurring over the household iron casting radiators produced by a company, and to compare the obtained numerical results with experimental results of the producer in terms of heat transfer rates, and also compare the numerically calculated convective heat transfer coefficient values with the results found by executing the correlations derived by researchers cited in Tables 1 and 2.

Figures 2a and 2b illustrate the comparison between the experimental data and the numerically obtained values with this study. At different radiator dimensions, as well as, inlet and outlet temperatures, it is observed that numerically acquired data predicts the experimental results of the radiator company within 5% and 10% deviation ratios for cornered end and chamfered end radiators, respectively. This comparison proves the correctness of the computational model performed in the present study.

Figures 3a and 3b demonstrate the increase in heat transfer rate through the radiators at different radiator sizes and water inlet and outlet temperatures, and at cornered end and chamfered end radiators, respectively. As expected, the more water velocities in radiators result in the rise in forced convection and thus heat transfer rate ascends. It is also obvious that the increase on radiator dimensions has led to a sharp increase in heat transfer rate. Very similarly, as shown in Figures 4a and 4b the increase in mass flow rate flowing in radiators at different dimensions and inlet/outlet temperatures accounts for the augmentation on radiator's heat transfer rate.

Figures 5a and 5b reveal that at the same radiator inlet and outlet water temperatures, as the radiator dimension rises from 500 mm to 900 mm, the heat transfer rate of the radiator shows a notable increase, for cornered end and chamfered end radiators, respectively. Moreover, it is evident that at the same radiator dimensions, with ascending values of water inlet temperatures, the heat transfer rate values illustrate a remarkable advance. Additionally, the data illustrated in Figures 6a and 6b demonstrate the augmentation in heat transfer rate of radiators with increasing values of temperature difference between surface and surrounding air temperature, at cornered end and chamfered end radiators, respectively.

Figures 7a and 7b shows the convective heat transfer characteristics occurring over the radiators and their variation with inlet and outlet temperature difference values. It can be observed that at different radiator dimensions (500 mm and 900 mm) and at the same radiator inlet and outlet temperatures, the convective heat transfer coefficient draws very slight variations, whereas, it shows a more remarkable change as the abovementioned temperature difference rises, when dimensions are the same.

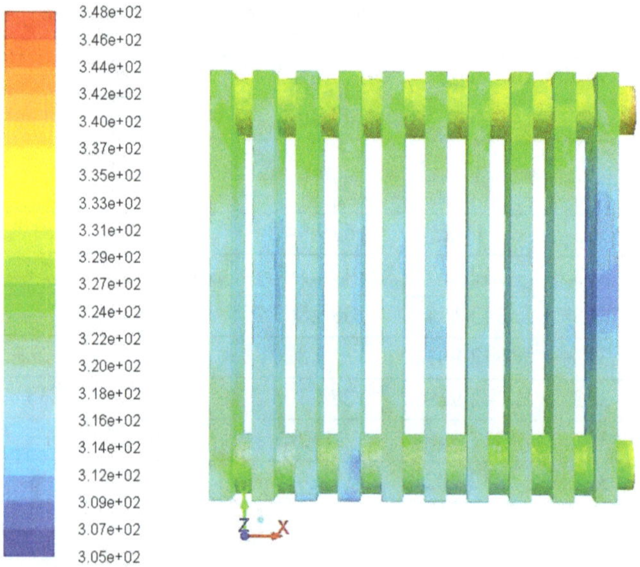

Figure 12. Temperature distribution on the surface of 500 mm – 160 mm radiator in 26 °C ambient temperature

Figures 8a and 8b illustrate the alteration of convective heat transfer coefficients over radiator surfaces with water velocity inlet values to the radiator. Similar to the Figures 3a and 3b demonstrating the change of heat transfer rate with velocity inlet values, the convective heat transfer coefficient values also rise with increasing velocity inlet values, since higher values of velocity have induced forced convection at higher levels within the radiator and therefore causing higher values of convective heat transfer coefficients over the radiator. In a very similar trend, Figures 9a and 9b draw the variation of the convective heat transfer coefficient with changing values of mass flow rates in radiators, for cornered end and chamfered end radiators, respectively. It is clear that with an increase in mass flow rate, in all case studies, the convective heat transfer coefficient demonstrates a noticeable growth over the radiator surface.

In Figures 10a and 10b show the change in convective heat transfer coefficient with the temperature difference between surface and surrounding air temperatures. From the figures, it can be seen that as the aforementioned temperature difference grows, due to increasing air movements over the radiator, the convective heat transfer coefficient ascends. Also, as expected, this coefficient illustrate a clear increase as the radiator water inlet temperature rises from 90°C to 75°C. Moreover, it should be noted that a very slight increase on this coefficient could be observed while one dimension of the radiators descends from 900 mm to 500 mm.

In addition to the validations implemented with experimental data in Figures 2a and 2b, the results obtained through the present work were compared with the data found utilizing correlations explored by some other researchers in the literature. In Figures 11a and 11b, for a 900 mm/160 mm cornered end, and for a chamfered end 500 mm/160 mm radiator, respectively, this comparison has been made. It is evident through the figures that the results of the present study lie within the range of the data acquired by executing the correlations cited in these figures. It should be noted that the sole successful correlation predicting the results of the current study which has been derived for free plates, is Hottinger's correlation, with an average deviation ratio of 4.08%. On the other hand, the correlations explored for enclosures have enabled much more consistent outputs with the results of this study. While the correlations proposed for heated floor and vertical walls by Min et al. deviate by 0.72% and 12.55% at average, respectively; the correlation of Li et al. obtained for an occupied office room under normal working conditions, deviates at an average value of 10.5%.

Figure 12 illustrates an image via Ansys, displaying the temperature distribution in one of the tested radiators which was abovementioned and presented in Figure 1. The influence of heat loss from hot water to surrounding air is noticed through this figure obviously.

CONCLUSION

As well as the thermal capacities of household radiators, convective heat transfer coefficient values occurring over their surfaces comprise the most crucial issues in terms of energy economy and thermal comfort. From this point of view, in the present study, a numerical model investigating household radiators at different dimensions, water inlet temperatures, inlet velocities and geometric shapes was generated by means of the program Ansys 17.

To obtain a The radiators were simulated using wide range of input ambient temperature conditions from 10 °C to 26 °C and water velocity ranging from 3.06 m/s to 7.51 m/s for 500 mm/160 mm type radiator and 4.82 m/s to 10.54 m/s for 900 mm/160 mm type radiator. The deductions obtained were summed up as follows:

The heat transfer capacity and the convective heat transfer coefficient over the radiator increase with increasing values of water inlet velocity and also mass flow rate, due to growing values of forced convection within radiators.

Using a producer's catalogue, found thermal capacity values and also the model developed at different dimensions, geometric shapes and water inlet temperatures, have been validated within the deviation values of 5%, and 10%, for cornered end, and chamfered end radiator types, respectively.

The results regarding the convective heat transfer coefficient values were compared with the data obtained using the correlations found by some other researchers, and noticed that a favorable consistency has been achieved.

ACKNOWLEDGMENTS

The authors are grateful to Turk Demirdokum A.Ş. the leading refrigerator company of Turkey for the experimental data. The first author wishes to thank King Mongkut's University of Technology Thonburi (KMUTT) for providing him with a Post-doctoral fellowship. The forth author is indebted to the Thailand Research Fund and the National Research University Project for supporting this study.

NOMENCLATURE

h_{num} convective heat transfer coefficient, W/m^2K
m_{in} mass flow rate, kg/s
Q_{exp} experimentally found thermal capacity, W
Q_{num} thermal capacity, W
T_{in} inlet temperature, K
T_{out} outlet temperature, K
$T_{surface}$ radiator surface temperature, K
$T_{surround}$ surrounding air temperature, K
v_{in} velocity inlet, m/s
exp experimental
in inlet
num numerical
out outlet

REFERENCES

[1] Calisir, T., Yazar, H. O., Baskaya, S. and Yucedag, S (2016). Experimental and numerical prediction of flow field around a panel radiator. International Scientific Journal Journal of Environmental Science 5.

[2] Embaye, M., Al-Dadah, R. K., and Mahmoud, S. (2016). Numerical evaluation of indoor thermal comfort and energy saving by operating the heating panel radiator at different flow strategies. Energy and Buildings, 121, 298-308.

[3] R.J. Ladumor, V. Y Gajjar, and K.K.Araniya, (2014). A review paper on analysis of automobile radiator. International Conference on Multidisciplinary Research & Practice 1 388-393.
[4] Calisir, T., Baskaya, S., Yazar, H. O., and Yucedag, S. (2015). Parametric numerical investigation of heat transfer from convectors to improve efficiency of panel radiators. In ICHMT Digital Library Online. Begel House Inc.
[5] Kayastha, K. S. (2015). CFD simulation of heat transfer analysis of automobile radiator using helical tubes. International journal of engineering research and development, 11(1), 24-35.
[6] Embaye, M., Al-Dadah, R. K., and Mahmoud, S. (2015). Thermal performance of hydronic radiator with flow pulsation–Numerical investigation. Applied Thermal Engineering, 80, 109-117.
[7] Johansson, P. O., and Wollerstrand, J. (2010). Heat output from space heating radiator with add-on-fan blowers. In Excerpt from Proceedings the COMSOL Conference.
[8] Shi, H. L., Liu, Y., Shao, Y. Z., and Jin, Y. A. (2015, June). Optimization design of plate-type radiator. In International Conference on Computer Information Systems and Industrial Applications. Atlantis Press.
[9] Myhren, J. A., and Holmberg, S. (2009). Design considerations with ventilation-radiators: Comparisons to traditional two-panel radiators. Energy and buildings, 41(1), 92-100.
[10] Sarbu, I., and Sebarchievici, C. (2015). A study of the performances of low-temperature heating systems. Energy Efficiency, 8(3), 609-627.
[11] Aydar, E., and Ekmekci, I. (2012). Thermal efficiency estimation of the panel type radiators with CFD analysis. Journal of Thermal Science and Technology, 32, 63-71.
[12] Salvio Chacko, D., Shome, B., Kumar, V., Agarwal, A. K., and Katkar, D. R. Numerical Simulation for Improving Radiator Efficiency by Air Flow Optimization.
[13] Sevilgen, G., and Kilic, M. (2011). Numerical analysis of air flow, heat transfer, moisture transport and thermal comfort in a room heated by two-panel radiators. Energy and Buildings, 43(1), 137-146.
[14] Shati, A. K. A., Blakey, S. G., & Beck, S. B. M. (2011). The effect of surface roughness and emissivity on radiator output. Energy and buildings, 43(2-3), 400-406.
[15] Arslanturk, C., and Ozguc, A. F. (2006). Optimization of a central-heating radiator. Applied energy, 83(11), 1190-1197.
[16] Menéndez-Díaz, A., Ordóñez-Galán, C., Bouza-Rodríguez, J. B., and Fernández-Calleja, J. J. (2014). Thermal analysis of a stoneware panel covering radiators. Applied energy, 131, 248-256.
[17] Brady, L., Abdellatif, M., Cullen, J., Maddocks, J., and Al-Shamma'a, A. (2016). An investigation into the effect of decorative covers on the heat output from LPHW radiators. Energy and Buildings, 133, 414-422.
[18] Kılıç, M., Sevilgen, G., and Mutlu, M. (2014). Three-Dimensional Numerical Analysis of Thermal Output of a Steel Panel Radiator. In Progress in Exergy, Energy, and the Environment (pp. 585-593). Springer, Cham.
[19] Jahanbin, A., and Zanchini, E. (2016). Effects of position and temperature-gradient direction on the performance of a thin plane radiator. Applied Thermal Engineering, 105, 467-473.
[20] Beck, S. B., Blakey, S. G., and Chung, M. C. (2001). The effect of wall emissivity on radiator heat output. Building Services Engineering Research and Technology, 22(3), 185-194.
[21] Khalifa, A. J. N. (2001). Natural convective heat transfer coefficient–a review: I. Isolated vertical and horizontal surfaces. Energy conversion and management, 42(4), 491-504.
[22] Khalifa, A. J. N. (2001). Natural convective heat transfer coefficient–a review: II. Surfaces in two-and three-dimensional enclosures. Energy Conversion and Management, 42(4), 505-517.
[23] Demir, H., Dalkilic, A. S., Kürekci, N. A., Duangthongsuk, W., and Wongwises, S. (2011). Numerical investigation on the single phase forced convection heat transfer characteristics of TiO2 nanofluids in a double-tube counter flow heat exchanger. International Communications in Heat and Mass Transfer, 38(2), 218-228.
[24] Incropera, F. P., and DeWitt, D. P. (1996). Fundamentals of Heat and Mass Transfer. 4th edn John Wiley & Sons. New York.

10

A PROPER SELECTION OF HOT AND COLD UTILITIES IN A PLANT CONTAINING MULTIPLE HEAT EXCHANGER NETWORKS

F. Pourfayaz[1], A. Kasaeian[1,*], M. Mehdizadeh Fard[1]

ABSTRACT

In this paper, the hot and cold utilities of a complex natural gas refinery containing multiple heat exchanger networks (HENs) were properly selected. Following using pinch analysis for designing the HENs of the natural gas refinery, this selection could be done by considering available utilities in the plant under consideration. The plant network was divided into the five heat exchanger networks for better and precise estimation of utility requirements in each smaller network. By comparing the type and temperature level of the utility required in each network, the best utility could be selected, in order to use as the heating and cooling media in the plant network with a considerable reduced operating cost of the system. The results show that in the new network design, there was about 38 and 100% reduction in cooling air and BFW respectively, but 45% increase in cooling water, as cold utilities and almost 64.4, 30.0, 90.9 and 100% reduction in HP steam, LP steam, hot flue gas and electric heater as hot utilities, respectively relative to the existing network.

Keywords: *Practical Pinch Analysis, Heat Exchanger Network, Hot and Cold Utilities, Appropriate Placement Principle*

INTRODUCTION

Because of increasing energy costs in recent years, the energy saving methods are usually considered at the early stage of any process design, especially in high energy demand industries such as complex oil and gas refineries. The other important reason for the efficient design of any energy system is mainly due to the environmental regulations, forcing to reduce the amount of pollution gases to the atmosphere [1-3]. Especially by the more efficient design of the utility system in a huge chemical processing plant, the amount of energy used in the plant life cycle as heating and cooling media will be reduced significantly. In this regard, a variety of optimization methods considering analytical or experimental techniques are available in the literature, showing how it would be possible to design more and more efficient energy systems for various industries [4-7].

As one of the conceptual methods in process integration, the pinch analysis is widely used as a suitable technique to evaluate and improve the energy systems. The term "Pinch Technology" was introduced by Linnhoff in 1979 to represent a new set of thermodynamically based methods that guarantee minimum energy levels in the design of heat exchanger networks. It is a systematic methodology to achieve utility savings by better process heat integration, maximizing heat recovery and reducing the external utility loads as heating or cooling media in the system. [8] This means in any process integration technique considered for total energy analysis of the systems, the targets for energy saving must be usually set before the design of the heat exchanger network running in the processing plant [9]. In the previous work done by the same authors, it was presented an energy analysis of a complex natural gas refinery by the super-targeting method, showing a considerable saving in yearly energy consumption of the plant [10]. There are a lot of works done in literature and research papers about energy saving techniques using pinch analysis methods, some of them were shown in [10] to [16].

The proper design of the utility system in any processing industry plays an important role in the operational cost of the system, by increasing or decreasing the heating and cooling requirements of the processing units running in the plant. The most usual types of heating media as hot utilities may include HP or LP steam, hot flue gases from furnaces or fired heaters, hot oil or water, electrical heating and some other less common media. Likewise, cold

This paper was recommended for publication in revised form by Regional Editor Tolga Taner
[1]Department of Renewable Energies and Environment, Faculty of New Sciences and Technologies, University of Tehran, Iran
**E-mail addresses: akasa@ut.ac.ir*

utilities, which remove heat from the processes by cooling media, may include cooling water or air, chilled water, refrigeration fluids and so on. Usually, the costs of cooling and heating system together with its operational cost are two most important factors for selection of the best utility media in the system. Especially in a huge natural gas refinery as a high energy demand industry, the efficient design of the utility system will considerably affect the energy saving and operational cost of the plant. It is obvious that it provides the key to choosing the correct levels and loads for the various hot and cold utilities. As a case study, the result of pinch analysis in a complex gas refinery which provided by the same authors in [10], was used to analyse the heat exchanger network running in the plant and consider opportunities to select the best heating and cooling media as the utility requirement of the system, using Appropriate Placement principle.

METHODOLOGY

The method is based on the concept of integration processes and the basic tools which used in the analysis are: Hot and Cold Composite Curves (HCC/CCC), and Grand Composite Curve (GCC). Pinch technology combines heat and mass balance information into simple charts and graphs to show the potential of energy recovery of the heat exchanger network (HEN).

In this analysis, the following steps were considered:

- Extraction of stream data from different processing units, considering the heat and mass balances of all process streams in the network, in order to calculate total heat exchanged required by the cold and hot streams, provided by hot and cold utilities, respectively.
- Drawing the hot and cold composite curves (HCC & CCC) as well as grand composite curve (GCC) for all the hot and cold streams in a T-H diagram, to analyze the minimum utility quantities required to solve the heat recovery problem.
- Determine the type and temperature level of utility required as the amount of energy consumption, such as fuel, water, LP and HP steam, refrigerants, air and so on in the network.
- Applying the steps 1 through 3 for all HENs running in the plant under consideration. In the case study of this manuscript, the entire system divided into the five smaller networks (see Table 3 for detail description), each consists of some processing units, so the procedure of utility selection could be applied for all sub-networks in the plant.
- By superposing all the utilities calculated in the previous step for all HENs, the total amount of energy consumption in the utility system of the plant can be determined, and the type and amount of heating and cooling media will be finalized.

CHOICE OF MULTIPLE UTILITY LEVELS BY THE APPROPRIATE PLACEMENT PRINCIPLE

The Appropriate Placement principle, originally introduced by Townsend and Linhoff in 1983 [8]. It is obvious that it provides the key to choosing the correct levels and loads for the various hot and cold utilities. Normally, because of cost differentials between utilities, it needs naturally to maximize the use of the least expensive ones. This usually led to maximize the use of the coldest hot utility and the hottest cold utility. The shape of the grand composite often dictates the most appropriate choice of levels and loads. Usually, the trade-offs from using multiple utilities are as follows:

- Steam can often be raised more efficiently at a lower temperature level, as more heat can be recovered from boiler flue gases;
- Power can be generated if the steam is let down from a higher pressure through a turbine;
- Lower-temperature steam is at lower pressure and hence the capital cost of the system is less;
- The driving forces between utilities and process are reduced, so that the surface area and capital cost of heaters/coolers will increase;
- Each extra level increases the complexity of design and incurs an additional capital cost for boilers and pipework.

Hence, according to the above consideration, the designer needs to balance the reduction in running cost against the increased capital cost brought about by increasing the number of levels. Based on the types of utility available in the plant, the external heating and cooling requirements are supplied as a hot and cold utility, at a specified temperatures sufficient to fulfill the duty. Referring to [17], in this case study, the heating media are high and low-pressure steam from steam boilers and hot flue gases from furnaces. The saturated HP steam at 45 bar and 274°C is generated in six gas-fired boilers, followed by letdown stations, in order to produce LP steam at 6.5 bar and 185°C by letting saturated high-pressure steam down to the low-pressure level followed by de-superheating. The hot flue gas delivered by gas-fired heaters, as the heating media with the temperature around 250-400 °C. Cooling media are air, chilled and cooling water and low-temperature propane and ethane as refrigerants. The cooling water is recirculated to a cooling system and returned to the cycle, which works in the range 20–40°C. Air as a cooling medium with the design outdoor temperature of 45°C, is derived by electrical fans in air cooled heat exchangers. The refrigeration system uses propane and ethane between -4.0°C to -90°C.

THE CASE STUDY: A BRIEF DESCRIPTION OF THE PLANT

The details of a general gas processing plant were shown in [18]. As the case study of this manuscript, a natural gas refinery designed for refining 2 billion standard cubic feet of natural gas feed per day, delivers 50 million cubic meters per day refined gas injecting to national grid pipeline, 2 million cubic meters per day Ethane, 135 tonnes per day NGL products (propane and butane liquids), 80,000 barrels per day gas condensate and 400 tons per day sulfur solids. The main processing units of the natural gas refinery have been indicated in Table 1. Detail information about this refinery was shown in [10], [17] and [19-22].

Table 1. Main process units of a complex natural gas refinery

UNIT Name	UNIT No.	DESCRIPTION
The gas receiving facilities	Unit 100	Reception Facilities
Gas Refining and Liquid Processing	Unit 101	Gas Treating
	Unit 104	Dehydration and Mercury Guard
	Unit 105	Ethane Recovery
	Unit 106	Export Gas Compression and Metering
	Unit 107	NGL Fractionation
	Unit 109	Sour Water Stripping
	Unit 111	Propane Refrigeration Unit
	Unit 113	Caustic Regeneration Unit for C3 and C4 SULFREX
	Unit 114	Propane Treatment and Drying
	Unit 115	Butane Treatment and Drying
	Unit 116	Ethane Treatment and Drying
	Unit 146	Chemicals Storage
	Unit 147	Propane Storage
	Unit 148	Butane Storage
Condensate stabilization	Unit 103	Condensate Stabilization
	Unit 110	Back-up Stabilization
	Unit 112	Demercaptinize Condensate (DMC)
Sulfur recovery and storage	Unit 108	Sulfur Recovery
MEG regeneration and re-injection	Unit 102	MEG regeneration
Utility services	Unit 120-132	The utility equipment and accessories (HP/LP Steam, Cooling air and water, Refrigerants and Power generation)

By extraction of all the process and utility data from engineering documents of the plant, the total energy requirements due to the utility consumptions of the plant were determined and summarized in Table 2. According to the Table 2, the cold and hot utilities of the existing plant were about 213.7 MW and 171.4 MW respectively. Meanwhile, there was 163.5 MW as the available heat recovery in the plant network [10]. According to the results, there was a huge energy consumption to operate the plant, with about 385 MW for the heating and cooling requirements of the process units. Therefore, any improvement in energy saving of the HEN running in the plant, will significantly reduce the annual operational cost, accordingly.

Table 2. Hot and cold utility consumptions in the existing plant

Service	Type of Utility	Consumption (KW)	Total (KW)
Cold Utility	Cooling Air	167,653	213,721
	Cooling Water	46,068	
Hot Utility	HP/LP Steam	160,398	171,422
	Fuel Gas	11,024	
Total Utility Consumption			385,143
Heat Recovery			163,527

RESULTS AND DISCUSSION

In this plant, the total numbers of hot and cold flow streams were 66 and 44 respectively and the total heat load released by all hot streams is 377,248 KW, whereas total heat load gained by all cold streams is 334,950 KW. The graph of hot and cold composite curves can be developed by drawing all stream temperatures versus cumulative heat loads of hot and cold streams at each temperature level, respectively (Fig. 1).

Referring to the Fig.1, the pinch temperature difference ΔT_{min} in the chart was determined around 66 °C, with the pinch at 97.8 °C shifted temperature, corresponding to 130.8°C for HCC and 64.8 °C for CCC. The Figure shows that total heat recovered by the network is about 163.5 MW, with energy targets of 171.4 MW hot utility and 213.7 MW cold utility.

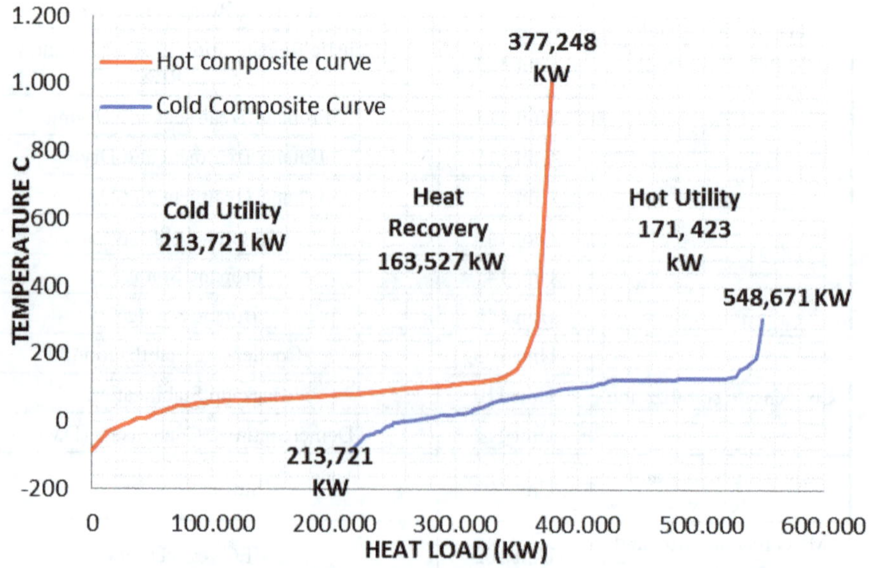

Figure 1. Hot and cold composite curves (HCC/CCC) of the plant network

The plant was divided into the five main areas, each consisting of some neighbor processing units. The nomination of the five areas and the contribution of the process units are shown in Table 3. Therefore, there is a Heat

Exchanger Network in each area, consisting of all the heat exchangers located in that area. For more detail information about the plant description and process data, refer to Mehdizadeh Fard et al. [10].

Table 3. Five divided areas in the plant and the relevant assigned units

Zone Number	Unit Number
AREA 1	Units 100/103/109/110
AREA 2	Units 101/104/105/106/111
AREA 3	Units 107/113/114/115/116
AREA 4	Unit 108
AREA 5	Units 146/147/148

The total numbers of heat exchangers in the network are equal to 104 units with total 208 fluid streams, consisting of 66 hot process streams, 44 cold process streams and 98 utility streams such as HP and LP steams, hot flue gases, cooling water and air. For the details of the hot and cold flow streams, inlet and outlet temperatures, and the heat loads refer to Mehdizadeh Fard et al. [10].

AREA No. 1 (Units 100/103/109/110):

By drawing hot and cold composite curves and considering the hot and cold utility requirements from heat balance of the streams in this network, the hot and cold composite curves and grand composite curve were shown in the Figures 2 and 3, respectively. The Figures show that this case was a special type of the pinch analysis, known as threshold problem. In this area, total numbers of hot and cold flow streams were 10 and 13 respectively. The total heat recovered by the network was about 58,707 KW, with energy targets of 33,458 KW hot utility and no cold utility (Fig. 2).

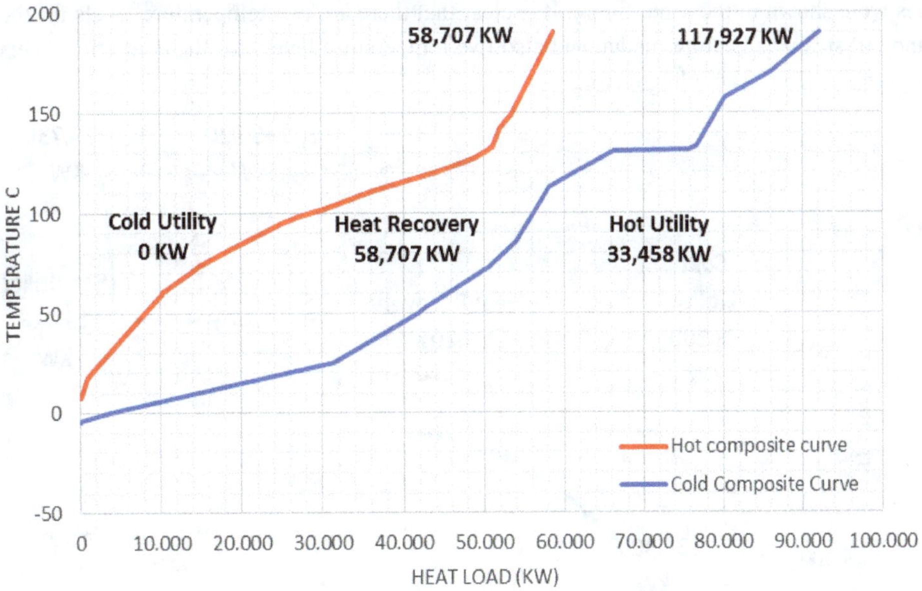

Figure 2. Hot and cold composite curves in Area 1

Referring to the GCC of the network as shown in the Figure 3, the selection of best utility could be done by the Appropriate Placement principle [8]. The type and temperature level of the appropriate utilities were shown in the Figure 3. It is clear that there are two temperature levels for HP steam as heating media, at 210 C and 160 C, and no cold utility.

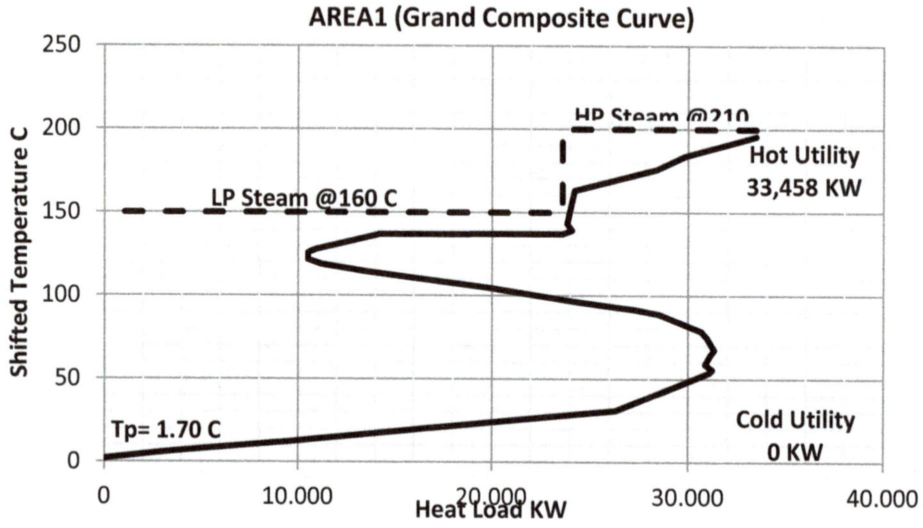

Figure 3. Grand composite curve in Area 1

AREA No. 2 (Units 101/104/105/106/111):

By drawing hot and cold composite curves and considering the hot and cold utility requirements from heat balance of the streams in this network, the hot and cold composite curves and grand composite curve were shown in the Figures 4 and 5, respectively. In this area, total numbers of hot and cold flow streams were 13 and 15 respectively. Hot and cold utility requirements of the existing network in this area were about 19,818 KW and 36,300 KW, respectively, with total heat recovered by the network about 99,620 KW (Figure 4)

Referring to the GCC of the network as shown in the Fig. 5, the selection of best utilities could be done by the same procedure explained for Area1 at previous steps. The type and temperature level of the appropriate hot and cold utilities were shown in the same figure. It is clear that there are two temperature levels for the hot utility, i.e. hot flue gas, and HP steam as heating media, and also two temperature levels for the cold utility, i.e. cooling air at 45 C and cooling water at 35 C as cooling media.

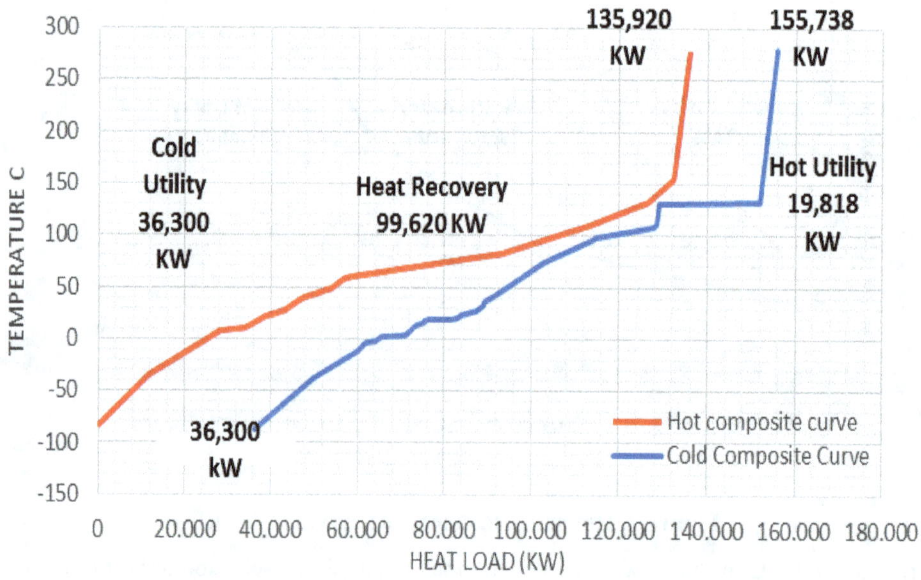

Figure 4. Hot and Cold composite curves in Area 2

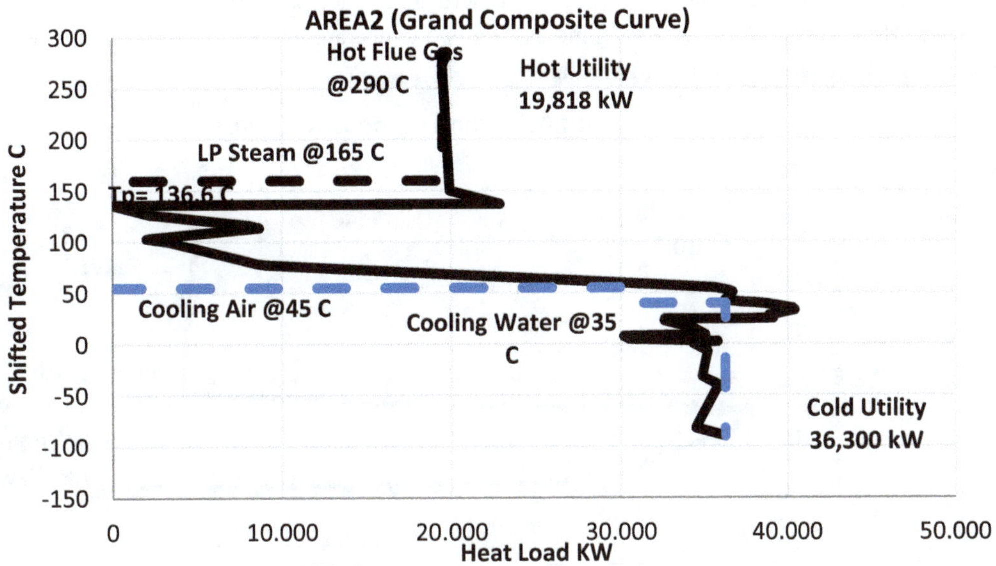

Figure 5. Grand composite curve in Area 2

AREA No. 3 (Units 107/113/114/115/116):

By drawing hot and cold composite curves and considering the hot and cold utility requirements from heat balance of the streams in this network, the hot and cold composite curves and grand composite curve were shown in the Figures 6 and 7, respectively. In this area, total numbers of hot and cold flow streams were 18 and 9 respectively. Hot and cold utility requirements of the existing network in this area were about 26,072 KW and 31,850 KW, respectively, with total heat recovered by the network about 11,837KW (Fig. 6).

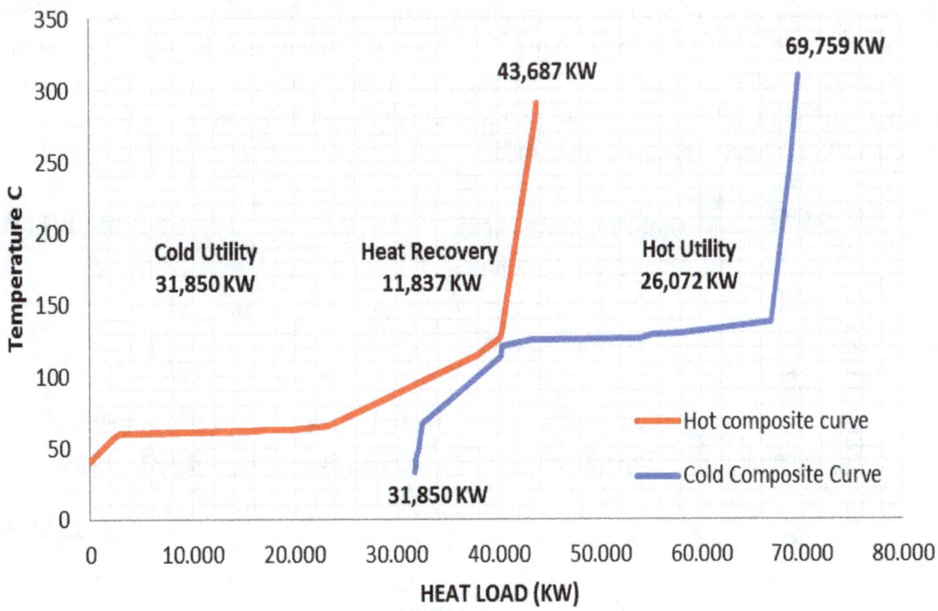

Figure 6. Hot and Cold composite curves in Area 3

Referring to the GCC of the network as shown in the Fig. 7, the selection of best utilities could be done by the same procedure explained for previous networks. The type and temperature level of the appropriate hot and cold utilities were shown in the same figure. It is clear that there are two temperature levels for the hot utility, i.e. hot flue gas at 320 and HP steam at 160 C as heating media, and two temperature levels for the cold utility, i.e. cooling air at 45 C and cooling water at 30 C as cooling media.

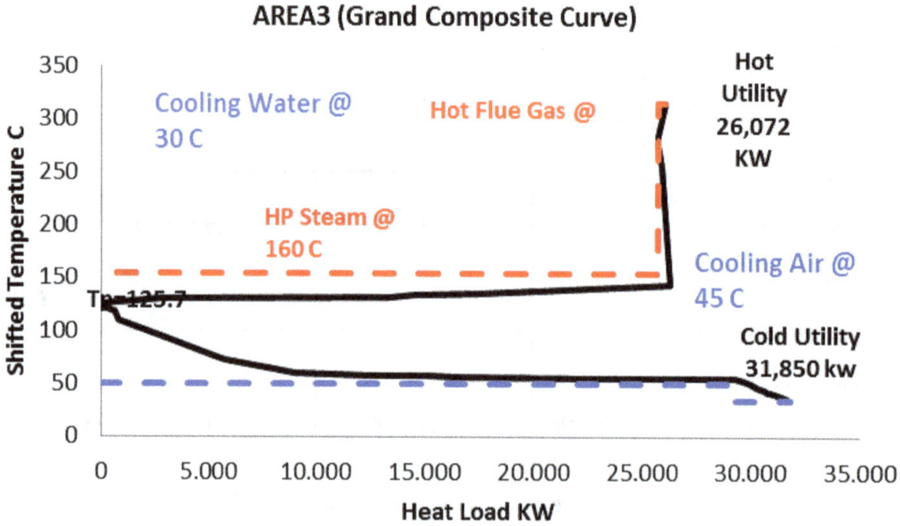

Figure 7. Grand composite curve in Area 3

AREA No. 4 (Unit 108)

By drawing hot and cold composite curves and considering the hot and cold utility requirements from heat balance of the streams in this network, the hot and cold composite curves and grand composite curve were shown in the Figures 8 and 9, respectively. In this area, total numbers of hot and cold flow streams were 12 and 3 respectively. Hot and cold utility requirements of the existing network in this area are about 23,675 KW and 54,348 KW, respectively, with total heat recovered by the network about 49,843 KW (Fig. 8).

Referring to the GCC of the network as shown in the Fig. 9, the selection of best utilities could be done by the same procedure explained in previous steps. The type and temperature level of the appropriate hot and cold utilities were shown in the same figure. It is clear that there is only one temperature level required as the hot utility, i.e. LP steam at 160 C as heating media, but two temperature level as the cold utility, i.e. cooling air and cooling water at 45 C and 30 C respectively, as cooling media.

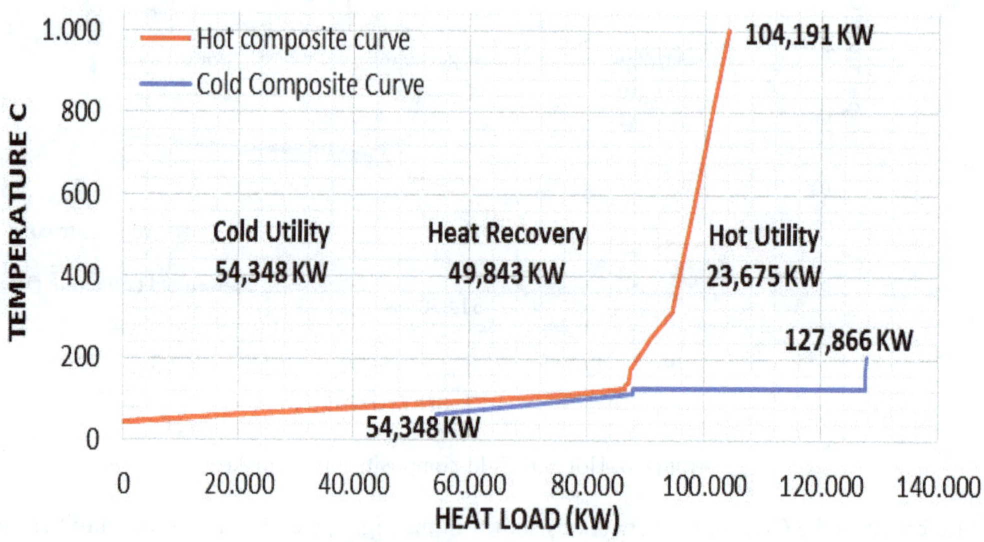

Figure 8. Hot and Cold composite curves in Area 4

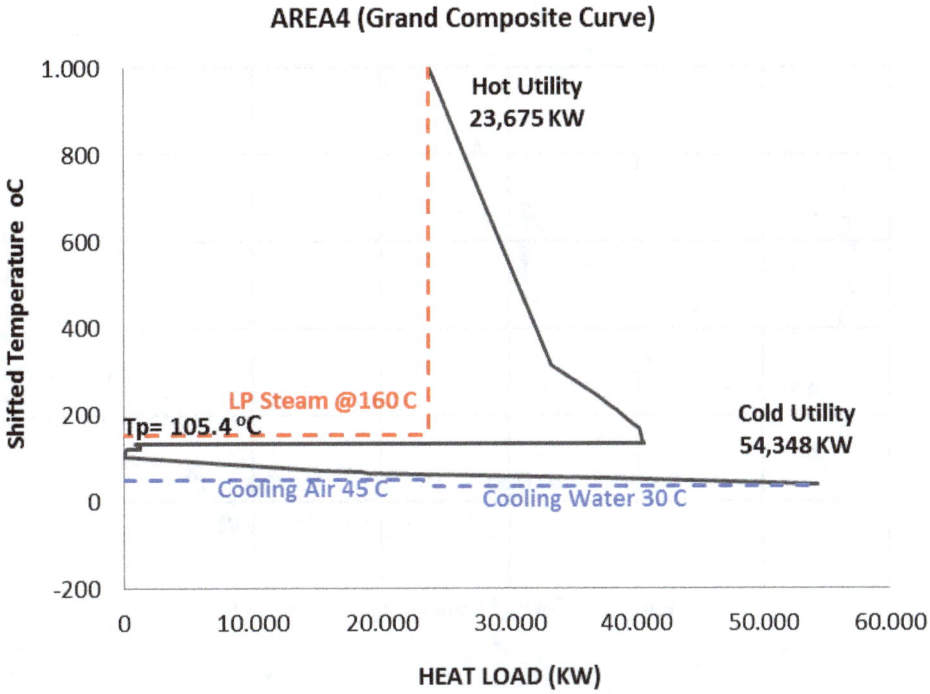

Figure 9. Grand composite curve in Area 4

AREA No. 5 (Units 146/147/148):

By drawing hot and cold composite curves and considering the hot and cold utility requirements from heat balance of the streams in this network, the hot and cold composite curves and grand composite curve are shown in the Figs. 10 and 11, respectively.

In this area, total numbers of hot and cold flow streams are 10 and 7 respectively. Cold utility requirement of the existing network in this area is about 22,823 KW, and no need to heating media as the hot utility. The graph shows that the heat recovery is equal to 11,920 KW, which is exactly equal to the heat gained by the cold composite curve. Therefore, in this special case, the heat recovery is limited by the total heat received by the cold composite curve (Fig. 10).

Referring to the GCC of the network as shown in the Fig. 11, the selection of best utilities could be done by the same procedure explained in previous steps. The type and temperature level of the appropriate hot and cold utilities were shown in the same figure. It is clear that there is only one temperature level required as the cold utility, i.e. cooling water at 35 C as cooling media, and no need to the hot utility.

Summary of the results from all five HENs

By the combination of the results obtained previously in each area of the refinery, the total energy consumption as the hot and cold utilities could be determined for all networks and the results were shown in Table 4. Meanwhile, the available hot and cold utility of the existing plant networks were summarized in Table 5. The comparison of the results shows that the utility consumption in the new network design would be much better than the existing one, and there is about 35.5% energy saving by applying the pinch analysis of the system. Also, the results show that there is a considerable saving in both heating and cooling media. There is about 32% reduction in cooling air and water, and almost 40% saving in steam consumption, relative to the existing network. In addition, the type and temperature level of the heating and cooling media were more reasonable by economic consideration.

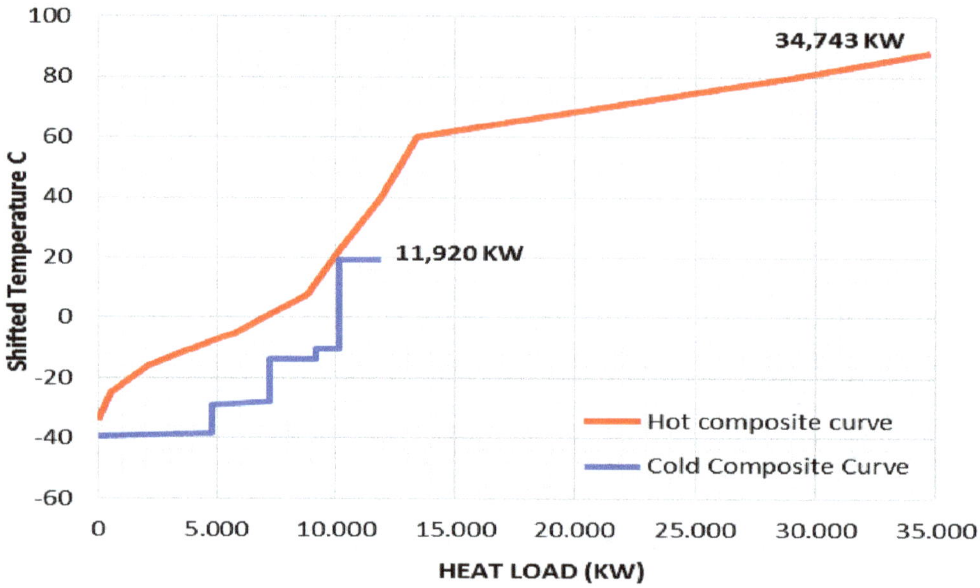

Figure 10. Hot and Cold composite curves in Area 5

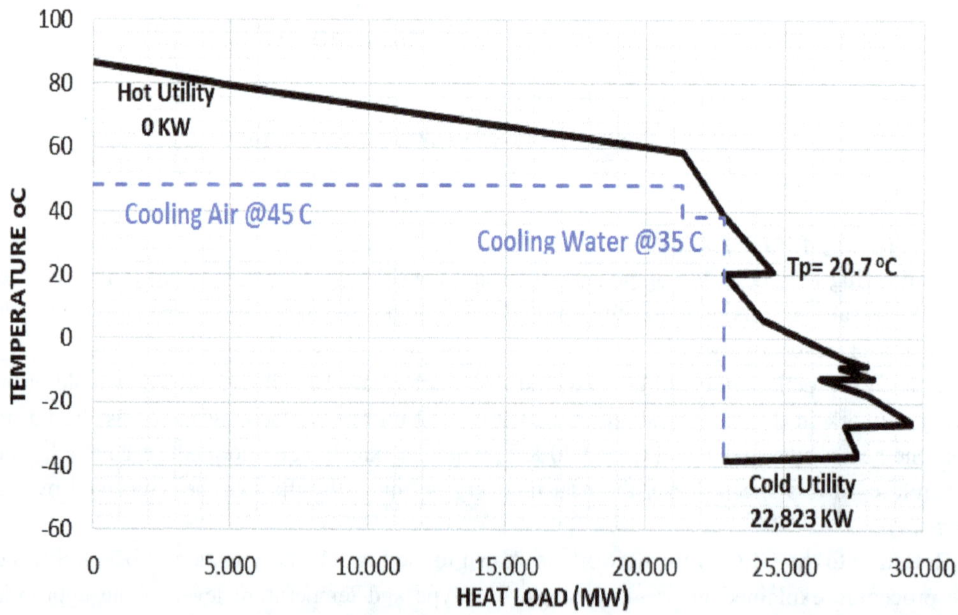

Figure 11. Grand composite curve in Area

The type and temperature level of cooling media were summarized and compared with the existing ones in Table 6. Referring to the table, it was clear that there was about 38 and 100% saving in the total cooling air and BFW consumption respectively, but 45% increase of water consumption as cold utilities.

Meanwhile, the type and temperature level of heating media were summarized and compared with the existing ones in Table 7. Referring to the table, it was clear that there was almost 64.4, 30.0, 90.9 and 100% saving in HP steam, LP steam, hot flue gas and electric heater usage as hot utilities, respectively.

Table 4. Total energy consumption and utility requirement in five networks

TOTAL ENERGY CONSUMPTION IN FIVE HENs (KW)					
Zone No.	COLD UTILITY		HOT UTILITY		TOTAL UTILITY
	KW	Type	KW	Type	KW
Area1	0	-	9,833	HP Steam @210 C	33,458
			23,625	LP Steam @160 C	
Area2	30,190	Cooling Air @45 C	308	Hot Flue Gas @290 C	56,118
	6,110	Cooling Water @35 C	19,510	LP Steam @165 C	
Area3	29,325	Cooling Air @45 C	381	Hot Flue Gas @320 C	57,922
	2,525	Cooling Water @30 C	25,691	LP Steam @160 C	
Area4	23,230	Cooling Air @45 C	23,675	LP Steam @160 C	78,023
	31,118	Cooling Water @30 C			
Area5	21,323	Cooling Air @45 C	0	-	22,823
	1,500	Cooling Water @ 35 C			
All Networks	**145,321**		**103,023**		**248,344**

Table 5. Total energy consumption and utility requirement in five networks (Existing network)

TOTAL ENERGY CONSUMPTION IN FIVE HENs (KW) (Existing Network)					
Zone No.	COLD UTILITY		HOT UTILITY		TOTAL UTILITY
	KW	Type	KW	Type	KW
Area1	23,215	Cooling Air @ 45 C	27,373	HP Steam @ 260 C	79,888
			29,300	LP Steam @ 155 C	
Area2	53,963	Cooling Air @ 45 C	6,426	Hot Flue Gas @ 290 C	103,789
	6,173	Cooling Water @ 35 C	37,227	LP Steam 155 C	
Area3	33,914	Cooling Air @ 45 C	1,125	Hot Flue Gas 350 C	67,954
	2,952	Cooling Water @ 35 C	3,473	Electric Heater	
			26,490	LP Steam @ 155 C	
Area4	36,738	Cooling Air @ 45 C	246	HP Steam @ 255 C	110,689
	16,325	Cooling Water @ 30 C	39,762	LP Steam @ 145 C	
	17,618	BFW @ 120 C			
Area5	19,823	Cooling Air @ 45 C	0	-	22,823
	3,000	Cooling Water @ 30 C			
All Networks	**213,721**		**171,422**		**385,143**

Table 6. Total cold utility requirement in five networks

Comparison of total cold utilities in exiting network & network after utilities selection following pinch analysis			
Zone No.	**Exiting network**	**Network after the utilities selection following pinch analysis**	**Percent Change in Cold Utility**
Area1	23,215 KW Cooling Air @ 45 C	0	100%
Area2	53,963 KW Cooling Air @ 45 C	30,190 KW Cooling Air @ 45 C	44.1%
	6,173 KW Cooling Water @ 35 C	6,110 KW Cooling Water @ 35 C	1.0%
Area3	33,914 KW Cooling Air @ 45 C	29,325 KW Cooling Air @ 45 C	13.5%
	2,952 KW Cooling Water @ 35 C	2,525 KW Cooling Water @ 30 C	14.5%
Area4	36,738 KW Cooling Air @ 45 C	23,230 KW Cooling Air @ 45 C	36.8%
	16,325 KW Cooling Water @ 30 C	31,118 KW Cooling Water @ 30 C	-90.6%
	17,618 KW BFW @ 120 C	0	100%
Area5	19,823 KW Cooling Air @ 45 C	21,323 KW Cooling Air @45 C	-7.6%
	3,000 KW Cooling Water @ 30 C	1,500 KW Cooling Water @ 35 C	50%
All Networks (KW)	167,653 KW; Cooling Air @ 45 C	104,068 KW; Cooling Air @ 45 C	**37.9%**
	28,450 KW; Cooling Water @ 30 & 35 C	41,253 KW; Cooling Water @ 30 & 35 C	**-45.0%**
	17,618 KW; BFW @ 120 C	0 KW; BFW @ 120 C	**100%**

Table 7. Total hot utility requirement in five networks

Comparison Of hot utilities in exiting network & network after utilities selection following pinch analysis			
Zone No.	**Exiting network**	**Network after the utilities selection following pinch analysis**	**Percent Reduction in Hot Utility**
Area1	27,373 KW HP Steam @ 260 C	9,833 KW HP Steam @ 210 C	64.1%
	29,300 KW LP Steam @ 155 C	23,625 KW LP Steam @ 160 C	19.4%
Area2	6,426 KW Hot Flue Gas @ 290 C	308 KW Hot Flue Gas @ 290 C	95.2%
	37,227 KW LP Steam @ 155 C	19,510 KW LP Steam @ 165 C	47.6%
Area3	1,125 KW Hot Flue Gas@ 350 C	381 KW Hot Flue Gas @ 320 C	95.2%
	3,473 KW Electric Heater	0	100%
	26,490 KW LP Steam @ 155 C	25,691 KW LP Steam @ 160 C	3.0%
Area4	246 KW HP Steam @ 255 C	0	100%
	39,762 KW LP Steam @ 145 C	23,675 KW LP Steam @ 160 C	40.5%
Area5	0	0	-
All Networks (KW)	27,619 KW HP Steam @ 255 & 260 C	9,833 KW HP Steam @ 210 C	**64.4%**
	132,779 KW LP Steam @ 145 & 155 C	92,501 KW LP Steam @ 160 & 165 C	**30%**
	7,551 KW Hot Flue Gas @ 290 & 350 C	689 KW Hot Flue Gas @ 290 & 320 C	**90.9%**
	3,473 KW Electric Heater	0	**100%**

CONCLUSION

As one of the conceptual methods in process integration, the pinch analysis is widely used as a suitable technique to evaluate and improve the energy systems. As a case study, the natural gas refinery of the South Pars Gas field in the Persian Gulf was chosen analyzed for energy requirement, in order to identify the type and temperature of appropriate utility as heating and cooling media in the total system.

The new network design improves total energy consumption about 35.52% relative to the existing one. Hot and cold utilities in the new network were 103,023 and 145,321 KW, respectively while the hot and cold utilities in the exiting network were 171,422 and 213,721 KW, respectively. Also, the results show that there was a considerable saving in both heating and cooling media. In this case study, there was about 38 and 100% reduction in cooling air and BFW respectively, but 45% increase in cooling water, as cold utilities, and almost 64.4, 30.0, 90.9 and 100% reduction in HP steam, LP steam, hot flue gas and electric heater as hot utilities, respectively relative to the existing network.

In many cases, this method can be targeted for both energy and cost savings achieved at the same time, by selection of the best utilities as heating and cooling media. The result shows that in addition to the energy saving in the new network design, the distribution of utility media in heating and cooling processes were much better than the existing network.

REFERENCES

[1] Aramesh M., Kasaeian A., Pourfayaz F., Wen D. (2017). Energy analysis and shadow modeling of a rectangular type salt gradient solar pond. Solar Energy, 146, 161–171.

[2] Kasaeian A., Barghamadi H., Pourfayaz F. (2017). Performance comparison between the geometry models of multichannel absorbers in solar volumetric receivers. Renewable Energy, 105, 1-12.

[3] Aghaie M., Mehrpooya M., Pourfayaz F. (2016). Introducing an integrated chemical looping hydrogen production, inherent carbon capture and solid oxide fuel cell biomass fueled power plant process configuration. Energy Conversion and Management, 124, 141–154.

[4] Almutairi MF., Bourisli RI. (2017). Optimum Orientation of a Mutually-Shaded Group of Buildings with Respect to External Solar Radiation. Journal of Thermal Engineering, 3(1), 1065-1070.

[5] Ahmadi MH., Mehrpooya M., Pourfayaz F. (2016). Thermodynamic and exergy analysis and optimization of a transcritical CO2 power cycle driven by geothermal energy with liquefied natural gas as its heat sink. Applied Thermal Engineering, 109, 640-652.

[6] Khir T., Hafdhi F., Ben Yahia A., Ben Braim A. (2017). Exergetic Optimization of Phosphoric Acid Factory Power Plant. Journal of Thermal Engineering, 3 (5), 1428-1441.

[7] Ahmadi MH., Mehrpooya M., Pourfayaz F. (2016). Exergoeconomic analysis and Multi objective optimization of performance of a carbon dioxide power cycle driven by geothermal energy with liquefied natural gas as its heat sink. Energy Conversion Management, 119, 422-434.

[8] Kemp, Ian C. (2007). Pinch Analysis and Process Integration, A User Guide on Process Integration for the Efficient Use of Energy, Second ed., Elsevier Ltd., UK.

[9] Quijera, J.A.; García, A.; Alriols, M.G.; Labidi, "J. (2013). Heat integration options based on pinch and exergy analyses of a thermo solar and heat pump in a fish tinning industrial process. Energy, 55, 23-37.

[10] Mehdizadeh Fard, M., Pourfayaz, F., Kasaeian, A.B., Mehrpooya, M. (2017). A Practical Approach to Heat Exchanger Network Design in a Complex Natural Gas Refinery. Journal of Natural Gas Science and Engineering, 40C, 141-158.

[11] Khorrammanesh, M., Amidpoura, M., Nasr, M.R.J. (2007). Application of process decomposition in multi-stream plate fin heat exchangers design to use in heat recovery networks. Chemical Engineering and Processing, 46, 941–954.

[12] Tabatabaiyan, S., Rohani, A.A., Bahmanpour, H. (2014). Energy recycling in a gas refinery using a thermal exchanger system: The case of Masjed Soleiman gas refinery. European Journal of Experimental Biology, 4(3), 402-406.

[13] Kazempour, H., Pourfayaz, F., Mehrpooya, M. (2017). Modeling and multi-optimization of thermal section of Claus process based on kinetic model. Journal of Natural Gas Science and Engineering, 38, 235-244.

[14] Kianfar, E., Tayebinejad, E. (2014). Energy Optimization of Ilam Gas Refinery Unit 100 (stabilization) by using HYSYS REFINERY Software. International Research Journal of Applied and Basic Sciences, 8 (9), 1265-1270.

[15] Joe, J.M., Rabiu, A.M. (2013). Retrofit of the Heat Recovery System of a Petroleum Refinery Using Pinch Analysis. Journal of Power and Energy Engineering, 1, 47-52.
[16] Keshavarzian, S., Verda, V., Colombo, E., Razmjoo, P. (2015). Fuel saving due to pinch analysis and heat recovery in a petrochemical company. Proceedings of ECOS 2015-The 28th International Conference on Efficiency, Cost, Optimization, Simulation and Environmental Impact of Energy Systems, Pau, France.
[17] Hamedni, N., Maddah, R. (2011). Description of Process and Utility Units of South Pars Gas Plants. University of Tehran Press, First Edition, Tehran.
[18] Kidnay, A.J., Parrish, W.R. (2006). Fundamentals of Natural Gas Processing, Taylor and Francis Group, CRC Press, NY.
[19] Hafeznia H., Pourfayaz F., Maleki A. (2017). An assessment of Iran's natural gas potential for transition toward low-carbon economy. Renewable and Sustainable Energy Reviews, 79, 71-81.
[20] Mehdizadeh Fard, M., Pourfayaz, F., Mehrpooya, M., Kasaeian, A.B. (2018). Improving energy efficiency in a complex natural gas refinery using combined pinch and advanced exergy analyses. Applied Thermal Engineering, 137, 341-355.
[21] Mehdizadeh Fard, M., Pourfayaz, F. (2018). A simple method for estimating the irreversibly in heat exchanger networks. Energy, 144, 633-646.
[22] Mehdizadeh Fard, M., Pourfayaz, F. (2019). Advanced exergy analysis of heat exchanger network in a complex natural gas refinery. Journal of Cleaner Production, 206, 670-687.

11

COST EFFECTIVE OPERATION OF HVAC SYSTEM UNDER THERMAL DISTURBANCE

O. Al-Tamimi[1], M. Kassas[1], W. M. Hamanah[1], B. S. Yilbas[1,*]

ABSTRACT

Varying cooling load requirements in HVAC system become a major concern to many consumers because of the additional cost of electricity. The cooling load variation of HVAC system usually occurs when the domestic house door is frequently kept open for prolonged durations, which is particularly important in hot seasons. The present study, addresses a comparison of energy consumption due to the thermal disturbance during domestic house door keep opened for various duration when the Variable Frequency Driver (VFD) and ON/OFF controllers are installed in HVAC system. The experiments are carried out in two identical houses during six warm days in a hot season where the outdoor temperatures have similar patterns throughout the days. The energy response of HVAC system is analysed incorporating the thermal and the electricity consumption data because of the change of the indoor temperature during opening of the house main door. The cost analysis is carried out and it is extended to include several repeats of door opening durations. It is the first time demonstrated that the energy consumption for HVAC system increases dramatically when the house main door is kept opened for several times in a day. The energy consumption and additional cost increase significantly for the case when the period of the main door opening is extended. In this case, VFD A/C controller unit achieves more energy savings than that of ON/OFF A/C controller unit.

Keywords: *HVAC, Energy Consumption, Variable Frequency Drives VFD, ON/OFF Cycle, Cost Analysis*

INTRODUCTION

Air conditioning (A/C) is mainly utilized for maintaining thermal comfort in indoor environments, especially for hot and humid climate. Nowadays, air conditioning has become a necessity in residential area, commercial buildings and industrial processes. It is considered to be one of the fastest growing energy sectors in Saudi Arabia. A/C system is directly participating in increasing energy consumption due to the Saudi Arabia weather which is characterized with high temperature, humidity, and dust storms. A temperature in Saudi Arabia follows the pattern of the desert_climate, with the exception of the southwest. The average summer temperature is about 45° C, and the highest temperature is reaching 55° C are not unusual [1-2]. In the following years, the climate change could make temperature raise high in summer days in the Arabian Gulf [1]. As a consequence, the higher outdoor temperature increases the demand for cooling over the years and the energy consumption. Although several attempts have been made to improve thermal performance of cooling systems, such as air conditioning (A/C) units [3-7], it is the first time to consider the efficient operation of cooling systems under temporal thermal disturbance due to house main door opening for several durations.

Minimizing the energy consumption in residential areas is always the main concern and retains a lot of research attention. Energy consumptions in the Kingdom of Saudi Arabia (KSA) have grown substantially over the previous years. Electricity consumption in the KSA is considered one of the highest per capita consumption rates in the world around 8.161 kWh /capita which is approximately three times more than the world's average [8]. The energy demand in correspondence has grown at a much higher pace of approximately 5-8% per year [8]. Moreover, A/C systems are considered to be the major energy consumptions in the residential area in KSA and it is accounted for approximately 60% [9-10]. The significant focus in evaluating the performance of HVAC systems is ensuring the thermal comfort and reducing the energy consumption.

On the other hand, most of the occupants have bad energy consumption habits. They frequently kept the main door of the house open while the A/C unit is in operation; this leads the hot and moist ambient air mixing with

This paper was recommended for publication in revised form by Regional Editor Ahmet Selim Dalkilic
[1]Department of Electrical Engineering, KFUPM, Dhahran, Saudi Arabia
**E-mail address: bsyilbas@kfupm.edu.sa*

the indoor cool air. Since the door opening behavior significantly affects the thermal comfort level, this may have an impact on energy consumption patterns. However, it does matter how frequent the bad habit repeats. If the main door of the house remained opened frequently and for long periods of time, large amount of energy might be wasted per a year. Therefore, the door opening/closing effect is truly a major concern to many consumers. Fabi et al. [11] discussed that the window opening behavior might have direct effects on energy consumption for cooling by varying the indoor air flow rate inside the houses. Pan et al. [12] had conducted an experimental test in an office building in Beijing, China and monitoring of occupants' window opening behavior, together with important indoor and outdoor environmental variables. They found that window opening behavior influences the performance of buildings significantly, and it was required to be improved to promote the energy efficiency of the building. Rijal et al. [13] developed an adaptive window opening algorithm to predict the thermal comfort, energy use and overheating in buildings. It was found that when the window was open the mean indoor and outdoor temperatures were higher than when closed, and they used adaptive algorithm to achieve more comfortable, lower energy buildings while avoiding overheating. Brown et al. [14] investigated a large scale energy surveys in the UK retail. They conducted the survey to examine differences in energy use behavior and the effect of open doors in retailers, between winter and summer season. The estimates of heating and cooling wastage have been made using computer simulation and empirical equations to model air flow through open doors with an inside to outside temperature difference. The results showed that the use of energy saving measures during the cooling season, such as closing external doors or using air curtains was minimal compared to the warm season.

Since the main door opening influences the energy consumption, duration of the house main door opening becomes critical for air-conditioning (A/C) unit operation. Although the effect of the door air flow on A/C energy consumption was studies previously [9-15], the temporal effects of the cooling load demand on energy consumption were not included in the studies. The temporal demands for the domestic cooling load can occur when the house door is frequently kept open in buildings. The energy consumption increases to overcome the need for the cooling load. This, in turn, influences the cooling system performance and the cost. Consequently, in investigation of outdoor air flow on energy consumption of A/C unit in hot environments becomes essential. Two experiments are conducted to investigate the energy consumption of HVAC systems, due to the warm and humid air transfer from the surroundings to the house, through the opening of the house main door for selected periods of time. In the present study, two different units of air conditioning systems are incorporated and each system is installed in two identical houses. Two cases are considered during six warm days where the outdoor temperatures have similar patterns throughout the days of the study. In the first case, the house main door is opened during the periods of 15 minutes, 30 minutes and 60 minutes for three days when A/C unit is active. The time of start for the main door opening is at 12:30 p.m. local time in Kingdom of Saudi Arabia. The second case is carried out for the other three days and the door is kept closed for all periods while A/C unit remains functional. Energy consumption and cost analyses are carried out to assess the performance of HVAC systems when the main door remained opened for selected durations. It should be noted that the duration for keeping the main door open is recorded over the period of one year in the local area where the tests are carried out. It is observed that the most frequent period of keeping the main door open is within 15, 30, and 60 minutes. The frequent door openings are mainly associated with the children in and outdoor activities. However, 60 minutes opening of is mainly related to forgetting to close the door.

EXPERIMENTAL SETUP

The performance and energy consumption of two A/C systems are assessed incorporating the weather conditions inside and outside of two identical houses considered. The monitoring and measurement hardware comprise of four major blocks: two 5-ton Al-Zamil rooftop A/C units, a Data Acquisition (DAQ) chassis with National Instrument modules (Lab-View), sensors, and a host computer. The National Instrument DAQ-chassis monitoring system has several modules such as voltage measurements, current measurements, and thermocouple input. In measurements, four temperature sensors (three sensors for indoor and one for outdoor), two humidity sensors (indoor and outdoor), an irradiation sensor, barometric pressure sensor, and three air flow sensors are used. The host computer has the National Instrument (NI) software, which is the main communicator with DAQ-chassis. The host computer initiates the execution commands, store the data, and display them on the monitor.

Experiment Procedures

The experiments were carried out in two identical houses at the KFUPM campus, Dhahran, Saudi Arabia. The floor plan is shown in Figure 1. Each house consists of a living room, one bedroom, a kitchen and a bathroom. All details of the houses which are integrated with ON/OFF and VFD HVAC systems are described in this section and followed by details of the HVAC monitoring and measurement systems [16-17].

Houses Details

General project information for two houses, location and environment parameters are presented in Table 1.

Table 1. Houses details and environment parameters for project

Weather reference city	Dhahran, Saudi Arabia
Project location	KFUPM, Guest-Houses #3305 and # 3307
House # 3305	ON/OFF HVAC unit
House # 3307	VFD HVAC unit
Barometric pressure	1007.652
Altitude	22.86 meters
Latitude and longitude	26.288768 and 50.114103

Sensors Location in the Houses

The measurement system and sensors are placed in both houses to monitor the required data for the performance assessment of the A/C units. Figure 2, shows the location of each sensor in the house. Three thermocouples are located in the bedroom wall (close to the door), one in the corridor wall (close to the house main door) and one in the living room wall. Temperatures are measured from three positions and later average temperature is estimated for the house. Airflow sensors are located in the middle of each duct exit to measure the A/C unit air mass flow rate. The pressure sensors are located inside the house to monitor indoor pressure variation. Thermocouples, outdoor humidity sensors, wind speed sensors, and the solar irradiation sensor are placed on the roof of the house to measure the ambient temperature, wind speed, and daily solar radiation.

Monitoring and Measurement Systems

The monitoring and measurement systems for both units are developed to achieve different tasks such as displaying several physical and electrical characteristics, and environmental conditions. Furthermore, the data acquisition systems incorporate signals, sensors, actuators, signal conditioning, data acquisition devices, and application software. The data acquisition system enables to measure electrical or physical parameters such as voltage, current, temperature, irradiation, humidity, pressure, and wind speed. In addition, the PC-based data acquisition system uses a combination of modular hardware, and application software (Lab-View). Figure 3 shows the chart of the monitoring and measurement system [18].

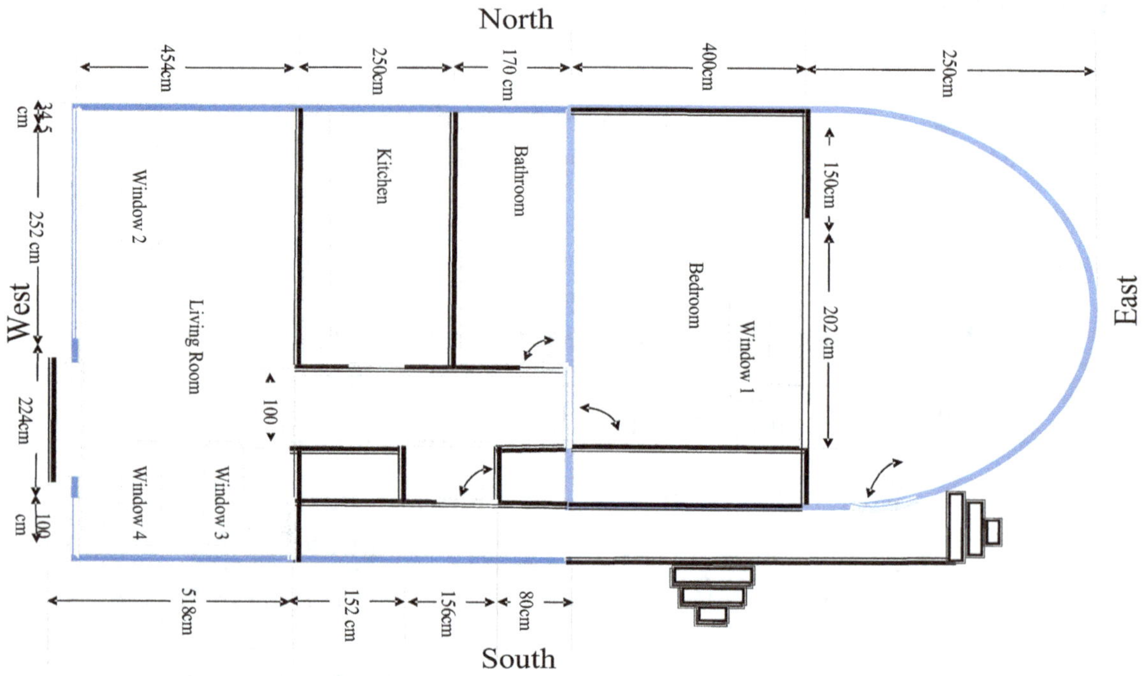

Figure 1. Floor plan for houses #3305& #3307

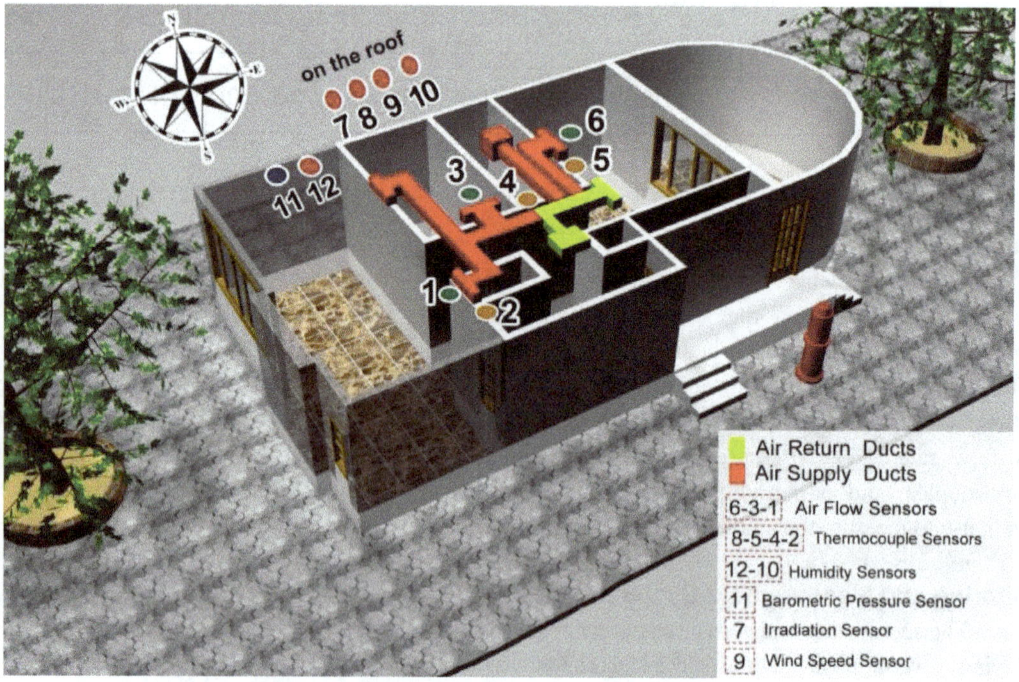

Figure 2. Ducts plan and sensor's locations in the houses

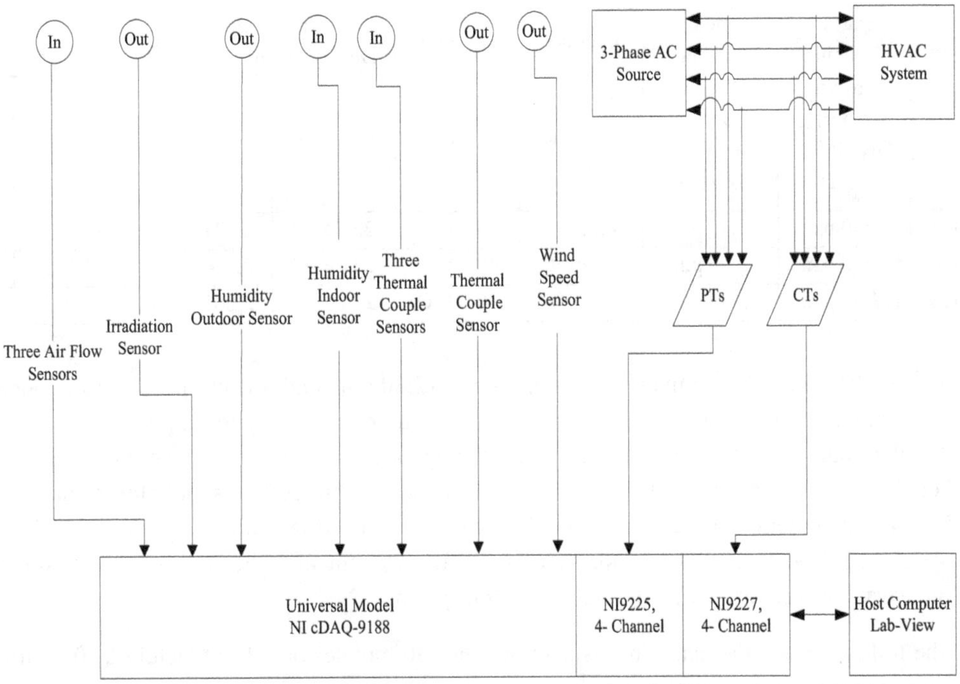

Figure 3. HVAC monitoring and measurement system

All the equipment and measuring devices are calibrated prior to the experiments. Table 2 gives the calibration data.

Table 2. The extra experimental tools for calibration and measurements

Instrument	Multi-Function Ventilation Meter VELOCICALC (9565)	Digital Handheld Multi-meter (PeakTech-3315)	Power Quality Analysis (PQA-HIOKI 3197)
Measurements	Temperature, humidity, pressure, air velocity, and wind speed	Current, frequency, and voltage, etc.	Validate the measured power by Lab-View platform
Accuracies	±3% of reading or ± 0.015 m/s, ± 0.3°C and ± 3% RH	± 0.1% reading + 0.2% range	±0.3 % rdg . ±0.01% f.s within ±0.5°C, ±3% at 40Hz to 5kHz

RESULTS AND DISCUSSION

The experiments are conducted to investigate the effect of opening the main doors of two identical houses on the cooling performance of A/C units for selected time periods. The study is performed during six warm days (8th to 13th of October, 2016) when the outdoor temperatures have similar patterns throughout the days.

Opening Main Door for House Equipped With VFD HVAC Unit

Figure 4 shows the outdoor air temperatures for the house equipped with VFD HVAC unit during the selected three days from 12:00 p.m. to 15:00 p.m. The indoor air temperature is taken as an average of three thermocouple readings placed at three different locations in the house. The indoor and outdoor temperatures are listed in Table 3 during the starting and ending of door opening operation.

Table 3. Outdoor and indoor temperature for both units on 11th, 12th and 13th of October, 2016

Date	Start of Opening (Main Door) ON/OFF& VFD	End of Closing (Main Door) ON/OFF &VFD	Outdoor Temp During Opening Time	Outdoor Temp During Closing Time	Indoor Temp During Opening Time		Indoor Temp During Closing Time	
					ON/OFF	VFD	ON/OFF	VFD
11/10/2016	12:30 p.m.	12:45 p.m.	42.23 °C	43.22 °C	22.74 °C	23.31 °C	26.17 °C	26.17 °C
12/10/2016	12:30 p.m.	13:00 p.m.	41.96 °C	40.98 °C	23.37 °C	22.14 °C	26.49 °C	26.49 °C
13/10/2016	12:30 p.m.	13:30 p.m.	40.54 °C	39.03 °C	24.06 °C	23.07 °C	25.41 °C	25.41 °C

On 11th of October, 2016 the main door is opened at 12:30 p.m. and it kept open for 15 minutes. The indoor temperature increases after one minute from the start of the main door opening period, which is associated with the hot air, which enters into the house. In this case, indoor temperature increases to reach 28 °C. However, it causes increasing of cooling air temperature, and gives rise to the operation of the A/C to supply low temperature cool air in the house. The increased temperature is recovered by the cool air flow until it reaches the desired set point temperature (23.5 °C). The main house door is closed after 15 min and the power reading for A/C unit and temperature recovery time are recorded via Data accusation system.

On the following day, the main door is kept open for 30 minutes on 12th of October, 2016, from 12:30 p.m. to 13:00 p.m. Indoor temperature variation is recorded staring at the time of the main door opening and the maximum temperature increases to reach 28 °C at the end of door opening period. However, temperature fluctuates in between 24.5 °C and 28 °C during the test because of non-uniform cooling. Once the main door is closed, the minimum indoor temperature becomes less than the set point temperature and it reduces around 21.5 °C. This variation occurs to recover the heat loss for the whole house to reach the set point temperature. On the last day, the main door is kept open for 60 minutes from 12:30 p.m. to 13:30 p.m.; similar changes takes place for the indoor temperature and some small temperature variation occur after the time of main door opening. It is noticed that the response of the VFD HVAC unit on the indoor temperature during the experiment is faster than the ON/OFF HVAC system. However, VFD unit has the ability to recover and match the desired temperature in a short time period.

Opening Main Door for House Equipped With On/Off HVAC Unit

The similar test of variable frequency HVAC unit is conducted for the house equipped with ON/OFF HVAC unit. The door is opened for three different periods 15 minutes, 30 minutes and 60 minutes on 11th, 12th and 13th of October, 2016. Figure 5 shows the outdoor temperature pattern and the indoor air temperature variation for each day. The switching ON/OFF band of the air conditioner is set as –1.5 °C and +1.5 °C. On the first day the main door is kept open for 15 minutes starting from 12:30 p.m. to 12:45 p.m. The variation on the indoor air temperature indicating that during the operation of opening the main door, the indoor temperature is lower than the set point and it takes approximately 4 minutes to pass the set point temperature and reaches 27 °C, then it continue fluctuating between 26 °C and 28 °C. The indoor temperature drops to 22 °C after 5 minutes from the door closing moment, and then it increases again to reach the desired set point. In the following day the door is opened for 30 minutes. According to Figure 6 the indoor temperature is varied from 24 °C to 28 °C during main door opening period, and due to the higher indoor temperature, the air conditioning system has run for a bit longer period of time, after the door closing to match the indoor temperature with the set point temperature.

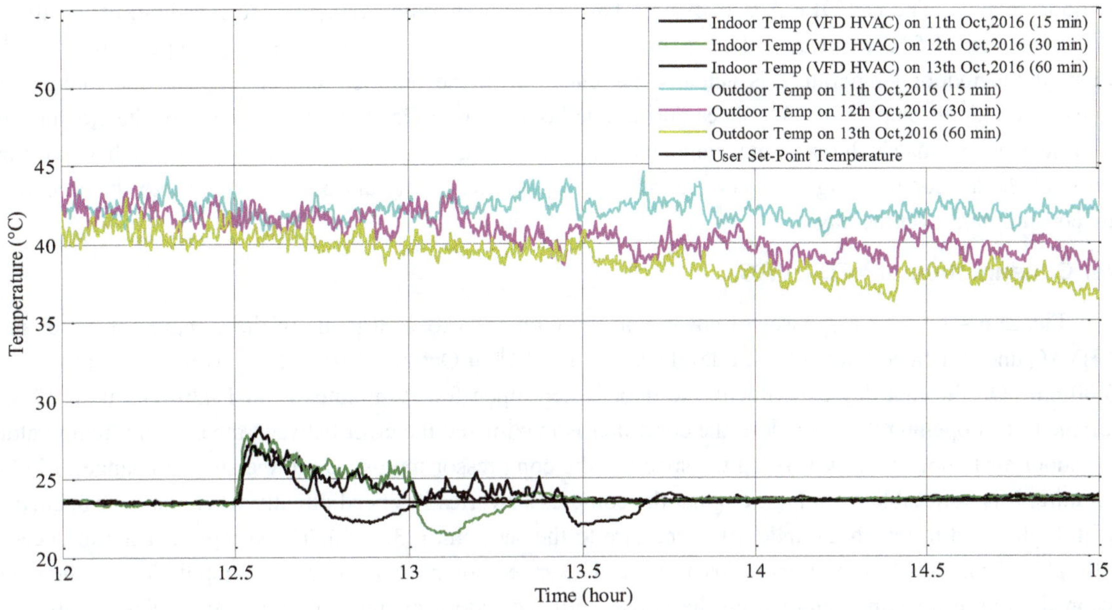

Figure 4. The outdoor air temperatures and indoor air temperatures responses during the door opening experiment in house #3307 VFD HVAC

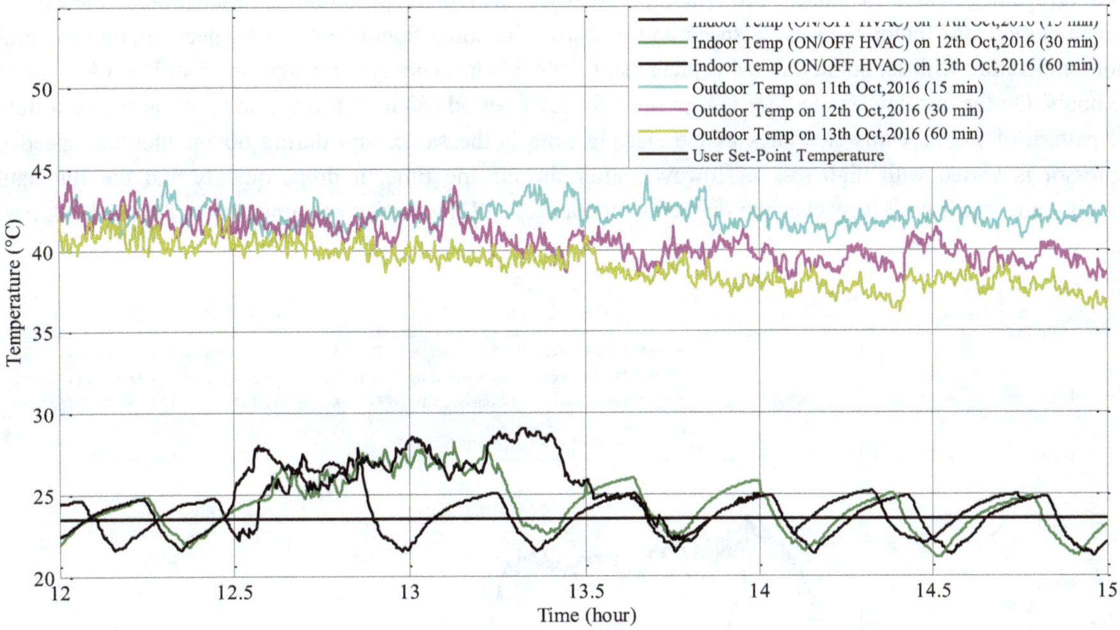

Figure 5. The outdoor air temperatures and indoor air temperatures responses during the door opening experiment in house #3305 ON/OFF HVAC

Finally, on the 13th of October, 2016, the door is kept open for 60 minutes starting at 12:30 p.m. and closed at 13:30 p.m. It can be observed From Figure 5 that the indoor temperature is nearly matching the set point temperature during the test of the main door opening and it is directly increases to 28 °C, which means there are some factors can contribute to increase the house indoor temperature in a short period of time, such as the wind speed and outdoor humidity. Moreover, at the time of door opening, the measured temperature is 24.07 °C, and after few second it increases to 25.5 °C. This behavior occurs because the A/C unit was OFF before and during the action of the door opening. The ON/OFF A/C is cycling based on the operation range or the threshold of the thermostat which set between –1.5 °C to +1.5 °C. The starting temperature is 25 °C and the stopping temperature is 22 °C. The variation of the indoor temperature is actually associated with the outdoor temperature and the internal houses activities. The indoor temperature fluctuates during one hour between 25 °C to 28 °C and until the closing time. It starts to decrease to match the set point temperature. In addition, opening the main door of the house during the operation of the air conditioning can vary the indoor temperature greatly and it is reflected into the amount of the energy consumption over the day.

Power Consumption of VFD HVAC

Figure 6 shows the responses of power consumption at various variation of the compressor motor speed of VFD HVAC unit, for three hours of each day 11th, 12th and 13th of October,2016, respectively, staring at 12:00 p.m. to 15:00 p.m. On the first day experiment, the door is kept open for 15 minutes from 12:30 p.m. until 12:45 p.m. During the test of opening the main door the controller is maximized the error between the set point temperature and actual indoor air temperature by varying the speed of the compressor motor, due to the sudden change of indoor air temperature. It is noticeable from Figure 6 that the compressor varies its speed directly when the door opened, and it run with high speed to match the indoor temperature to the set point (23.5 °C). The compressor motor speed drops abruptly after 3 minutes from the door closing time. The speed compressor motor is manipulating by the controller based on the return air temperature from the house until the indoor air temperature is at or close to its set point temperature. Table 4 gives the energy consumption by VFD HVAC unit for the first day when the door is kept open for 15 minutes. 5.9153 kWh is energy consumptions for three hours from 12:00 p.m. to 15:00 p.m. for the first day test. On second day, the power response of the VFD follows the pattern of the previous day as the starting time of main door opening is also the same. Furthermore, it continues with high speed until the main door is closed at 13:00 p.m. The compressor motor reduces its speed and it drops with some fluctuation in the speed attempt to adjust the indoor temperature with set point. Table 4 depicts that 6.7069 kWh is energy consumption of VFD HVAC unit during three hours. On the last day, as the door is kept open for long period (60 minutes) the compressor motor follows the same pattern of the previous two days as the starting time is the same, and during 60 minutes the speed of the compressor is varied with high speed. However, after the closing time, it drops quickly and the fluctuation is relatively quite small until it reaches the desired temperature, and the energy consumption is 8.8365 kWh per three hours.

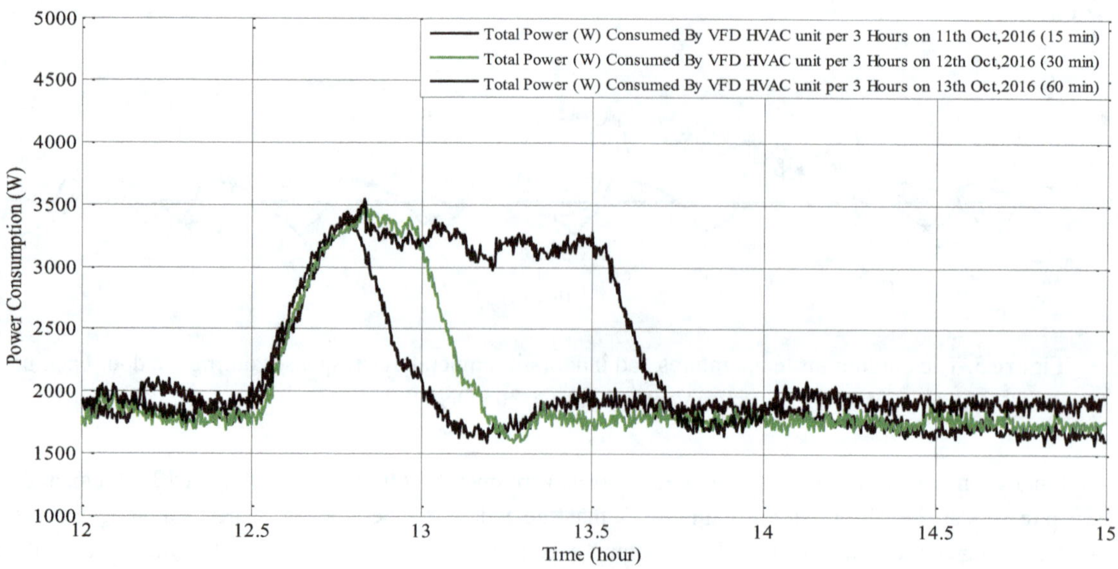

Figure 6. Power consumptions for 3 hours by VFD unit on 11th, 12th and 13th of Oct .2016

As can be concluded at the end of this section, the VFD HVAC responds faster once the door is opened. This leads to the level of user comfort can be matched shortly. In addition, as the door is kept open for three different periods of time. It can be noticed, that after the door closing time of each period the curves looks similar to each other and the power consumption has its steady state value around 1.65 kWh, beyond which, the power becomes nearly constant or just fluctuating in the narrow range.

Table 4: Energy consumption by VFD & ON/OFF unites on 11th, 12th and 13th of Oct. 2016

Date	Time of the Experiment (Door Opened)	Energy Consumption Time (Hours)	Time (3 hours)	VFD Unit Energy Consumption	ON/OFF Unit Energy Consumption	Difference in Energy Consumption VFD &ON/OFF
11/10/2016	15 Min	3 hours	From 12:00 p.m. To 15:00 p.m.	5.9153 kWh	9.393 kWh	3.478 kWh
12/10/2016	30 Min	3 hours	From 12:00 p.m. To 15:00 p.m.	6.7069 kWh	10.884 kWh	4.1772 kWh
13/10/2016	60 Min	3 hours	From 12:00 p.m. To 15:00 p.m.	8.8365 kWh	12.923 kWh	4.0868 kWh

Power Consumption of On/Off HVAC

ON/OFF control is one of the oldest techniques that is practiced in buildings for the purpose of energy saving and occupant thermal comfort level. Figure 7 shows the power consumptions of ON/OFF HVAC system for three hours of each day 11th, 12th and 13th of October, 2016, respectively; staring at 12:00 p.m. to 15:00 p.m. On the first day, the main door is kept open for 15 minutes from 12:30 p.m. to 12:45 p.m. The compressor motor is OFF during the staring time of the main door opening test. The thermostat ON/OFF controller takes around 2 minutes to minimize the error between the set point temperature and the indoor air temperature by turning ON the compressor motor. It is clearly observed from Figure 7 that the controller turns OFF after 13 minutes from the door closing time. This action is taken to allow a faster heat loss recovery to the house until the indoor air temperature reaches the lower limit of the set point temperature. Under the ON/OFF condition, the compressor motor speed fluctuates from equivalent 635 Watt to full speed and stops several times as the blower motor is set on the fixed mode and it is working continuously during the whole day. From Table 4, 9.3933 kWh is the energy consumption of ON/OFF HVAC unit for three hours of the first test.

Based on the experiment on the second day, the main door of the house is kept open for 30 minutes from 12:30 p.m. to 13:00 p.m. Figure 8 shows the compressor motor is already ON during the main door opening operation. After 12:30 p.m. the compressor motor is running continuously and turns OFF after 12 minutes from the door closing moment. Moreover, the compressor motor resumes its ON/OFF cycling until the end of the day. Table 4 gives the energy consumption 10.8841 kWh during the three hours, and compares to the three hours of the first day the energy consumption is higher by 1.491 kWh. On the other hand, the main door of the house is kept open for 60 minutes in the third day of test, and Figure 8 shows that after 2 minutes from the main door opened the compressor motor turns ON. The compressor motor runs continuously and exceeds the door closing time by 18 minutes, before resuming its ON/OFF cycling. The reason that the compressor runs continuously for a longer period of time is that main door is kept open for one hour, much of the surrounding warm and moist air mixes with house cool indoor air temperature which increases the indoor air temperature above the desired temperature rapidly. According to Table 4, 12.9233 kWh is the energy consumption of the three hours when the door kept opened for 60 minutes.

Frequently opening the main door causes the increase of energy consumption. The performance of the ON/OFF system is running under one fixed compressor speed without any capacity control. In addition, the ON/OFF is using the full speed of the electric compressor that is equivalent to 6200 Watt. It can be claimed that energy consumption for the ON/OFF unit is higher than the VFD unit. The main reason for this is the difference between the operation modes as the ON/OFF unit is running under constant speed, and the VFD unit has the ability to vary its speed.

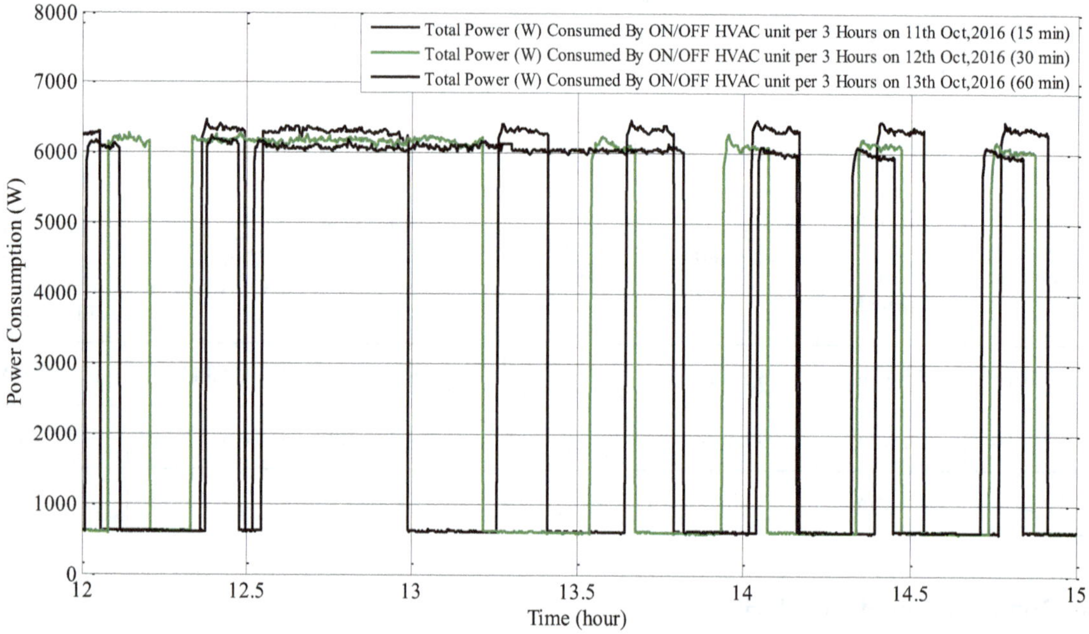

Figure 7. Power consumptions for 3 hours by ON/OFF unit on 11th, 12th and 13th of Oct. 2016

From the technical side of both HVAC units, the ON/OFF unit is controlling by the thermostat device and the VFD unit is controlling by using two types of Proportional, Integral and Derivative (PID) control integrated with variable frequency drive. However, one PID controller use to control on the startup of the unit and the second PID use to control during unit operation. The startup control must prevent an excess of requested power. Since at startup the status of the load is not known but only the temperature is, the power must be entered little by little, waiting for the reaction of the system. It can regulate on the value of the return temperature using a wide proportional band (2-3 times the nominal thermal gradient) and a large enough integral time that is greater than the system time constant. The control during operation must be quick in order to follow any load variations and maintain the supply temperature as close to the set point value as possible. In this case, the time constant is given by the reaction of the compressor - evaporator system and is in the order of a few tens of seconds (slower with shell and tube evaporators, faster with plate evaporators). The PID controllers integrate the "anti-windup" function that limits the integral action when the request has reached the maximum and minimum limits.

It can be concluded that VFD A/C unit has the ability to control the cooling load by varying its speed (compressor speed) according to the temperature of the air return from the house. The response of speed controlling of the compressor is fast which has an influence in the amount of the energy variation and energy consumption. The VFD A/C unit can achieve the desired temperature to the house without costing significant energy consumption. ON/OFF A/C unit is cycling based on the temperature obtained by the thermostat. The time response of ON/OFF unit to reach the desired temperature is very long comparing to the time response of VFD unit. The longer time taking for response and achieve the set point temperature is the huge energy consumption. In addition, the condenser motor of the VFD A/C unit is running on the same mode of the compressor motor (Auto Mode), and the condenser motor of the ON/OFF unit has different mode comparing to the compressor motor mode (Fixed Mode). The condenser motor is running continuously without any stop throughout day. The motors of ON/OFF unit are running under full power load about (6000 W) and the motors of VFD A/C unit are running under different power load based on the required air return.

Closing of House Main Door Equipped With VFD HVAC Unit

Figure 8 shows the indoor and outdoor temperatures for three days from 12:00 p.m. to 15:00 p.m. on 8th, 9th and 10th of October, 2016, while the main door of the house is kept closed. The outdoor temperatures vary between 37 °C to 45 °C, and the indoor air temperature is taken as an average of three thermocouple sensors located at three different places in the house #3307. It is clearly shown that the indoor air temperatures have similar patterns throughout the days, and it is almost similar to the set point temperature. A variable frequency compressor could distribute conditioned air at different temperature. Consequently, closing the door during the operation of air conditioning gives the occupants a very comfortable level of satisfaction, with minimum energy consumption.

The indoor and outdoor temperatures during the starting and ending of door closing tests are given in Table 5.

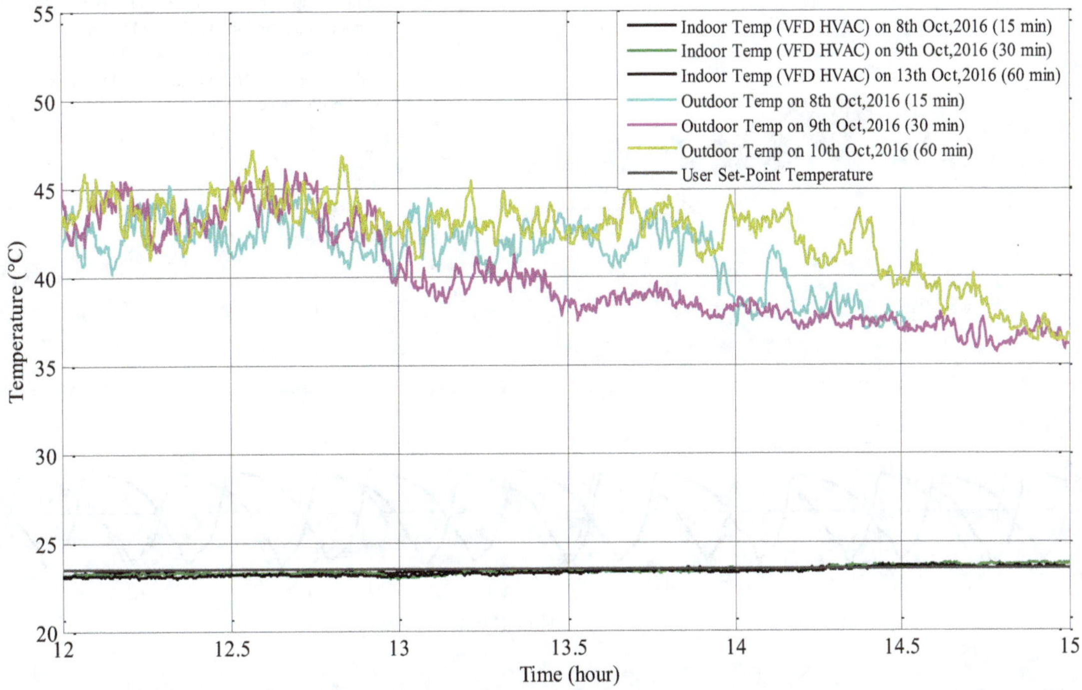

Figure 8. Indoor and Outdoor temperature for 3 hours for house #3307 on 11th, 12th and 13th of Oct. 2016

Table 5: Outdoor and indoor temperature for both unites on 11th, 12th and 13th of Oct. 2016

Date	Start of Closing (Main Door) ON/OFF &VFD	End of Closing (Main Door) ON/OFF &VFD	Outdoor Temp During the Start of Closing Time	Outdoor Temp During the End of Closing Time	Indoor Temp During the Start of Closing Time		Indoor Temp During the Closing Time	
					ON/OFF	VFD	ON/OFF	VFD
08/10/ 2016	12:30 p.m.	12:45 p.m.	41.36 °C	43.11 °C	24.46 °C	22.80 °C	24.87 °C	21.85 °C
09/10/ 2016	12:30 p.m.	13:00 p.m.	42.37 °C	42.10 °C	24.63 °C	22.79 °C	22.17 °C	22.6 °C
10/10/ 2016	12:30 p.m.	13:30 p.m.	41.85 °C	41.12 °C	23.98 °C	22.9 °C	23.65 °C	22.87 °C

Closing of House Main Door Equipped With On/Off HVAC Unit

Figure 9 shows indoor and outdoor temperatures for three days from 12:00 p.m. to 15:00 p.m. on 8th, 9th and 10th of October, 2016, as the door is closed of house #3305. It is clearly shown that the ON/OFF HVAC system turns on when indoor air temperature is greater than that of the set point temperature 23.5 °C plus a band range +1.5 °C, and then it turns off at the lower limit –1.5 °C of set point temperature. It is noticeable that keeping the door closed maintains the air conditioning system cycling during the day and the indoor temperature fluctuates in the range of ±1.5 °C without any extra variation.

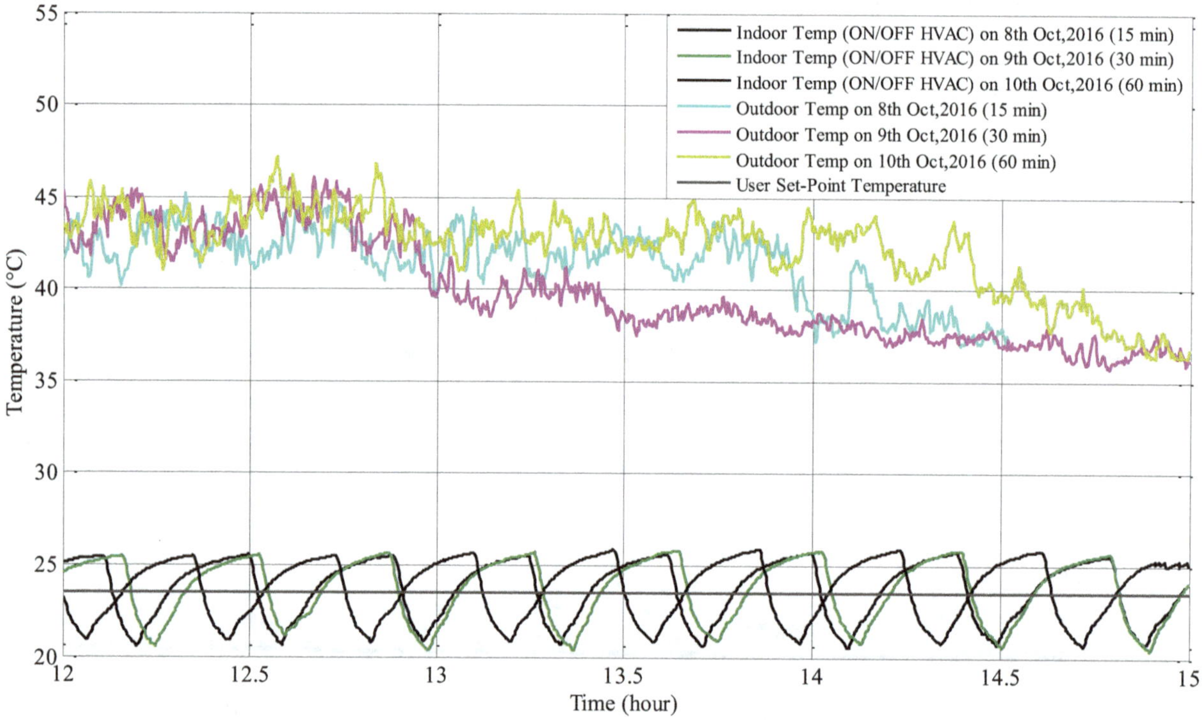

Figure 9. Indoor and outdoor temperatures for 3 hours for house #3305 on 8th, 9th and 10th of Oct .2016

Keeping the main door of the house closed during the operation of the air conditioning systems maintains the indoor air temperature on the range of the set point temperature, and provides the desired thermal comfort throughout the day with minimum energy consumption.

Power Consumption of VFD HVAC Unit

Figure 10 shows the power consumptions of VFD HAVC unit for three days without main door opening from 12:00 p.m. to 15:00 p.m. The compressor capacity of the air conditioning system is matched to the load by regulating the speed of the compressor motor, for example with low load, the compressor runs at low speed, and hence consumed less energy. It is shown in Figure 10 that a considerable amount of power consumptions vary between 1500 watts to 2000 watts. There is a small difference in the energy consumptions of VFD HVAC during the three days, and that is due to the changes in the house activities and the outdoor temperature. The difference is approximately 0.5 kWh. As the compressor motor is reduced, a significant reduction in power requirement is achieved. Consequently, power consumptions could be reduced by closing the main door of the house during the operation of the air conditioning system. Without a huge variation in the power consumptions, Table 6 gives the energy consumptions by the VFD HVAC unit within three hours on 8th, 9th and 10th of October, 2016.

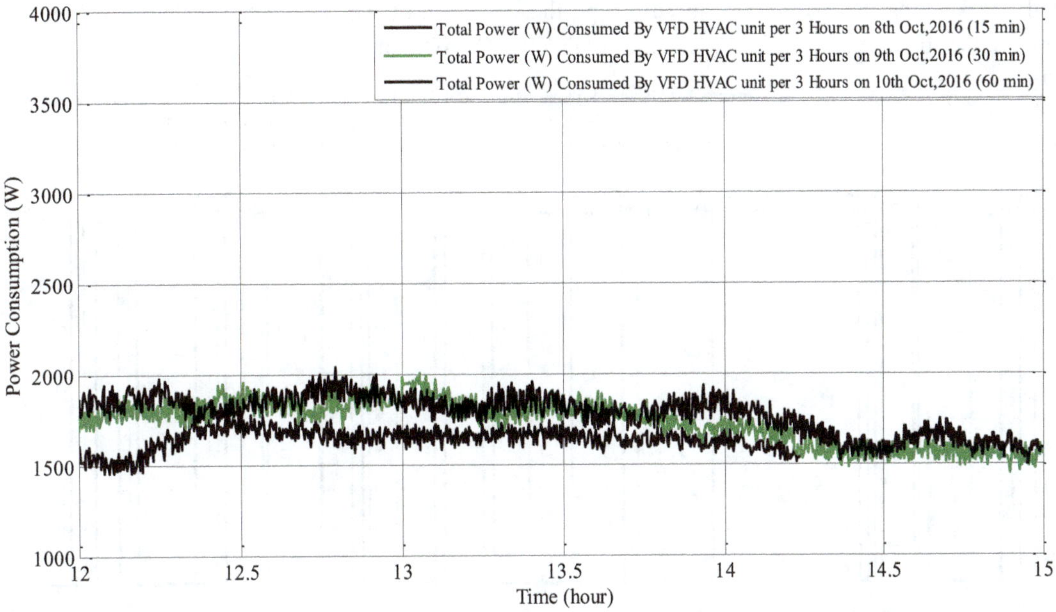

Figure 10. Power consumptions for 3 hours by VFD unit on 8th, 9th and 10th of Oct. 2016

Table 6. Energy consumed by ON/OFF & VFD unite on 8th, 9th and 10th of Oct.2016 (Door Closed)

Date	Time of the Experiment (Door Closing)	Energy Consumption Time (Hours)	Time (3 hours)	VFD Unit Energy Consumption	ON/OFF Unit Energy Consumption	Difference in Energy Consumption VFD&ON/OFF
08/10/2016	15 Min	3 hours	From 12:00 p.m. To 15:00 p.m.	4.8976 kWh	5.8708 kWh	0.9732 kWh
09/10/2016	30 Min	3 hours	From 12:00 p.m. To 15:00 p.m.	5.2179 kWh	6.1616 kWh	0.9437 kWh
10/10/2016	60 Min	3 hours	From 12:00 p.m. To 15:00 p.m.	5.3268 kWh	6.3131 kWh	0.9863 kWh

Power Consumption of On/Off HVAC Unit

Figure 11 shows the power consumptions for 3 hours from 12:00 p.m. to 15:00 p.m. by ON/OFF unit on 8th, 9th and 10th of October, 2016. In addition, Table 6 lists the energy consumption by the ON/OFF HVAC system when the main door is closed within three hours. Closing the main door of the house during the operation of A/C system has great effect on the power consumption. If the door is closed for a long period of time, energy consumption decreased obviously. The ON/OFF HVAC unit mode is triggered by the thermostat at a specific set point temperature. As a result the thermostat triggers the compressor in OFF mode quickly which ultimately increases the compressor ON/OFF cycling. The power consumption of the compressor motor varies between equivalent 640 Watt to the full speed of the compressor which to 6200 Watt and stops several times. In addition, the blower motor consumes 635 Watt and it sets on a fixed mode. Due to the main door closing condition, the indoor temperature drops below the set point, and the ON/OFF HVAC turns OFF and it repeatedly turns itself ON and OFF throughout the day.

On the other hand, it is well established from experimental study in [9], that the compressor ON/OFF cycle causes significant amount of efficiency and capacity loss. In addition, it is experimentally found that efficiency loss of the system is 9% and the capacity loss is 11% due to compressor motor ON/OFF cycling. The frequent sharp pulse for compressor always generates noisiness and affects the lifetime of the motor seriously. Based on the experiments, it is found that keeping the main door closed during the operation of the A/C system helps to reduce the energy consumption either by using the VFD HVAC unit or ON/OFF HVAC unit.

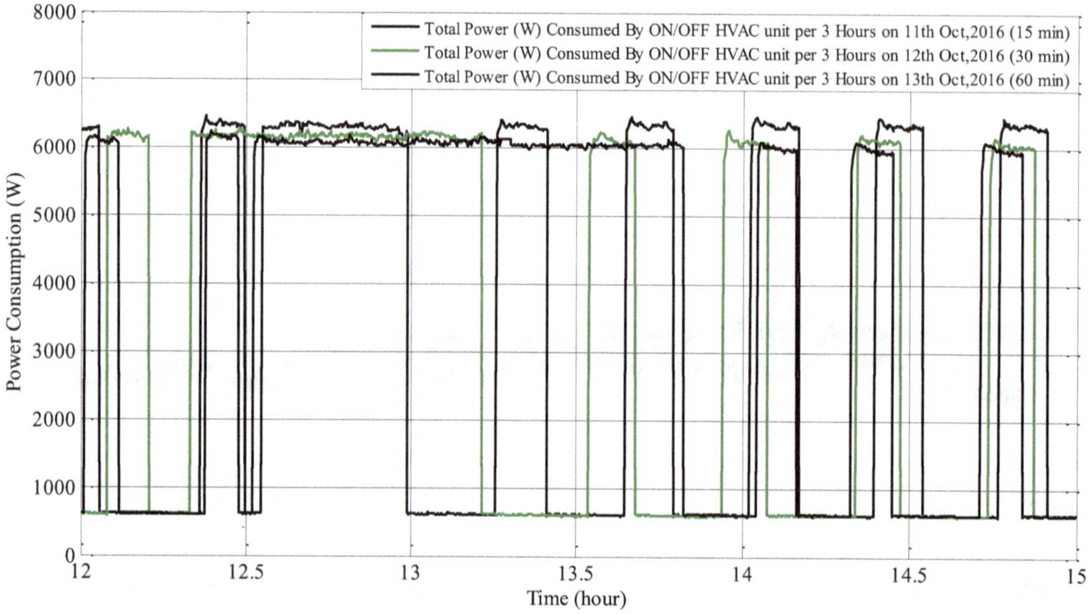

Figure 11. Power consumptions for 3 hours by ON/OFF unit on 8th, 9th and 10th of Oct. 2016

Difference of Door Opening/Closing Energy Consumptions

Table 7 gives the energy consumptions of the VFD HVAC unit during the main door opening /closing conditions. Form the experiment results, energy consumption of VFD HVAC unit when the main door kept open for 15 minutes increases by 1.0177 kWh compared to the same unit without door opening. The energy consumption when the main door kept open for 30 minutes is greater by 1.489 kWh compared to the energy consumption of the closing door test. In addition, when the main door opening for 60 minutes, the VFD HAVC unit consumes 3.5097 kWh compared to closing door condition for the same unit.

Table 7: Difference in energy consumption VFD door opened and VDF door closed

Date (Door Opened)	Date (Door Closed)	(Door Opening Time& Door Closing Time	Energy Consumption Time (3 hours)	VFD Unit Energy Consumptio n (Door Opened)	VFD Unit Energy Consumption (Door Closed)	Difference in Energy Consumption for VFD unit
08/10/2016	08/10/2016	15 Min	From 12:00 p.m. To 15:00 p.m.	5.9153 kWh	4.8976 kWh	1.0177 kWh
09/10/2016	09/10/2016	30 Min	From 12:00 p.m. To 15:00 p.m.	6.7069 kWh	5.2179 kWh	1.489 kWh
10/10/2016	10/10/2016	60 Min	From 12:00 p.m. To 15:00 p.m.	8.8365 kWh	5.3268 kWh	3.5097 kWh

Table 8 gives the energy consumptions of the ON/OFF HVAC unit during the main door opening/ closing tests. From the test result of door opening condition for 15 minutes, the energy consumptions increases by 3.5225 kWh compared to the closed door condition. When the duration of opening the main door increases to 30 minutes, the differences of energy consumption of opening and closing the main door increases to 4.7225 kWh. Furthermore, opening the main door for 60 minutes has a strong influence on the energy consumption for the ON/OFF HVAC unit. It increases by about 6.6102 kWh compared to the door closing condition. Consequently, it is noticeable from the experiment results that a significant amount of energy consumption increases with increasing of the duration of the door opening test. Moreover, the energy consumption difference between the door opening/closing tests for the ON/OFF HVAC unit is greater than the difference of the energy consumption by the VFDHVAC unit. The reason behind that is the ON/OFF HVAC is cycling during the air conditioning operation which reflected to consume more energy.

Table 8: The energy consumptions of the ON/OFF HVAC unit during the door opening/closing conditions

Date (Door Opened)	Date (Door Closed)	(Door Opening Time& Door Closing Time)	Energy Consumption Time (3 hours)	ON/OFF Unit Energy Consumption (Door Opened)	ON/OFF Unit Energy Consumption (Door Closed)	Difference in Energy Consumption for ON/OFF unit
11.10.2016	08.10.2016	15 Min	From 12:00 p.m. To 15:00 p.m.	9.3933 kWh	5.8708 kWh	3.5225 kWh
12.10.2016	09.10.2016	30 Min	From 12:00 p.m. To 15:00 p.m.	10.8841 kWh	6.1616 kWh	4.7225 kWh
13.10.2016	10.10.2016	60 Min	From 12:00 p.m. To 15:00 p.m.	12.9233 kWh	6.3131 kWh	6.6102 kWh

Cost Analysis

Table 9 gives the cost analysis of electric energy consumption for both VFD & ON/OFF HVAC units. The energy consumption by ON/OFF HVAC is greater than the energy consumption by the VFD HVAC. Therefore, opening the house main door, which is equipped with a conventional ON/OFF unit, causes the additional cost to the occupants due to the increasing of the electricity bill. When comparing to VFD and ON/OFF HVAC units in terms of energy consumptions, there appears to be small difference in energy consumptions. In addition, if the occupants are conservative and aware of the result for opening/closing the main door during the operation of air conditioning units, a significant amount of energy could be saved. This can be observed in Table 9. The additional cost due to energy consumption is presented by US Dollars. It should be noted that the average cost of electric energy in the world is $ 0.11683 per kWh. [19].

The cost analysis is carried out for a several repeating days of main door opening conditions. 10, 30, and 60 days are selected in one year period to conduct the overall cost analysis and forecasting the cost of energy consumption due to the main door opening. The difference of the energy consumption of each time duration of door opening 15 min, 30 min, and 60 min is calculated with all durations of the selected days in a year. Table 9 illustrates the difference in energy consumption, energy increases per day, and additional cost in US Dollars.

Based on the calculation performed in Table 9, in the first case, when the main door is kept open for 15 minute within 10, 30 or 60 days, the additional cost difference between the ON/OFF and the VFD units in the period of these days are $ 2.8091, $ 8.4695, and $ 16.9606, respectively. It is observed that the conventional ON/OFF unit costs more than the VFD unit for the same conditions. On the other hand, the cost different due to frequent keeping the main door opened for 30 minutes of 10, 30, or 60 days, is having $ 3.7774, $ 11.3316, and $ 22.665 addition to the regular cost of electricity bill. In this case ON/OFF unit consumed more energy and cost the occupants some extra money than that of the VFD unit. In the third case, if the main door opened for one hour throughout 10, 30 or 60 days, ON/OFF unit charges about $ 3.6223, $ 10.867, and $ 21.736 more than VFD unit from the additional cost. Overall, it can be noticed that both unites are cost the occupants some extra money, when they open the main door for period of time during the operation of A/C systems. If the duration of door opening increases, the additional cost increases, however, the ON/OFF unit charges more money than the VFD unit.

Table 9: Cost analysis of energy consumptions during the door opening/closing conditions

Door Opening Time	Repeating of Door Opening/Day	Difference in Energy Consumption		Energy Increased Per Day		Additional Cost kWh = $ 0.11683 U.S. Dollar	
		VFD (kWh)	ON/OFF (kWh)	VFD (kWh)	ON/OFF (kWh)	VFD ($)	ON/OFF ($)
15 Minutes	10 days	1.017	3.422	10.177	34.225	1.1889	3.998
	30 days			30.177	102.67	3.5255	11.995
	60 days			60.177	205.35	7.0304	23.991
30 Minutes	10 days	1.489	4.722	14.892	47.225	1.7398	5.5172
	30 days			44.676	141.67	5.2194	16.551
	60 days			89.352	283.35	10.438	33.103
60 Minutes	10 days	3.509	6.610	35.097	66.102	4.1003	7.7226
	30 days			105.29	198.30	12.301	23.168
	60 days			210.58	396.61	24.60	46.336

CONCLUSIONS

The energy consumption and corresponding cost analysis of HVAC systems due to opening/closing of the main door of the house are investigated, for the first time, for the selective periods in hot season of Saudi Arabia. The tests are carried out in two identical houses during three warm days where the outdoor temperatures have similar patterns. However the main door is kept open for 15 min, 30 min and 60 min respectively for the tests and analysis in line with regular practices of hot house of local area in Saudi Arabia. The tests due to door closing are carried out on the other three days. In addition, experimental set up is developed incorporating the Lab-View based data acquisition monitoring and measurement system for each house. House #3305 is equipped with a conventional ON/OFF A/C controller while house #3307 is equipped with VFD A/C controller. The findings revealed that keeping the house main door opened for considerably prolonged time, electrical energy consumption increases significantly to maintain the thermal comfort level in the house in terms of the desired temperature.

The specific conclusions derived from the present work are as follows:

- In the case of VFD A/C controller, it is found that energy consumption increases by 1.10 kWh, 1.4 kWh and 3.5 kWh when the door is kept open for 15 minutes, 30 minutes, and 60 minutes, respectively. In the case of the conventional ON/OFF A/C controller, the energy consumption increases by 3.52 kWh, 4.72 kWh, and 6.61 kWh when the house door is kept opened for 15 minutes, 30 minutes, and 60 minutes, respectively. `
- Indoor temperature fluctuation increases with increasing door opening duration, which makes indoor air temperature level uncomfortable. The energy consumption increases for both HVAC controllers used because of the excessive operation of the A/C compressors to meet and recover the desired indoor temperature.
- The additional cost difference between the ON/OFF and the VFD controller unit is found to be about $9.41, when the main door is kept open for 15 minutes. On the other hand, when the main door is kept open for 30 minutes, this cost increases to approximately $12.6. The additional cost difference becomes about $12, when the main door is kept open for 60 minutes. Therefore, increasing the duration of the door opening results in is excessive additional to the cost to the electricity bill.

ACKNOWLEDGMENT

The author acknowledges the support by the DSR funded project RG1205-1&2 at King Fahd University of Petroleum and Minerals, Dhahran, Saudi Arabia for this research work.

REFERENCES

[1] Pal, J. S., and Eltahir, E. A. (2016). Future temperature in southwest Asia projected to exceed a threshold for human adaptability. Nature Climate Change, 6(2), 197-200.

[2] Belhadj, C. A., Hamanah, W. M., and Kassas, M., LabVIEW based real time monitoring of HVAC system for residential area. 2017 IEEE International Conference on Computational Intelligence and Virtual Environments for [1] Measurement Systems and Applications (CIVEMSA).

[3] Afram, A., and Janabi-Sharifi, F. (2015). Gray-box modeling and validation of residential HVAC system for control system design. Applied Energy, 137, 134-150.

[4] Du, Z., Jin, X., Fang, X., and Fan, B. (2016). A dual-benchmark based energy analysis method to evaluate control strategies for building HVAC systems. Applied Energy, 183, 700-714.

[5] Radhakrishnan, N., Su, Y., Su, R., and Poolla, K. (2016). Token based scheduling for energy management in building HVAC systems. Applied Energy, 173, 67-79.

[6] Fiorentini, M., Wall, J., Ma, Z., Braslavsky, J. H., and Cooper, P. (2017). Hybrid model predictive control of a residential HVAC system with on-site thermal energy generation and storage. Applied Energy, 187, 465-479.

[7] Alibabaei, N., Fung, A. S., Raahemifar, K., and Moghimi, A. (2017). Effects of intelligent strategy planning models on residential HVAC system energy demand and cost during the heating and cooling seasons. Applied Energy, 185, 29-43.

[8] Syed, H. (2009). Modeling, Analysis and Optimization of the Thermal Performance of Air Conditioners (Doctoral dissertation, King Fahd University of Petroleum and Minerals).

[9] Al-Shaalan, A., Ahmed, W., and Alohaly, A. (2014). Design guidelines for buildings in Saudi Arabia considering energy conservation requirements. In Applied Mechanics and Materials (Vol. 548, pp. 1601-1606). Trans Tech Publications.

[10] Kassas, M. Modeling and simulation of Residential HVAC Systems Energy Consumption. The 5th International Conference on Sustainable Energy Information Technology, Procedia Computer Science 52 (2015) 754-763.

[11] Fabi, V., Andersen, R. V., Corgnati, S., and Olesen, B. W. (2012). Occupants' window opening behaviour: A literature review of factors influencing occupant behaviour and models. Building and Environment, 58, 188-198.

[12] Pan, S., Xu, C., Wei, S., Hassan, T. M., Xie, L., Xiong, Y., and de Wilde, P. (2016). Improper window use in office buildings: findings from a longitudinal study in Beijing, China. Energy Procedia, 88, 761-767.

[13] Rijal, H. B., Tuohy, P., Nicol, F., Humphreys, M. A., Samuel, A., and Clarke, J. (2008). Development of an adaptive window-opening algorithm to predict the thermal comfort, energy use and overheating in buildings. Journal of Building Performance Simulation, 1(1), 17-30.

[14] Brown, N., Wright, A. J., Caeiro, J. A., Altan, H., and Summerfield, A. J. (2006, December). Large scale energy surveys in the UK retail sector. In Proc. of the RICS Annual Conference Cobra.

[15] Khan, M. I. H., and Afroz, H. M. (2014). An experimental investigation of door opening effect on household refrigerator; the perspective in Bangladesh. Asian Journal of Applied Sciences, 7(2), 79-87.

[16] Hamanah,W.M., 2016, "Modeling, Simulation and Energy Performance of VFD and ON/OFF Cycle HVAC Systems," M.S. thesis, KFUPM, Az Zahran, Saudi Arabia.

[17] W. M. A. Hamanah, M. Kassas, E. M. A. Mokheimer, C. B. Ahmed, and S. A. M. Said, "Comparison of Energy Consumption for Residential Thermal Models With Actual Measurements," J. Energy Resour. Technol., vol. 141, no. 3, p. 032002, 2018.

[18] Belhadj, C. A., Hamanah, W. M., and Kassas, M., 2017, "LabVIEW based real time Monitoring of HVAC System for Residential Load," IEEE International Conference on Computational Intelligence and Virtual Environments for Measurement Systems and Applications (CIVEMSA), Annecy, France, June 26–28, pp. 66–71.

[19] Dincer, I., M. M. Hussain, and I. Al-Zaharnah. "Analysis of sectoral energy and exergy use of Saudi Arabia." International Journal of Energy Research 28.3 (2004): 205-243.

12

INVESTIGATION OF LORENTZ FORCE EFFECT ON STEADY NANOFLUID FLOW AND HEAT TRANSFER THROUGH PARALLEL PLATES

A. T. Akinshilo[1,*], A.O. Ilegbusi[2]

ABSTRACT

In this paper Lorentz force effect on steady fluid flow and heat transfer of nanofluid is examined. The nanofluid is transported through horizontal parallel plates with magnetic flux of uniform density acting perpendicular to the plates. The effects of thermo-fluidic parameters such as Schmidt number, viscosity and magnetic parameter on flow and heat transfer are presented. Other important heat and mass transfer parameters such as Nusselt and Sherwood numbers practically relevant were also studied. Obtained results from analytical solutions shows quantitative increase of Magnetic parameter varied within the range of 1-4 depicts increasing temperature distribution. Also results when compared with past literatures forms good agreement. Therefore study provides a good emphasis for the advancements of Nano fluidics such as micro mixing, friction reduction, energy conservation, and biological samples.

Keywords: *Steady Flow, Nanofluid, Horizontal Plates, Lorentz Force, Homotopy Perturbation Method*

INTRODUCTION

The transport of fluid through parallel surfaces has become increasingly high in most industrial and manufacturing applications today such as power transmission, polymer processing and heat exchangers. Owing to its efficient flow and heat transfer of fluids to its required destination. As energy pricing has become a significant factor in determining the mode of fluid transport, the need to manage heat transfer efficiently as led to renewed effort by scientist and engineers to research into more efficient ways of managing energy. In efforts to enhance heat transfer Abu Nada et al. [1] studied heat transfer under natural convection in horizontal concentric annuli using nanoparticles to increase heat transfer effect on the fluid flow. Heat transfer and flow over stretching sheet was investigated by Cortell [2] where study outcome reveals the influence of radiation on flow while Ellahi et al. [3] presented heat transfer of non-Newtonian fluid through two coaxial cylinders with porous media using variable viscosity, effect of heat transfer on nanoparticle under the influence of magnetism is analyzed. Mixed convection of nanofluid flow through square cavities where analyzed by Garoosi et al. [4-5] under two phase simulation without and with external heating where results show the effect of buoyancy on cavity flow. Shortly after Garoosi et al. [6] Investigated nanofluid flow in heat exchangers adopting the Buongiorno model of numerical technique, they discovered that at low Rayleigh number heat transfer decreases. Brownian motion and thermophoresis effects on slip flow of alumina/water were investigated by Malvandi and Ganji [7] where they showed effects of Nanoparticle concentration on heat transfer. Heat transfer over stretching surface was studied by Mehmood and Ali [8] presenting a three dimensional analytic solution employing the homotopy analysis method. Rashidi et al. [9] analyzed entropy generation in a rotating porous disk under steady flow condition considering the influence of magnetism with results revealing effect of entropy on temperature distribution. Rashidi et al. [10] in the bid to control wake and vortex shedding later considered heat transfer around a porous obstacle. Convective surface boundary condition was used by Shehzad et al. [11] to study three dimensional flow of Jeffery fluid presenting velocity and temperature distribution using series solutions obtained from non-linear equations. Later Shehzad et al. [12-13] investigated thermal radiative effects on mixed convection of thixotropic fluid, results reveal numerical values of wall shear and heat transfer rate. Flow and heat transfer in a rotating system was investigated by Sheikholeslami et al. [14] between stretching sheet and porous surface, discussing the effects of thermo-fluidics on flow and heat transfer. Two phase simulation of nanofluid was also investigated by Sheikholeslami et al. [15] using heat line model analyzing the effect of magnetism on heat and mass transfer.

This paper was recommended for publication in revised form by Regional Editor Jaap Hoffman Hoffman
[1]Mechanical Engineering Department; University of Lagos, Akoka-Yaba, Nigeria
[2]Mechanical Engineering Department; Yaba College of Technology, Yaba, Nigeria
**E-mail address: ta.akinshilo@gmail.com*

In recent years the need to improve the thermal conductivity of fluids as led to the addition of nanoparticles into base fluid, owing to the fact that nanoparticles have higher thermal conductivities, which enhances the overall energy transport during fluid flow. Therefore this creative approach has been widely adopted by researchers in the study of flow and heat transfer [16-39].Nanofluid displays characteristics of high heat transfer and enhanced flow character due to high thermal conductivity of nanoparticles which makes them potentially useful in medical, fuel cells and microelectronic processes.

Approximate analytical methods of solutions applied by researchers in study of the fluid flow include the pertubation method (PM),homotopy analysis method (HAM), homotopy pertubation method (HPM), adomian decomposition method (ADM) and Differential transformation method (DTM) [39-43] .Methods such as PM are limited owing to the problems of weak nonlinearities and artificial pertubation parameter which may be non existent in practical sense . The need to find an initial condition to satisfy the boundary condition makes methods such as DTM requires computational tools such as Matlab, Maple or Mathematica in handling a solution of large parameters resulting to large computational cost and time. Lack of rigorous theories for determining the initial approximation of the HAM , its auxilliary function and parameter restricts the HAM. Also the problem of finding the adomian polynomials makes the ADM not attractive to researchers. However the homotopy pertubation method been a relatively simplistic method of solving nonlinear, coupled equations due to its highly successive and accurate approximation making it a favourable analytical technique to researchers.

Motivated by past research works, the homtopy pertubation method (HPM) is used to investigate Lorentz force effect which is a resistive magnetic force type on steady nanofluid flow and heat transfer coneyed through horizontal parallel plates .

MODEL DEVELOPMENT AND ANALYTICAL SOLUTION

Here nanofluid flows through parallel plates held horizontally against each other under steady flow condition. The plates are held at a distance h, where the x axis is along the plate and the y axis is perpendicular which is described in the physical model of problem Figure 1. The nanofluid flow along the x axis which is normal to the plate with an angular velocity. A uniform magnetic flux acts perpendicular to the axis of flow, such that the plates is held by equal but opposite force which fixes its position at reference (0, 0, 0) i.e. its position remains the same. The formulation of the model development of the nanofluid is developed with respect to the above conditions following the assumptions that the fluid is incompressible, radiation heat transfer is negligible and two component mix is in thermal equilibrium. Therefore the governing equations of the system can be described as:

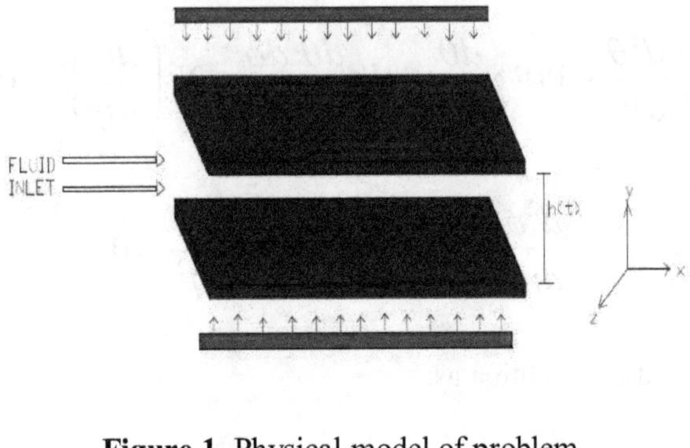

Figure 1. Physical model of problem

$$\frac{\partial u}{\partial x}+\frac{\partial v}{\partial y}=0 \tag{1}$$

$$\rho_f\left(u\frac{\partial u}{\partial x}+v\frac{\partial u}{\partial y}\right)=-\frac{\partial p^*}{\partial x}+\mu\left(\frac{\partial^2 u}{\partial x^2}+\frac{\partial^2 u}{\partial y^2}\right)-\sigma B_0^2 u \quad (2)$$

$$\rho_f\left(u\frac{\partial v}{\partial x}+v\frac{\partial T}{\partial y}\right)=-\frac{\partial p^*}{\partial y}+\mu\left(\frac{\partial^2 v}{\partial x^2}+\frac{\partial^2 v}{\partial y^2}\right) \quad (3)$$

$$u\frac{\partial T}{\partial x}+v\frac{\partial T}{\partial y}=\alpha\left(\frac{\partial^2 T}{\partial x^2}+\frac{\partial^2 T}{\partial y^2}\right)+\frac{(\rho Cp)_p}{(\rho Cp)_f}x\left[D_{\mathbf{B}}\left\{\frac{\partial C}{\partial x}\frac{\partial T}{\partial x}+\frac{\partial C}{\partial y}\frac{\partial T}{\partial y}\right\}+(D_T/\mathrm{T_C})\left\{\left(\frac{\partial T}{\partial x}\right)^2+\left(\frac{\partial T}{\partial y}\right)^2\right\}\right] \quad (4)$$

$$u\frac{\partial C}{\partial x}+v\frac{\partial C}{\partial y}=D_B\left(\frac{\partial^2 C}{\partial x^2}+\frac{\partial^2 C}{\partial y^2}\right)+\left(\frac{D_T}{T_0}\right)\left\{\frac{\partial^2 T}{\partial x^2}+\frac{\partial^2 T}{\partial y^2}\right\} \quad (5)$$

Taking boundary conditions as:

$$u=ax, v=0, T=T_h, C=C_h \quad \text{at } y=0 \quad (6)$$

$$u=0, v=0, T=T_0, C=C_0 \quad \text{at } y=+h \quad (7)$$

With the dimensionless parameters stated in the nomenclature, the dimensionless equations are given as [37]:

$$\frac{d^4 f}{\partial\eta^4}-R\left(\frac{df}{d\eta}\frac{d^2 f}{d\eta^2}+f\frac{d^2 f}{d\eta^2}\right)-M^2\frac{d^2 f}{d\eta^2}=0 \quad (8)$$

$$\frac{d^2\theta}{d\eta^2}+\mathrm{PrRf}\frac{d\theta}{d\eta}+Nb\frac{d\theta}{d\eta}\frac{d\phi}{d\eta}+Nt\left(\frac{d\theta}{d\eta}\right)^2=0 \quad (9)$$

$$\frac{d^2\phi}{d\eta^2}+RScf\frac{d\phi}{d\eta}+\frac{Nt}{Nb}\frac{d^2\theta}{d\eta^2}=0 \quad (10)$$

Taking appropriate boundary condition as:

$$f=0, \frac{df}{d\eta}=1, \theta=1, \phi=1 \text{ at } \eta=0 \quad (11a)$$

$$f=0, \frac{df}{d\eta}=0, \theta=0, \phi=0 \text{ at } \eta=1 \quad (11b)$$

Important characteristics of flow, heat and mass transfer for practical relevance are reduced to skin friction, Nusselt and reduced Sherwood number which can be defined as:

$$C_{fr} = f''(1) \text{ where } C_{fr} = \frac{H^2\sqrt{1-\alpha t \operatorname{Re}_r}}{r^2} C_f \tag{12a}$$

$$Nur = -\theta'(1) \text{ where } Nur = \sqrt{1-\alpha t Nu} \tag{12b}$$

$$shr = -\phi'(1) \text{ where } \operatorname{shr} = \sqrt{1-\alpha t sh} \tag{12c}$$

where $\operatorname{Re}_r = \dfrac{r\alpha H}{2v}$ is the local squeeze Reynolds number.

Principles of Homotopy Perturbation Method

The following equation is considered in explaining the fundamentals of the homotopy perturbation method [38]:

$$A(u) - f(r) = 0 \quad r \in \Omega \tag{13}$$

Utilizing the boundary condition:

$$B(u, \frac{\partial u}{\partial \eta}) = 0 \quad r \in \Gamma \tag{14}$$

A is the general differential operator, B is the boundary operator, f(r) is the analytical function and Γ is the boundary domain of Ω. Separating A into two components of linear and nonlinear terms L and N respectively. The Eq. (13) is reconstructed as:

$$L(u) + N(u) - f(r) = 0 \quad r \in \Omega \tag{15}$$

Homotopy perturbation structure takes the form:

$$H(v,p) = (1-p)[\mathrm{L(v)} - \mathrm{L(u_0)}] + P[A(v) - f(r)] = 0 \tag{16}$$

where $v(r,p) : \Omega x[0,1] - R$

In Eq. (16) P ϵ (0, 1) is the embedding parameter and U_0 is taken as the initial term that satisfies boundary condition. The power series of Eq. (16) can be expressed therefore as:

$$v = v_0 + Pv_1 + P^2 v_2 + P^3 v_3 + \ldots\ldots \tag{17}$$

Most appropriate solution for the problem takes the form:

$$u = \lim_{p\to 1}(v_0 + Pv_1 + P^2v_2 + P^3v_3 + ...) \tag{18}$$

Application of the Homotopy Perturbation Method

The homptopy pertubation method (HPM) which is an analytical scheme for providing approximate solutions to the ordinary differential equations, is adopted in generating solutions to the coupled ordinary nonlinear differential equation .Upon constructing the homotopy, the Eqs. (8)- (10) can be expressed as:

$$H_1(\text{p},\eta) = (1-p)\left[\frac{d^4 f}{\partial\eta^4}\right] + p\left[\frac{d^4 f}{\partial\eta^4} - R\left(\frac{df}{d\eta}\frac{d^2 f}{d\eta^2} + f\frac{d^2 f}{d\eta^2}\right) - M^2\frac{d^2 f}{d\eta^2}\right] \tag{19}$$

$$H_2(p,\eta) = (1-p)\left[\frac{d^2\theta}{d\eta^2}\right] + p\left[\frac{d^2\theta}{d\eta^2} + \text{PrRf}\frac{d\theta}{d\eta} + Nb\frac{d\theta}{d\eta}\frac{d\phi}{d\eta} + Nt\left(\frac{d\theta}{d\eta}\right)^2\right] = 0 \tag{20}$$

$$H_3(p,\eta) = (1-p)\left[\frac{d^2\phi}{d\eta^2}\right] + p\left[\frac{d^2\phi}{d\eta^2} + RScf\frac{d\phi}{d\eta} + \frac{Nt}{Nb}\frac{d^2\theta}{d\eta^2}\right] = 0 \tag{21}$$

Taking the appropriate boundary condition as:

$$f = 0, \frac{df}{d\eta} = 1, \theta = 1, \phi = 1 \text{ at } \eta=0 \tag{22a}$$

$$f = 0, \frac{df}{d\eta} = 0, \theta = 0, \phi = 0 \text{ at } \eta=1 \tag{22b}$$

Taking power series of velocity, temperature and concentration fields yields:

$$f = P^0 f_0 + P^1 f_1 + P^2 f_2 + ... \tag{23}$$

$$\theta = P^0\theta_0 + P^1\theta_1 + P^2\theta_2 + ... \tag{24}$$

$$\phi = P^0\phi_0 + P^1\phi_1 + P^2\phi_2 + ... \tag{25}$$

Substituting Eq. (23) into (20) and selecting at the various order yields:

$$P^0 : \frac{d^4 f_0}{d\eta^4} \tag{26}$$

$$P^1 : \frac{d^4 f_1}{d\eta^4} - \frac{d^2 f_0}{d\eta^2} M + R \frac{d^2 f_0}{d\eta^2} f_0 - \frac{d^2 f_0}{d\eta^2} \frac{df_0}{d\eta} \tag{27}$$

$$P^2 : \frac{d^4 f_2}{d\eta^4} - \frac{d^2 f_1}{d\eta^2} M + R \frac{d^2 f_0}{d\eta^2} f_1 + \frac{d^2 f_1}{d\eta^2} f_0 - \frac{df_0}{d\eta} \frac{d^2 f_1}{d\eta^2} - \frac{df_1}{d\eta} \frac{d^2 f_0}{d\eta^2} \tag{28}$$

Substituting Eq. (24) into (21) and selecting at the various order yields:

$$P^0 : \frac{d^2\theta_0}{d\eta^2} \tag{29}$$

$$P^1 : \frac{d^2\theta_1}{d\eta^2} + N_t \left(\frac{d\theta_0}{d\eta} \right)^2 - N_b \frac{d\theta_0}{d\eta} \frac{d\phi_0}{d\eta} + \Pr R f_0 \frac{d\theta_0}{d\eta} \tag{30}$$

$$P^2 : \frac{d^2\theta_2}{d\eta^2} + 2N_t \frac{d\theta_0}{d\eta} \frac{d\theta_1}{d\eta} + N_b \frac{d\theta_0}{d\eta} \frac{d\phi_1}{d\eta} + N_b \frac{d\theta_1}{d\eta} \frac{d\phi_0}{d\eta} + \Pr R \frac{d\theta_0}{d\eta} f_1 + \Pr R \frac{d\theta_1}{d\eta} f_0 \tag{31}$$

Substituting Eq. (25) into (22) and selecting at the various order yields:

$$P^0 : \frac{d^2\phi_0}{d\eta^2} + \left(\frac{N_t}{N_b} \frac{d^2\theta_0}{d\eta^2} \right) \tag{32}$$

$$P^1 : \frac{d^2\phi_1}{d\eta^2} + \left(\frac{N_t}{N_b} \frac{d^2\theta_1}{d\eta^2} \right) - RSc \frac{d\phi_0}{d\eta} f_0 \tag{33}$$

$$P^2 : \frac{d^2\phi_2}{d\eta^2} + \frac{N_t}{N_b} \frac{d^2\theta_2}{d\eta^2} + RSc \frac{d\phi_0}{d\eta} f_1 + RSc \frac{d\phi_1}{d\eta} f_0 \tag{34}$$

Taking leading order boundary condition as:

$$f_0 = 0, \frac{df_0}{d\eta} = 1, \theta_0 = 1, \phi_0 = 1 \text{ at } \eta=0 \tag{35a}$$

$$f_0 = 0, \frac{df_0}{d\eta} = 0, \theta_0 = 0, \phi_0 = 0 \text{ at } \eta=1 \tag{35b}$$

Simplifying Eq. (26) applying the boundary condition Eq. (35) yields:

$$f_0 = \eta^3 - 2\eta^2 + \eta \tag{36}$$

Simplifying Eq. (29) applying the boundary condition Eq. (35) yields:

$$\theta_0 = 1 - \eta \tag{37}$$

Simplifying Eq. (32) applying the boundary condition Eq. (35) yields:

$$\phi_0 = 1 - \eta \tag{38}$$

Taking the first order boundary condition as:

$$f_1 = 0, \frac{df_1}{d\eta} = 1, \theta_1 = 1, \phi_1 = 1 \text{ at } \eta=0 \tag{39a}$$

$$f_1 = 0, \frac{df_1}{d\eta} = 0, \theta_1 = 0, \phi_1 = 0 \text{ at } \eta=1 \tag{39b}$$

Simplifying Eq. (27) and applying the first order boundary condition Eq. (39) yields:

$$f_1 = \frac{17R\eta^7}{420} - \frac{5R\eta^6}{36} - \frac{R\eta^8}{280} - \eta^4\left(\frac{M}{6} + \frac{R}{6}\right) + \eta^5\left(\frac{M}{20} + \frac{13R}{60}\right) - \frac{\eta^2}{2}\left(\frac{2M}{15} + \frac{R}{84}\right) + \frac{\eta^3}{6}\left(\frac{11M}{10} + \frac{23R}{210}\right) \tag{40}$$

Simplifying Eq. (30) applying the first order boundary condition Eq. (39) yields:

$$\theta_1 = \frac{\Pr R\eta^5}{20} - \frac{\Pr R\eta^4}{6} + \frac{\Pr R\eta^3}{6} + \left(-\frac{N_b}{2} - \frac{N_t}{2}\right)\eta^2 + \left(\frac{N_b}{2} + \frac{N_t}{2} - \frac{\Pr R}{20}\right)\eta \tag{41}$$

Simplifying Eq. (33) applying the first order boundary condition Eq. (39) yields:

$$\phi_1 = \left(\frac{RSc}{6} - \frac{N_t Pr\, R}{20N_b}\right)\eta^5 + \left(\frac{N_t Pr\, R}{6N_b} - \frac{RSc}{6}\right)\eta^4 + \left(\frac{RSc}{6} - \frac{N_t Pr\, R}{6N_b}\right)\eta^3 + \left(\frac{N_t}{2} + \frac{N_t^2}{2N_b}\right)\eta^2 + \left(\frac{N_t Pr\, R}{20N_b} - \frac{RSc}{20} - \frac{N_t}{2} - \frac{N_t^2}{2N_b}\right)\eta \tag{42}$$

The coefficient p^2 for $f(\eta), \theta(\eta) and \phi(\eta)$ in Eqs. (28), (31) and (34) were too long to be mentioned here but it is expressed graphically in all the results and in the result validation, Table 1. Therefore substituting eq. (36) and (40) into the power series Eq. (23) yields:

$$f = \eta^3 - 2\eta^2 + \eta + \frac{17R\eta^7}{420} - \frac{5R\eta^6}{36} - \frac{R\eta^8}{280} - \eta^4\left(\frac{M}{6} + \frac{R}{6}\right) + \eta^5\left(\frac{M}{20} + \frac{13R}{60}\right) - \frac{\eta^2}{2}\left(\frac{2M}{15} + \frac{R}{84}\right) + \frac{\eta^3}{6}\left(\frac{11M}{10} + \frac{23R}{210}\right) + \cdots \tag{43}$$

Similarly substituting the Eqs. (37) and (41) into the power series Eq. (24) can be expressed as:

$$\theta = 1 - \eta + \frac{\Pr R\eta^5}{20} - \frac{\Pr R\eta^4}{6} + \frac{\Pr R\eta^3}{6} + \left(-\frac{N_b}{2} - \frac{N_t}{2}\right)\eta^2 + \left(\frac{N_b}{2} + \frac{N_t}{2} - \frac{\Pr R}{20}\right)\eta + \ldots \tag{44}$$

Also the Eqs. (38) and (42) upon substituting into the power series (25) is expressed as:

$$\phi = 1 - \eta + \left(\frac{RSc}{6} - \frac{N_t Pr\,R}{20N_b}\right)\eta^5 + \left(\frac{N_t Pr\,R}{6N_b} - \frac{RSc}{6}\right)\eta^4 + \left(\frac{RSc}{6} - \frac{N_t Pr\,R}{6N_b}\right)\eta^3 + \left(\frac{N_t}{2} + \frac{N_t^2}{2N_b}\right)\eta^2 + \left(\frac{N_t Pr\,R}{20N_b} - \frac{RSc}{20} - \frac{N_t}{2} - \frac{N_t^2}{2N_b}\right)\eta + \cdots \tag{45}$$

Table 1. Comparison of values for temperature and concentration when M=R=1, Ec=Pr=Sc=N_t=N_b=0.1, Pr=10

	θ(η)			ϕ(η)		
η	Sheikholeslami et al. [37]	Present work	Error %	Sheikholeslami et al. [37]	Present work	Error %
0	1.0000	1.0000	0.0000	1.0000	1.0000	0.0000
0.1	0.8562	0.8566	0.0004	0.9426	0.9429	0.0003
0.2	0.7209	0.7213	0.0004	0.8776	0.8779	0.0003
0.3	0.5590	0.5993	0.0003	0.7994	0.7996	0.0002
0.4	0.4912	0.4916	0.0004	0.7069	0.7072	0.0003
0.5	0.3954	0.3960	0.0004	0.6024	0.6027	0.0003
0.6	0.3094	0.3097	0.0003	0.4888	0.4891	0.0003
0.7	0.2293	0.2296	0.0003	0.3691	0.3695	0.0004
0.8	0.1524	0.1527	0.0003	0.2462	0.2467	0.0003
0.9	0.0763	0.0767	0.0004	0.1225	0.1229	0.0004
1.0	-0.000	-0.000	0.0000	-0.000	-0.000	0.0000

RESULTS AND DISCUSSION

The result obtained from the analytical solutions using the HPM is discussed in this section. As observed the validity of results when compared with solutions obtained in literature forms satisfactory agreement as depicted in Table 1.0. Effect of parameters on flow, heat transfer and concentration are reported graphically. With thermal fluidic parameters at various values on the velocity, temperature and concentration profile are presented. The Figure 2 shows the effect of viscosity parameter (R) on the velocity distribution, as depicted as R increases the velocity distribution increases slightly due to increase in fluid activation energy causing an increase in fluid motion with maximum effect near the lower plate. While it is shown from Figure 3 that velocity distribution decreases at increasing numerical values of Lorentz force or magnetic parameter (M) which can be physically explained by the decrease in momentum boundary layer thickness caused by magnetic force field whose effect is maximum towards the lower plate and falls rapidly towards the upper plate.

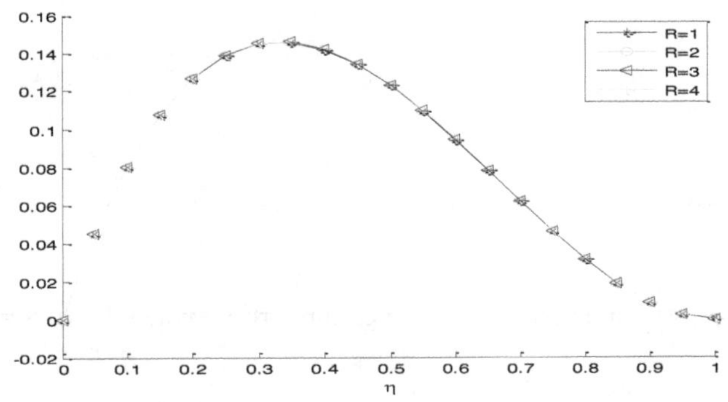

Figure 2. Effect of viscosity parameter (R) on velocity profile when Pr=10, Sc=N_t= N_b =0.1 and M=1

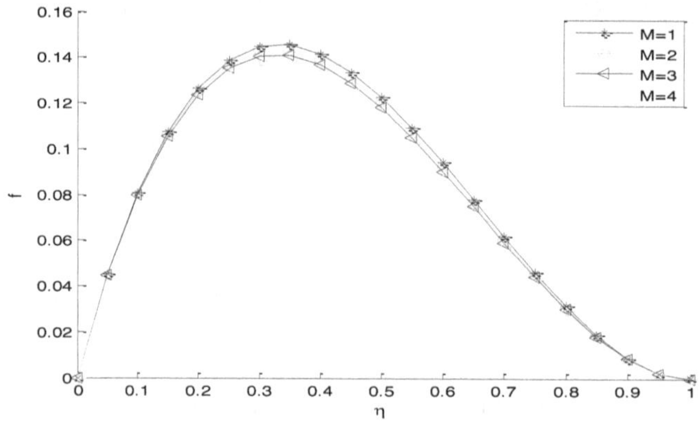

Figure 3. Effect of Magnetic parameter (M) on velocity profile when Pr=10,Sc=N_t= N_b =0.1 and R=1

The effect of increasing values of viscosity parameter (R) is demonstrated in Figure 4. It is observed that temperature distribution decreases at increasing R as a result of enhanced fluid motion causing rapid heat dissipation with the temperature profile maximum at the lower and minimum at the upper plate respectively. Magnetic parameter (M) effect on temperature distribution is clearly illustrated in Figure 5 which shows quantitative increase in M result in increasing thermal boundary layer thickness represented by slight increase in temperature. Increasing effect of thermophoretic parameter (N_t) on temperature distribution is seen in Figure 6 which depicts an increase in temperature profile while Brownian parameter (N_b) effect on temperature distribution is seen in Figure 7 where it is illustrated that increasing N_b gives a corresponding increase in temperature distribution.

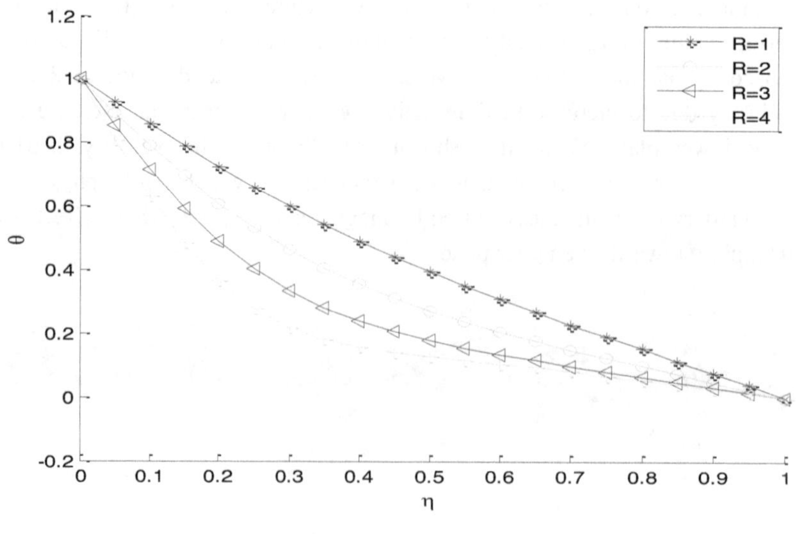

Figure 4. Effect of viscosity parameter (R) on Temperature profile when Pr=10,Sc=N_t= N_b =0.1 and M=1

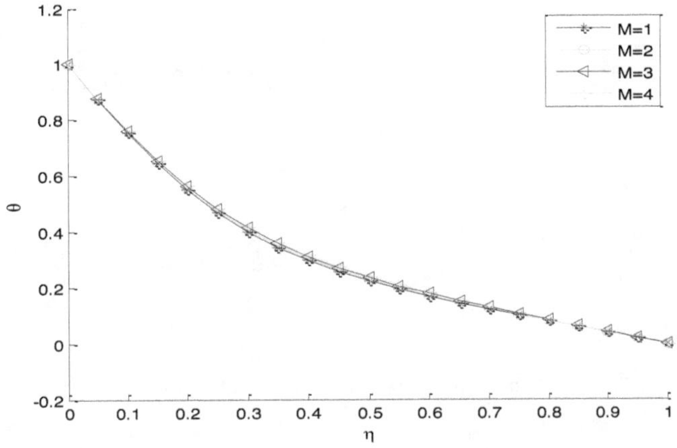

Figure 5. Effect of Magnetic parameter (M) on Temperature profile when Pr=10, Sc=N_t= N_b =0.1 and R=1

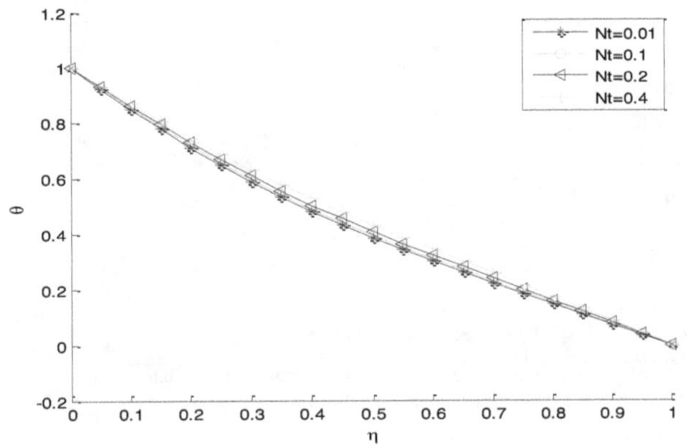

Figure 6. Effect of thermophoretic parameter (N_t) on Temperature profile when Pr=20, Sc= N_b =0.1 and M= R=1

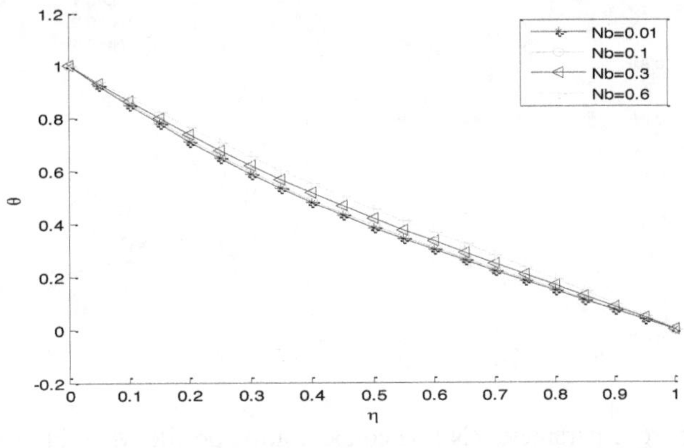

Figure 7. Effect of Brownian motion parameter (N_b) on Temperature profile when Pr=10, Sc= N_t =0.1 and M= R=1

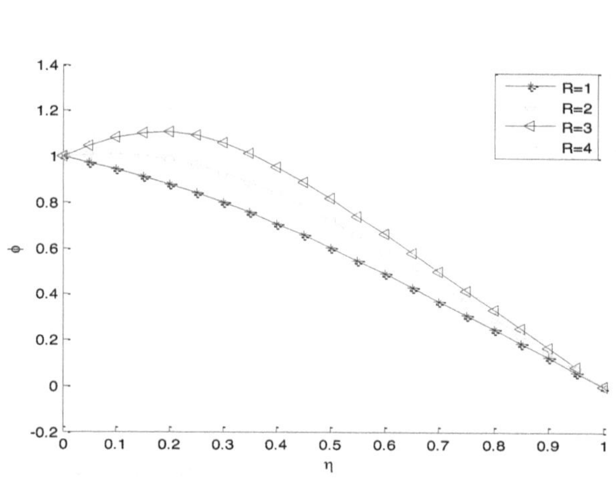

Figure 8. Effect of viscosity parameter (R) on concentration profile when Pr=10,Sc=N_t = N_b =0.1 and M=1

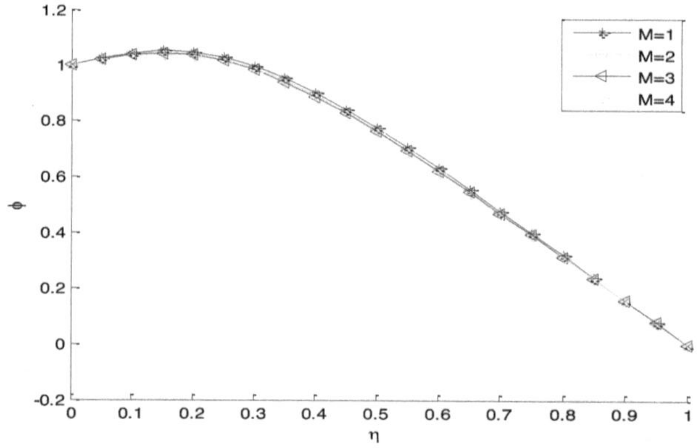

Figure 9. Effect of magnetic parameter (M) on concentration profile when Pr=10,Sc=N_t = N_b =0.1 and R=1

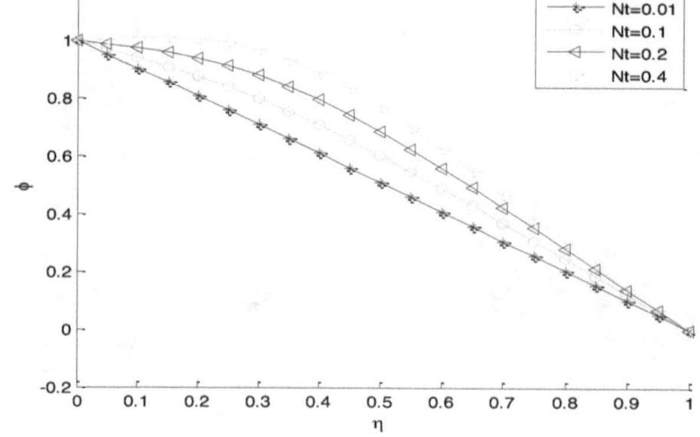

Figure 10. Effect of thermophoretic parameter (N_t) on concentration profile when Pr=10, Sc= N_b =0.1 and M= R=1

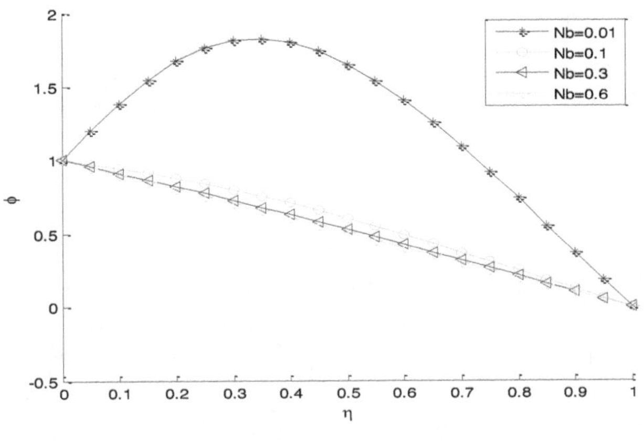

Figure 11. Effect of Brownian motion parameter (N_b) on concentration profile when Pr=10, Sc= N_t =0.1 and M= R=1

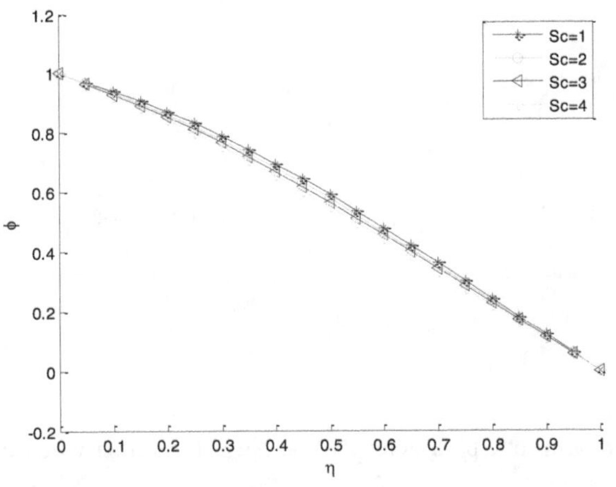

Figure 12. Effect of Schmidt number (Sc) on concentration profile when Pr=10, N_b= N_t =0.1 and M= R=1

Viscosity parameter (R) effect on Nano concentration is observed in Figure 8. It is shown that increasing R, concentration effect is maximum towards the lower plate due to increasing overall transport energy capacity. Magnetic parameter (M) effect on nanoparticle concentration is demonstrated in Figure 9 which depicts decreasing Nano concentration caused by increased magneto hydrodynamic boundary layer owing to higher thermal conductivity. It can be observed in Figure 10 that thermophoretic parameter (N_t) causes an increase in concentration profile as N_t increases quantitatively due to increasing fluid thermal diffusivity. Figure 11 depicts the effect of increasing values of Brownian motion parameter (N_b) on Nano concentration. As observed increasing values of N_b causes significant decrease in nano particle concentration owing to increased diffusion rate of reacting species. The effect of Schmit number (Sc) on concentration is observed in Figure 12. It is seen that increasing values of Sc leads to increased fluid viscosity causing a corresponding decrease in Nano concentration.

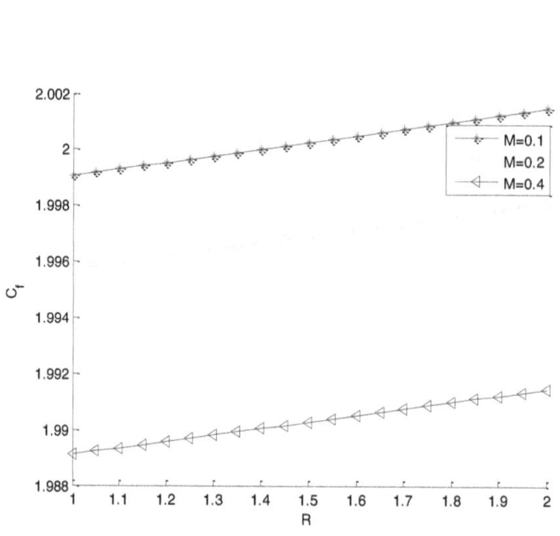

Figure 13. Effect of Magnetic parameter (M) on skin friction when Pr=10, $N_b = N_t = 0.1$

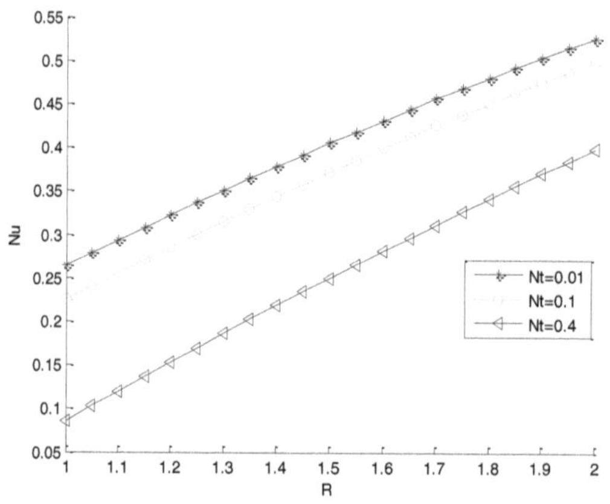

Figure 14. Effect of thermophoretic parameter (N_t) on Nusselt number when Pr=10, $N_b = 0.1$ and M=1

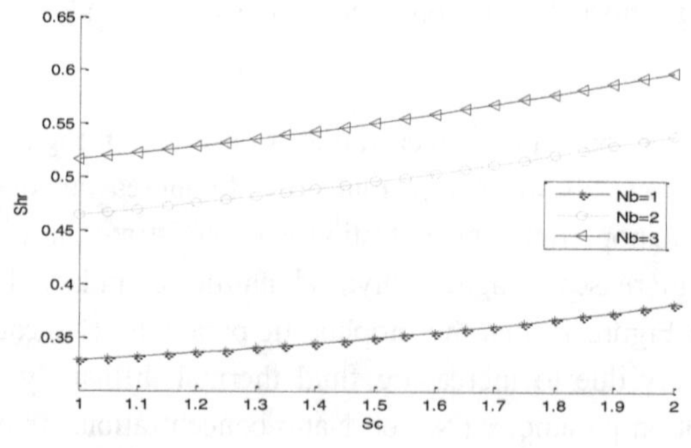

Figure 15. Effect of Brownian motion parameter (N_b) on Sherwood number when Pr=10, $N_t = 0.1$ and M=R=1

The effect of magnetic parameter (M) is illustrated on the skin friction in Figure 13, it is demonstrated that increasing M causes decrease in skin fluid friction which is higher at the upper plate wall whereas increasing values of thermophoretic parameter (N_t) as observed in Figure 14 causes a similar decrease on the Nusselt number i.e. the fluid heat transfer. Also the effect of Sherwood number on flow is observed in Figure 15. It is illustrated that increasing values of Brownian parameter (N_b) causes an increase on mass flow rate of fluid.

CONCLUSION

This present study analyses nanofluid flow through parallel plates arranged horizontally against each other, under the influence of uniform magnetic flux. The nanofluid flow described by nonlinear ordinary differential equations arising from the mechanics of the fluid are analyzed adopting the homotopy perturbation method (HPM). Various thermal-fluidic parameters such as viscosity parameter, Magnetic parameter, Schmidt number are investigated on flow, heat transfer and concentration. Result reveals skin friction and Nusselt number decreases while Sherwood number increases near walls during heat and mass transfer of Nano mix. This study proves useful in the advancement of transport and heat transfer processes in micro mixing, fuel cells, energy conservation, and pharmaceutical processes amongst other applications.

NOMENCLATURE

C	Concentration
c_p	Specific heat capacity of nanofluid
D_B	Diffusion coefficient of diffusing species
k	Thermal conductivity
$M=\dfrac{\sigma B_0^2 h^2}{\rho v}$	Magnetic parameter
$N_b=(\rho c)_p D_B(C_h)/(\rho c)_f\alpha)$	Brownian motion parameter
$N_t=(\rho c)_p D_T(T_H)/[(\rho c)_f \alpha T_c]$	Thermophoretic parameter
p^*	Nanofluid fluid pressure
$\Pr=\dfrac{\mu}{\rho_f \alpha}$	Prandtl's number
$R=\dfrac{ah^2}{v}$	viscosity variational parameter
$Sc=\dfrac{\mu}{\rho_f D}$	Schmidt number
T	Temperature
$u=axf'(\eta)$	x component velocity
$v=-ahf(\eta)$	y component velocity
μ	Dynamic viscosity
ρ_f	Base fluid density
$\eta=\dfrac{y}{h}$	Similarity variable
$\theta(\eta)=\dfrac{T-T_h}{T_0-T_h}$	Dimensionless Temperature
$\phi(\eta)=\dfrac{C-C_h}{C_0-C_h}$	Dimensionless Concentration

REFERENCES

[1] Abu-Nada, E., Masoud, Z., Hijazi, Z., (2008). Natural convection heat transfer enhancement in horizontal concentric annuli using nanofluids, International Communication Heat Transfer, 35, 657-665.
[2] Cortell, R., (2014). Fluid flow and radiative nonlinear heat transfer over a stretching sheet, Journal of King Saud University, 26, 161-167.
[3] Ellahi, R., Aziz,S., Zeshan, A., (2013) .Non-Newtonian nanofluid flow through porous medium between two coaxial cylinder with heat transfer and variable viscosity, Journal of Porous Media, 16 , 205-216.
[4] Garoosi. F., Bagheri,G., Rashidi,M.M.,(2015). Two phase simulation of natural convection and mixed convection of nanofluid in square cavity, Powder Technology, 275, 239-256.
[5] Garoosi,F., Rohani, B., Rashidi, M.M., (2015). Two phase modeling of mixed convection nanofluids in a square cavity with internal and external heating, Powder Technology, 275, 304-321.
[6] Garoosi, F., Jahanshaloo, L., Rashidi,M.M., Badakhsh, A. Alli, A. (2015). Numerical simulation of natural convection of the nanofluid in heat exchangers using a Buongiorno model, Applied Mathematics and Computation, 254, 183-203.
[7] Malvadi, A., Ganji, D.D., (2014). Brownian motion and thermophoretic effects of slip flow of alumina/water nanofluid inside a circular microchannel in the presence of magnetic field, International Journal of Thermal Science, 84, 196-206.
[8] Mehmood, A., Ali, A., (2008). Analytic solution of three dimensional viscous flow and heat transfer over a stretching surface by homotopy analysis method, American Society of Mechanical Engineers, 130 , 21701-21707.
[9] Rashidi, M.M., Abelman, S., Freidooni, S., Mehr, N., (2013). Entropy generation in steady MHD flow due to rotating porous disk in a nanofluid, International Journal of Heat and Mass Transfer, 62, 515- 525.
[10] Rashidi, S., Dehghan,M., Ellahi, R., Biaz, M., Jamal-Abad, M.T., (2015) .Study of streamwise transverse fluid with convective surface boundary condition, International Journal of Heat and Mass Transfer, 378 ,128-137.
[11] Shehzad, S.A., Alsaedi, A., Hayat, T., (2012). Three dimensional flow of Jeffery fluid with convective surface boundary condition, International Journal of Heat and Mass Transfer, 55, 3971-3976.
[12] Shehzad, S.A., Qasim, M., Alsaedi, A., Hayat, T., Alhuthali, M.S., (2013). Combined effects of thermal stratification and thermal radiation in mixed convection flow of thixotrophic fluid, European Physics Journal, 128-137.
[13] Shehzad, S.A., Alsaedi, F.E., Hayat, T., Monaquel, S.J., (2014). MHD mixed convection flow of thixotrophic fluid with thermal radiation, Heat Transfer Research, 45, 659-676.
[14] Sheikholeslami, M., Ashorynejad, H.R., Domairry, G., Hashim, I., Flow and heat transfer of Cu-water nanofluid between a stretching sheet and a porous surface in rotating system, Journal of Applied Mathematics. Article ID: 421320.
[15] Sheikholeslami, M., Gorji-Bandpy, M., Soleimani, S., (2013). Two phase simulation of nanofluid flow and heat transfer using heat analysis, International Communication of Heat and Mass Transfer, 47, 73-81.
[16] Akbar, N., Rahman, S.U., Ellahi, R., Nadeem, S., (2014). Nanofluid flow in tapering stenosed arteries with permeable walls, International Journal of Thermal Science, 85 ,54-61.
[17] Ashorynojad, H.R., Sheikholeslami, M., Pop, M., Ganji, D.D., (2013). Nanofluid flow and heat transfer due to stretching cylinder in the presence of magnetic field, Heat Mass Transfer Research, 49, 427-436.
[18] Ashorynojad, H.R., Mohamad, A.A., Sheikholeslami, M., (2013). Magnetic field effects on natural convection flow of a nanofluid in a horizontal cylindrical annulus using Lattice Boltzmann method, International Journal of Thermal Science, 64, 240-250.
[19] Domairry, D., Sheikholeslami, M., Ashorynejad, M.R., Suba Reddy Goria, R., (2012). Natural convection of flow of a non-Newtonian nanofluid between two vertical parallel plates, Proceeding of Institute of Mechanical Engineers Part N: Journal of Nanoengineering Nanosystem, 225, 115-122.
[20] Elllahi, R., (2013). The effect of MHD and temperature dependent viscosity on flow of non-Newtonian nanofluid in a pipe: analytical solutions, Applied Mathematical Model, 37, 2013, pp. 1451-1457.
[21] Ellahi, R., Reza, M., Vafai, K., (2012). Series solution of non-Newtonian nanofluids with Reynolds model and Vogel's model by means of homotopy analysis method, Mathematical Computational Model, 55, 1876-1891.
[22] Ellahi, R., Hassan, M., Zeeshan, A., (2015). Shape effect of nanosize particles on Cu-H20 nanofluid on entropy generation, International Journal of Heat and Mass Transfer, 81, 449-456.
[23] Hatami, M., Sheikholeslami, M., Hosseini, M., Ganji, D.D., (2014). Analytical investigation of MHD nanofluid flow in non-parallel walls, Journal of Molecular Liquids, 194, 251-259.
[24] Hatami, M., Shikholeslami, M., Ganji, D.D., (2014). Nanofluid flow and heat transfer in asymmetrical porous channel with expanding or contracting wall, Journal of Molecular Liquid, 195, 230-239.

[25] Hatami, M., Shikholeslami, M., Ganji, D.D., (2014). Laminar flow and heat transfer of nanofluid between contracting and rotating disks by least square method, Powder Technology. 253, 769-779.
[26] Jou, R.Y., Teng, S.C., (2006). Numerical research of natural convective heat transfer enhancement filled with nanofluid in rectangular enclosures, International Communications in Heat and Mass Transfer, 33, 727-736.
[27] Ketayari, G.H.R., (2013). Lattice Boltzmann simulation of natural convection in a nanofluid filled inclined cavity at presence of magnetic field, Science Iran, 20, 1517-1527.
[28] Ketayari, G.H.R., (2013). Lattice Boltzmann simulation of natural convection in nanofluid filled 2D long enclosures at presence of magnetic source. Theoretical Computational Fluid Dynamics, 27, 865-883.
[29] Ketayari, G.H.R., (2013). Lattice Boltzmann simulation of MHD natural convection in nanofluid filled cavity with sinousoidal temperature distribution, Powder Technology, 243, 171-183.
[30] Khan, W.A., Pop, I., (2010). Boundary layer flow of a nanofluid past a stretching sheet. International Journal of Heat Mass Transfer, 53, 2477-2483.
[31] Sheikholeslami, M., Ganji, D.D., (2013). Heat transfer of Cu-water nanofluid flow between parallel plates, Powder Technology, 235, 873-879.
[32] Chand, R., Rana, G., Hussein, A.K. (2015). On the onset of thermal Instability in a low Prandtl number nanofluid layer in a porous medium. Journal of Applied Fluid Mechanics, 8 (2), 265-272.
[33] Mohammed, H., Al-Aswadi, A., Abu-Mulaweh, H., Hussein, A.K., Kanna, P. (2014). Mixed convection over a backward-facing step in a vertical duct using nanofluids-buoyancy opposing case, Journal of Computational and Theoretical Nanoscience, 11, 1-13.
[34] Hussein, A.K., Ashorynejad, H., Shikholeslami, M., Sivasankaran, S. (2014). Lattice Boltzmann simulation of natural convection heat transfer in an open enclosure filled with Cu–water nanofluid in a presence of magnetic field. Nuclear Engineering and Design, 268, 10-17.
[35] Chand, R., Rana, G. a, Hussein, A.K. (2015). Effect of Suspended Particles on the Onset of Thermal Convection in a Nanofluid Layer for More Realistic Boundary Conditions. International Journal of Fluid Mechanics Research, 42 (5),375-390.
[36] Akinshilo, A.T., Olofinkua, J.O., Olaye, O., (2017). Flow and Heat Transfer Analysis of Sodium Alginate Conveying Copper Nanoparticles between Two Parallel Plates. Journal of Applied and Computational Mechanics, DOI:10.22055/jacm.2017.21514.1105.
[37] Sheikholeslami, M., Rashidi, M.M., Alsaad, D.M., Firouzi, F., Rokni, H.B., Domairry, G., (2015). Steady nanofluid flow between parallel plates considering thermophoresis and Brownian effect. Journal of King Saud, http.dx.doi.org/10.1016/j.jksus.2015.06.003.
[38] Hussein, A.K., Bakier, M., Ben Hamida, M., Sivasankaran, S. (2016). Magneto-hydrodynamic natural convection in an inclined T-shaped enclosure for different nanofluids and subjected to a uniform heat source, Alexandria Engineering Journal, 55, 2157-2169.
[39] Hussein, A.K., (2017). Mustafa, A. Natural convection in fully open parallelogrammic cavity filled with Cu-water nanofluid and heated locally from its bottom wall, Thermal Science and Engineering Progress, 1, 66-77.
[40] Kargar, A., Akbarzade, M., (2012). Analytical solution of Natural convection Flow of a non-Newtonian between two vertical parallel plates using the Homotopy Perturbation Method. World Applied Sciences Journal. 20, 1459-1465.
[41] Arslanturk, A., (2005). A decomposition method for fin efficiency of convective straight fin with temperature dependent thermal conductivity. International Communications in Heat and Mass Transfer, 32, 831-841.
[42] Aziz, A., Enamul-Huq, S.M., (1973). Perturbation solution for convecting fin with temperature dependent thermal conductivity. Journal of Heat Transfer, 97, 300-310.
[43] Zhou, J.K., (1986). Differential Transformation and its application for Electrical circuits, Huazhong University Press, China.

13

DETERMINATION OF CONVECTIVE, RADIATIVE, AND TOTAL HEAT TRANSFER CHARACTERISTICS OVER A RADIANT HEATED CEILING: A COMPUTATIONAL APPROACH

Ozgen Acikgoz[1*]

ABSTRACT

In the present study, a computational investigation is carried out within a model room of which dimensions has also been given in international standards for experimental approaches. In accordance with the experimental standards, the analyses are realized at the room dimensions of (L x H x W= 4 x 3 x 4 m), and at the thermal boundary conditions of (T_c= 35 – 50°C), (T_w= 15 – 24°C), (T_f= 15 – 24°C). As a result of iterations conducted using a computational fluid dynamics package program code, each corresponding to a case study, convective, radiative, and total heat fluxes from radiant heated ceiling are obtained, and afterward using relevant reference temperature proposed in the literature, heat transfer coefficients pertaining to convection, radiation, and total heat transfer have been found. In conclusion, average heat flux values of 95.83 Wm^{-2} and 17.49 Wm^{-2} are gained through the examined radiant heated ceiling, by means of radiation and convection, respectively. This denotes that roughly 85% of total heat transfer from the ceiling arises through radiation that is very compatible with the relevant literature. Finally, spanning the entire different case studies, average convective, radiative, and total heat transfer coefficients of 1.1 $Wm^{-2}K^{-1}$, 5.4 $Wm^{-2}K^{-1}$, 7.6 $Wm^{-2}K^{-1}$ are acquired.

Keywords: *Ceiling Heating, CFD, Radiation, Convection*

INTRODUCTION

On radiant heating and cooling applications, numerous studies exist in the relevant literature. While some of them are experimentally carried out investigations, significant portion of them are numerical studies. This arises from the fact that building up an experimental chamber providing stable thermal conditions which results in trustable outcomes, is not only difficult but also expensive. This stems from the difficulties in preparing hydronic circuits connected to a chamber equipped with well-protection from surrounding environment. Therefore, remarkable amount of experimental radiant heating works were conducted by means of electrically heated panels mounted on surfaces of enclosures, whereby, implementation of experiments was quite facilitated. Moreover, numerical studies via CFD (Computational Fluid Dynamics) programs were done to simulate a room of which surfaces had been heated/cooled by means of either constant heat flux or constant surface temperature boundary conditions.

With regard to reaching thermal comfort zone, the forth – coming point of radiant heating and cooling systems is that relatively lower radiant surface temperatures and higher radiant surface temperatures in heating and cooling modes bring about significant decrease in convective heat transfer and thus lower amount of air movements. To summarize the recently published papers, a short but dense literature review is given in the following paragraphs.
Wu et al. [1] progressed a new model to calculate surface temperatures and heat transfer for radiant systems, by using conduction shape factor. The data gained showed that the differences between found surface temperatures and heat transfer through the suggested model and the measured data were determined as 0.8°C and 8.1 W/m^2, while the average water temperature was within the range of 40°C and 60°C.

To ascertain the relation between the solar radiation utilization ratio and the supply heating temperatures, along with evaluating the efficiency of the radiant capillary module and the thermal comfort of the system for effective applications, Zhao et al. [2] have done a set of experiments in a residential building in a province of China. As a conclusion, they have deduced that by reducing the supply water temperatures, energy consumption of the system may be noticeably get better, in case the supply water temperature is below 35°C.

This paper was recommended for publication in revised form by Regional Editor Bekir Yilbas
[1] Department of Mechanical Engineering, Yildiz Technical University, Istanbul, Turkey
**E-mail address: oacikgoz@yildiz.edu.tr*

Evren et al. [3] carried out a variety of measurements in an experimental chamber whose surface temperatures and the dry-bulb air temperature were kept under control. The authors utilized two type of electric fan heaters, which have the same heating potentials, whereas different fan outputs, to employ it combined with a hydronic floor heating system. In the experiments, the floor system, and the fan heaters were performed at a number of heating outputs and hence radiation and convection ratios were varied while energy consumption data were being saved. As a result, it was concluded that radiative-convective ratio was lying within the range of 0.65 to 0.75.

Zhao et al. [4] examined the performance of a radiant floor heating and cooling application employed in an international airport. They indicated that the radiant floor application provided a heating output of 30 to 70 W/m^2, while the hot water temperature fluctuates between 35 and 40°C that may be evaluated as a trustable choice to cover heating requirements.

Zhang et al. [5] employed a model of a radiant ceiling cooling system used in association with an underfloor ventilation system. For this aim, they asserted a dynamic simplified resistance and capacity network of the mentioned combination. Via least square method, the authors determined thermal resistances, capacities and convective heat transfer coefficients of the model. To verify the accuracy of the developed model, seven case studies were made, while identification of the parameters were done via experimental data. As a result, authors have concluded that operative differences are less than 0.5 K, as inaccuracy of ceiling heat flux are within a range of 7.7%, between case studies and measured results.

By operating the program EnergyPlus, Bojic et al. [6] noted that the newly developed floor and ceiling radiant heating system had the most efficient performance, whereas, the conventional ceiling heating had the worst performance. On the other hand, it was estimated that in the upcoming days, with the decrease of primary energy consumption coefficient (R) the panel heating using ground source heat pump as heating source might have less total energy consumption than a system consuming natural gas.

Feng et al. [7] had a purpose of examining the effects of solar radiation on floor cooling capacity, as well as, forming a new and easy path to estimate the output of radiant floor cooling system, when direct solar radiation exists. They operated the program EnergyPlus to assess the influences of solar at a set of design conditions. As a conclusion, the results indicated that the actual cooling capacity were 1.44 times higher than the values calculated with ISO 11855 method, while the number is 1.2 times higher than that of ASHRAE method.

Bojic et al. [8] have individually investigated radiant heating systems used by means of all surfaces of rooms. It was demonstrated that the floor-ceiling had had the most favorable efficiency with regard to energy consumption, environmental fronts, and economy. Moreover, the least preferable system among different surface heating applications was found as the classical ceiling heating.

Shin et al. [9] generated design charts to evaluate heat flux, difference between maximum and minimum floor surface temperatures, and maximum floor surface temperatures in the design stage, through analyzing the correlation between heat flux and design parameters, alongside carrying out numerical case studies to determine temperature distributions on floor surfaces. By means of experimental investigations, the authors have validated whether the charts are providing true results.

Awbi [10] has reported the outcomes of a numerical investigation on convective heat transfer coefficients over a heated floor, wall, and ceiling. For this purpose, he employed two turbulence models: a standard k-ε model using wall functions and a low Reynolds k-ε model. The numerical outcomes were compared with those found by means of two different experimental rooms.

In addition to the numerical efforts in the field, Giovanelli and Bashir [11] have published a work on the design and analysis of a solar receiver. The main point of the work associated with the present investigation is that the authors to study the charge and discharge mechanism of the storage system that is combined with the solar receiver, using a CFD approach. The method implemented by the authors has composed an insightful path with regard to the way that should be followed through a CFD approach and a beneficial frame over the system they analyzed from a thermodynamics and heat transfer point of view.

In addition to the recently published articles summed up in this section, the articles and some international standards mainly focused on estimating heat transfer coefficients should be mentioned [12-22]. Either they are experimental or numerical, the significant part of these papers are related to convection. This is because, the radiative heat transfer in enclosures is calculated mainly by a method named net – radiosity method, and almost constant radiative heat transfer characteristics have been obtained in the extant literature. Therefore, the point still at some extent, has necessitated further investigation is the convective phenomenon which also determines the total heat transfer characteristics over a radiant heated/cooled surface.

To respond to this purpose, in the present study, the heat transfer characteristics pertaining to a radiant ceiling heating system are investigated in a numerically modeled room at the dimensions of LxHxW= 4x3x4 m, and at the thermal boundary conditions of (T_c= 35 – 50°C), (T_w= 15 – 24°C), (T_f= 15 – 24°C). The emissivity value of all surfaces was assumed 0.9 which is a reasonable, but also commonly measured value in experimental chambers. In the current investigation, using the programs DesignModeler and FLUENT in the package program ANSYS 14, the chamber is drawn and the governing equations related to the air flow within the chamber along with the radiation heat transfer problem between surfaces are solved in accordance with the prevalently acknowledged CFD solution approximation. Moreover, to generalize the behavior of convection and radiation over a heated ceiling in a room, a correlation which is a function of convective and radiative heat transfer coefficients found using the CFD approach, as well as of temperature difference between ceiling surface temperature and air temperature in the middle of the chamber, is derived and contributed to the literature for future researchers to compare their results.

THEORETICAL CALCULATIONS

Total heat transfer rate by the panels (Q_t); is composed of the radiative heat transfer emitted to the surrounding surfaces (Q_r), and convective heat transfer (Q_c) arising between radiant surface and the air. This energy equivalence is given in Equation 1.

$$Q_t = Q_r + Q_c \tag{1}$$

The total heat flux rate balance can also be gained with Equation 2.

$$h_t(T_s - T_{op}) = h_r(T_s - AUST) + h_c(T_s - T_a) \tag{2}$$

The radiative heat transfer coefficient is found with the help of AUST of which calculation way is presented in Equations 3-5.

$$AUST = \sqrt[4]{\sum_{j=1}^{n} (F_{s-j}\, T_j^4)} \tag{3}$$

$$F_{\varepsilon_{s-j}} = \frac{1}{[(1-\varepsilon_s)/\varepsilon_s] + (1/F_{s-j}) + (A_s/A_j)[(1-\varepsilon_j)/\varepsilon_j]} \tag{4}$$

$$F_{s-j} = \frac{1}{A_i}\int_{A_i}\int_{A_j} \frac{\cos\theta_i \cos\theta_j}{\pi R^2} dA_i dA_j \tag{5}$$

The radiant heat-flux ($\dot{q}_r$) from the heated surfaces to the other surfaces is obtained through Equation 6 and the coefficient symbolizing the pertaining heat transfer is attained by Equation 7.

$$\dot{q}_{rad} = \sigma \sum_{j=1}^{n} F_{\varepsilon_{s-j}} (T_s^4 - T_j^4) \tag{6}$$

$$h_r = \frac{q_r}{T_s - AUST} = \frac{\sigma \sum_{j=1}^{n} F_{\varepsilon_{s-j}} \left(T_s^4 - T_j^4\right)}{T_s - AUST} \tag{7}$$

For the entire surface emissivity values input in the program FLUENT, the value 0.9 has been employed. This value is prevalently used in the calculation method of radiative heat transfer between room surfaces, and also in experimentally found value of this parameter is roughly 0.9.

Equation 8 provides the total heat transfer coefficient (h_t), in which, the total heat-flux density ($\dot{q}_t$) is the sum of the radiative ($\dot{q}_r$) and the convective ($\dot{q}_c$) heat transfer heat-flux.

$$h_t = \frac{\dot{q}_t}{T_c - T_s} \tag{8}$$

The convective heat transfer coefficient between the heated surfaces and room – depends on the heat-flux densities, heated surface and air temperatures – was acquired with Equations 9-10.

$$\dot{q}_c = \dot{q}_t - \dot{q}_r \tag{9}$$

$$h_c = \frac{\dot{q}_c}{T_c - T_a} \tag{10}$$

CFD APPROACH

With the advancements in computer processors, the utilization of CFD programs for the determination of heat transfer characteristics in living areas has become reasonable. However, despite significant progression, reaching ultimate solution of iterations, in such large geometries, such as the office in the present study, has still taken notable amount of time. To mitigate this, generating cells close to the edges more frequent and preparing less frequent toward the center of the office have been applied, and thus enabling solutions in tolerable periods along with trustable accuracy has become possible. It should be noted that the present investigation provides results obtained within a wide range of thermal boundary conditions mentioned in Introduction Section.

A three dimensional model is generated to reach the findings of the present study. The governing equations are given for the steady-state air consisting of related models such as buoyancy influences, turbulent dispersion, and radiant heat transfer model. The equation of continuity in Equation 11, momentum in Equation 12, and energy equation in Equation 13 are given below for steady-state analyses:

$$\frac{\partial}{\partial x_i}(\rho u_i) = 0 \tag{11}$$

$$\frac{\partial}{\partial x_i}(\rho u_i u_j) = -\frac{\partial p}{\partial x_i} + \frac{\partial}{\partial x_j}\left[\mu\left(\frac{\partial u_i}{\partial x_j} + \frac{\partial u_j}{\partial x_i} - \frac{2}{3}\delta_{ij}\frac{\partial u_l}{\partial x_l}\right)\right] + \frac{\partial}{\partial x_j}\left(-\rho \overline{u_i' u_j'}\right) \tag{12}$$

$$\frac{\partial}{\partial x_i}\left(\rho u_i h - \lambda \frac{\partial T}{\partial x_i}\right) = u_i \frac{\partial p}{\partial x_i} + \tau_{ij}\frac{\partial u_i}{\partial x_i} \tag{13}$$

where, S_{ij} and τ_{ij} are mean rate of strain tensor and viscous stress tensor, respectively as shown in Equations 14 and 15.

$$\tau_{ij} = 2\mu S_{ij} - \frac{2}{3}\mu \frac{\partial u_l}{\partial x_l}\delta_{ij} \tag{14}$$

$$S_{ij} = \frac{1}{2}\left(\frac{\partial u_i}{\partial x_j} + \frac{\partial u_j}{\partial x_i}\right) \tag{15}$$

Solving governing equations (continuity, momentum, and energy) requires a package program to be employed. For this purpose, initially, the simulated office was drawn and meshed by means of the program DesignModeler within ANYSY 14, and then, thermal boundary conditions were input and necessary options such as the turbulence model to be used, was selected in the program FLUENT. Figure 1 illustrates the office meshed as it is depicted in the previous lines.

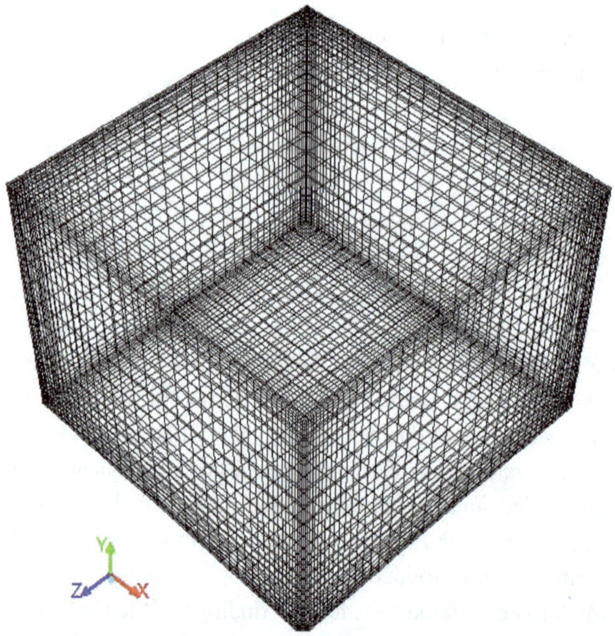

Figure 1. Meshing of modeled office (Lx H x W = 4 x 3 x 4 m)

The program's solution method is established on the control volume theory, ie. by means of turning relevant equations into algebraic equations for them to be solved [23]. Because, usage of mentioned plausible thermal boundary conditions result in turbulent flow over the ceiling, -due to Rayleigh numbers greater than 10^9- , a turbulence model has been chosen. The type of k-ε model was determined as the k-ε RNG model, in accordance with the proposal of Seyam et al. [24]. The under – relaxation factors pertaining to density, momentum, turbulence kinetic energy turbulence dissipation rate, turbulent viscosity and energy have been selected 1.0, 2.0, 0.8, 0.8, 1 and 0.9, respectively. It should also be noted that it took about 3 hours, the program to reach proposed residual values of 10^{-3} and 10^{-6}, for momentum/continuity and energy equations, respectively. Additionally, thermos – physical properties of air were entered for the air temperature value at the center of the enclosure, obtained from the air's thermos – physical tables of Incropera and DeWitt [25]. Also, the Boussinesq approach has been selected, because in natural convection it is strongly recommended by the program manual. Such as in hydronic radiant heating applications, a constant surface temperature has been preferred in the present work, to gain realistic outcomes. It should also be expressed that density – based coupled solver was determined as the solver. The gravitational force was also entered in the y direction as -9.81 m/s^2.

In addition to the points and selections made which should be considered while conducting a CFD simulation, one issue which should also be taken into account is independency of results from cell numbers within the simulated chamber. To ascertain whether the influence of meshing structure is sufficiently minimized or not, the Nusselt numbers are acquired through the program's area weighted average specialty. It should also be noted that the Nusselt number over the heated surface points out to convective heat transfer along this surface and can be assessed as a reference in this context. Altering interval counts at x, y, and z axes generates the diverse total cell numbers in the air, and this can be observed in Table 1.

It is clear that at the same thermal boundary conditions, in case the total cell number rose from 50625 to 90750, the Nusselt number deviated less than 1%. Moreover, since the results corresponding the abovementioned meshing numbers are so close, the interval combination next to the last one, corresponding to 50625 cells was determined as the appropriate option, and thus it was deduced that the effect of cell numbers on the results have been averted. From this point on, the case studies were carried out with selected interval count numbers.

To dispel the number of cells on results, at different cell number options, the iterations were done until the abovementioned residual values were reached and concluded that at the total cell number of 50625, the independency of results from total cell numbers had been realized. Hence, whole case studies have been implemented at this total cell numbers.

Table 1. The variation of Nusselt number as the number of cells within the cavity alters

Combinations of interval counts in x-y-and z axes	**Number of cells**	**Nusselt number**
20x20x20	8000	140.01
35x20x35	24500	141.87
45x25x45	50625	142.56
55x30x55	90750	142.94

RESULTS AND DISCUSSION

In case the relevant literature is investigated, one can notice that the most recent papers are related to radiant wall and floor heating and cooling. In addition to this, when the literature is rigorously scrutinized, a significant number of radiant ceiling cooling studies can be met. Nonetheless, it can clearly be observed that radiant ceiling heating systems comprise the least investigated portion in the field. The studies of Causone et al. [12], Koca et al. [13], and Rahimi and Sabernaeemi [26] are the most recent and detailed works in the literature. Nevertheless, by very nature of experimental studies, these works have also been carried out within confined thermal boundary conditions, and implemented in chambers of which dimensions are constant and cannot be altered. Therefore, carrying out case studies in computational fluid dynamics programs such as one used in the present work ANSYS 14, provides researchers and engineers to do these investigations in broader thermal boundary conditions and different room dimensions.

In the light of abovementioned considerations, the present study presents the results of a numerical work which were conducted within a wide range of thermal boundary conditions, and in a simulated room meshed as illustrated in Figure 1. As can be observed from the figure, meshes near edges are maintained highly frequent, while preferring them less frequent toward the center of the chamber. This choice can be accounted for by the fact that in the thermal boundary layers of heated and cooled surfaces, rapid temperature changes occur, and thus effect of these variations ought to be taken into account by frequent cell densities in these regions. Nonetheless, due to the fact that temperature variations out of boundary layer regions that are toward the center are rather negligible, mesh density can be selected much less frequent in this region. By this means, computation duration has also been diminished, otherwise, in large geometries such as that of the present investigation, acquiring a realistic result in reasonable periods would have been unfeasible.

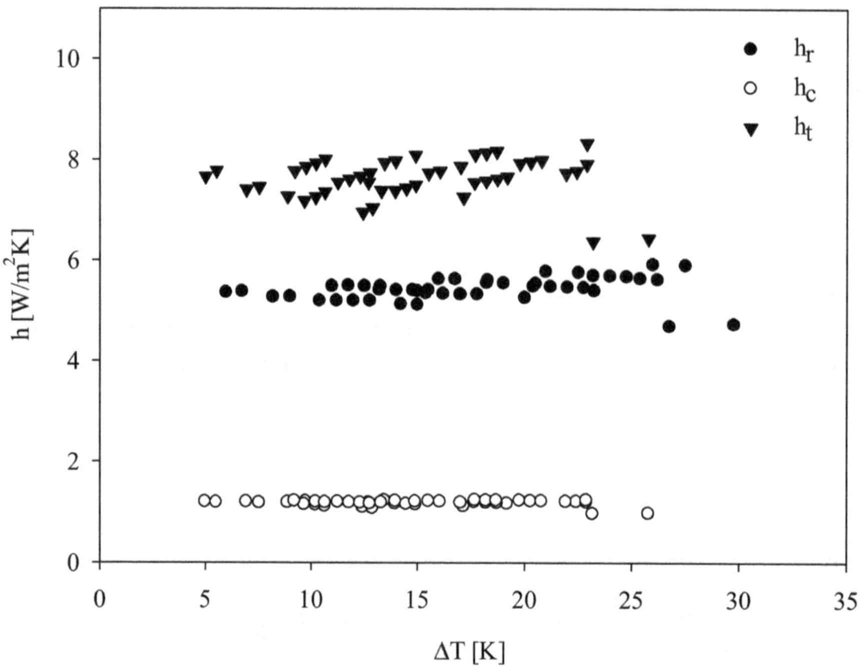

Figure 2. Radiative, convective, and total heat transfers' variation with corresponding temperature difference values

In Figure 2 the variation of radiative, convective, and total heat transfer coefficients with corresponding temperature differences (ΔT) are indicated. It can be observed from the figure that especially convective and radiative heat transfer coefficients mostly draw an unchanging trend with corresponding temperature values, (T_c-T_a), and (T_c-AUST), respectively. Consequently, it may be deduced that as a result of the present one is able to obtain an average radiative heat transfer coefficient value of 5.4 W/m^{2}K, and an average convective heat transfer coefficient value of 1.1 W/m^{2}K. Though total heat transfer coefficient values appears to be more scattered than that of radiative and convective heat transfer coefficients, the data cluster in the vicinity of 7.5 W/m^{2}K, while an average value of 7.6 W/m^{2}K is obtained, as a recommendation of the present investigation. Relatively scattered appearance of total heat transfer coefficients in the figure may be attributed to reference temperature selection pertaining to total heat transfer characteristics, which has been a major uncertainty in the relevant research field. In the meantime, it should be noted that in this work the air temperature in the middle of the chamber was used as the reference temperature for total heat transfer characteristics. Figure 3 presents a comparison between the total heat flux of the radiant ceiling system in the present work, and results obtained using the correlation proposed by EN 15377 [17]. As it can be noticed in Figure 2, the relatively scattered trend of the heat flux trend from the present study compared to the results of correlation suggested by EN 15377 [17], can also be accounted for the explanations noted for Figure 2.

In addition to this, between the outcomes of two studies presented in this figure, a slight deviation between the trendlines of studies are discerned. This may be attributed to unavoidable experimental errors in the corresponding study, numerical errors due to the method performed in the current examination, or thermal boundary conditions of surfaces other than the heated ceiling in the present investigation. Nevertheless, the deviation of the results of the correlation of EN 15377 [17] and of the results of this study are very close at lower temperature difference ($T_c - T_a$) values, which are more realistic values, while at higher temperature difference ($T_c - T_a$) values, which are not so realistic in pratice, the deviation grows. Thus, this also elucidates that the explained deviation does not denote that the numerical results of this study to be unrealistic.

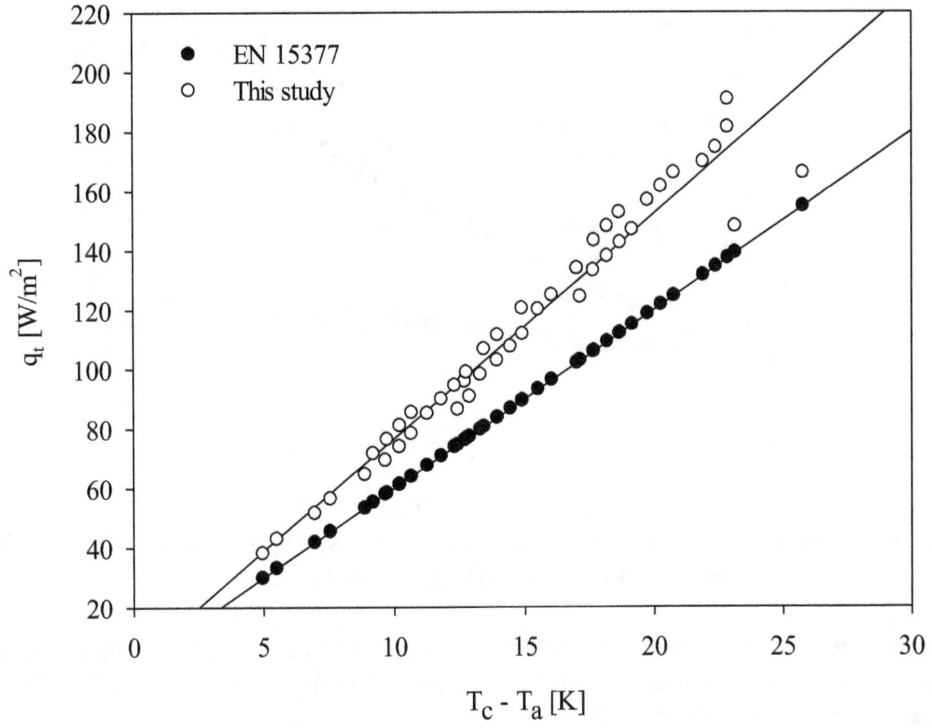

Figure 3. The comparison of total heat flux results of this study and results obtained employing the correlation proposed by EN 15377

Figure 4 illustrates a comparison of the results of study with of Koca et al. [13] and of Causone et al. [12] in both which radiant ceiling heating systems are experimentally analyzed. It should be noted that since in the paper of Causone et al. [12] no AUST value given, the radiative heat transfer characteristic results of this study have not been able to be compared with those of the corresponding work. In terms of convective heat flux, it is evident from the figure that the results of three study are considerably consistent with each other, while results of Causone et al. [12] give the lowest convective heat flux value, the present study results in the highest average, and Koca et al. [13] concludes as the value remaining in the middle. With regard to radiative heat flux, the results of this study, along with that of Koca et al. [13] draw almost the same trend line.

Thus, they are compatible with each other. Total heat flux values slightly deviate from each other possibly due to the differences in convective heat flux results, which is an ingredient composing total heat flux. It is clear from the figure that as in the convective heat flux remarks, despite the fact the differences are not so significant, this study results in highest heat flux outputs, while Causone et al. [12] the lowest, and Koca et al. [13] in the middle of two Figure 5 compares the results of this study and of Koca et al. [13] in terms of heat transfer coefficients. It is noticeable that similar to convective heat flux values, convective heat transfer coefficients found via the present study demonstrate a slightly higher tendency than those of Koca et al. [13]. Furthermore, the radiative heat transfer coefficient data from this investigation keep track of the results of Koca et al. [13] almost in the same points, while total heat transfer coefficients of the present work which involves convective heat transfer data illustrate a higher tendency line, because of the higher convective results both indicated with respect to convective heat transfer coefficients in Figure 5 and convective heat flux in Figure 4, of which deviation sources were accounted for in previous paragraph, as interpreting Figure 4.

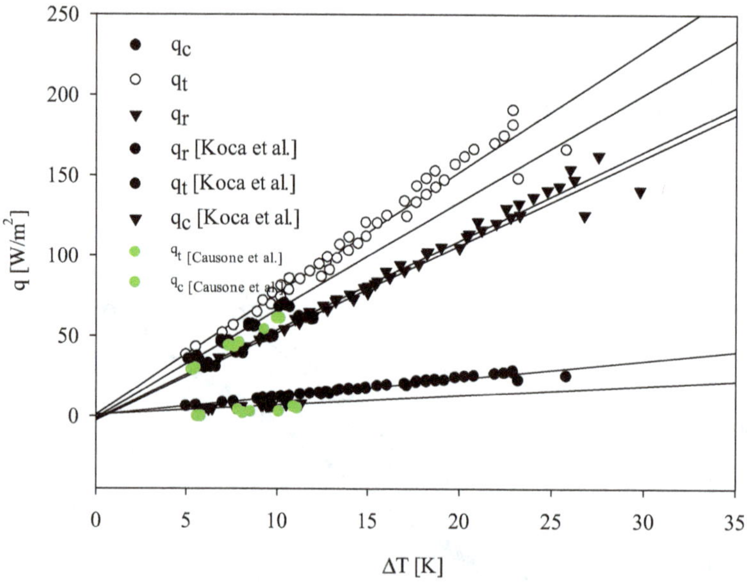

Figure 4. Comparison of convective, radiative, and total heat outputs of the present work, and of experimental results of Koca et al. [11] and Causone et al. [10]

Furthermore, to form a correlation between the radiative and convective findings of the present investigation, an equation valid within the thermal boundary conditions expressed in the last paragraph of Introduction section is produced and can be seen in Equation 16 as follows:

$$\frac{h_r}{h_c} = 4.11\,(T_c - T_a)^{0.043} \tag{16}$$

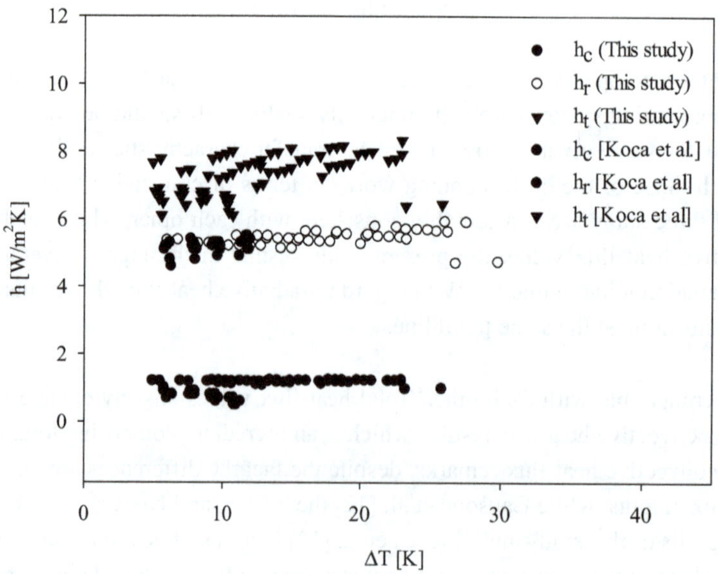

Figure 5. A comparison between heat transfer coefficient findings of the current study, and of experimental study of Koca et al. [11]

To demonstrate the h_r/h_c ratio results of this study alongside the outputs of the correlation expressed in Equation 16, Figure 6 is generated. As can be seen through the figure, a trend line symbolizing the outweighing behavior of the proportion of radiative and convective heat transfer coefficient, was drawn. It ought to be noted that the correlation derived which also symbolizes this line has a maximum deviation ratio of 6%, and an average deviation ratio of 1.2%, which are both satisfactory values, with regard to being a benchmark of obtaining an accurate equation.

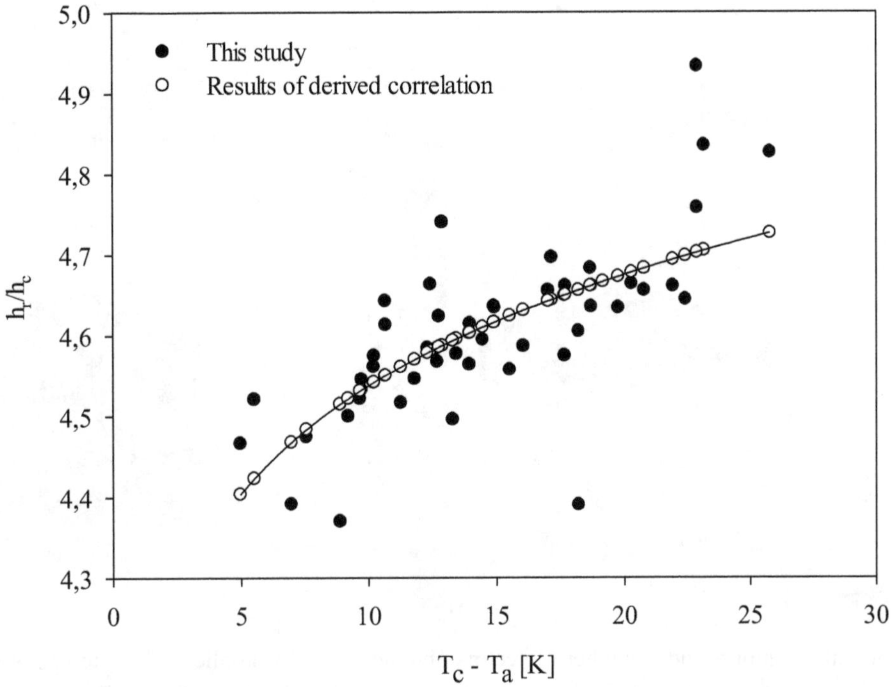

Figure 6. The results of the present study pertaining to the proportion of h_r and h_c, and the results obtained by the Eq. [16] derived for this ratio

Table 1. Comparison of the present work with some other radiant studies in the literature, in terms of heat transfer coefficients

Researchers	Heating / Cooling type	h_c (W/m²K)	h_r (W/m²K)	h_t (W/m²K)
Causone et al. [10]	Radiant ceiling heating	0,3	5,6	5,8
Koca et al. [11]	Radiant ceiling heating	0,7	5,2	6,6
Cholewa et al. [12]	Radiant floor heating	2,2 - 3,4	5,6	8,5 - 10,4
Andres - Chicote et al. [13]	Radiant ceiling cooling	4,2	5,6	13,2
EN 1264 – 5 [14]	Radiant heated floor	-	-	10,8
EN 15377-1 [15]	Radiant heated floor	-	5,5	-
This study	Radiant heated ceiling	1,1	5,4	7,6

In Table 1, the findings of the present paper are compared with studies covering heat transfer coefficients over radiant surfaces. The most similar studies in the literature were compared with those of the present investigation by means of Figures 2 – 6, however, some other researchers' findings were also assembled in Table 1 for future researchers to have a better idea over various radiant heating and cooling systems in terms of heat transfer characteristics.

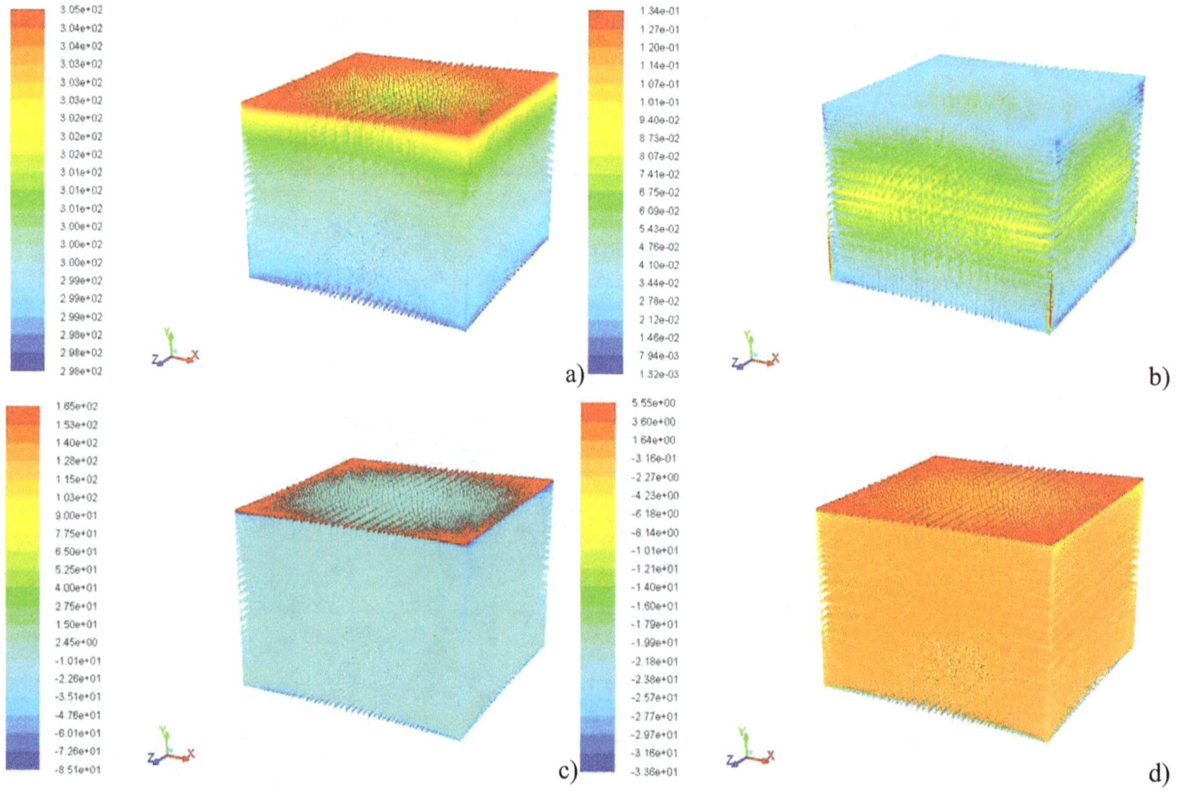

Figure 7. Temperature contours and a number of vectors obtained via CFD application (a) temperature contour, (b) velocity vectors, (c) radiant heat flux vectors, and (d) surface heat transfer coefficient vectors

Figures 7 (a-d) demonstrate a set of contours and vectors along with heat transfer characteristics, in case the thermal boundary conditions of (T_s= 50°C, T_w= 24°C, and T_f= 18°C) are applied to the numerically modeled room whose dimensions are given in Section 1. It is obvious from Figure 7a that radiant heated ceiling results in an gradual thermal stratification next to the heated ceiling, and a less gradual one beside the floor region where air temperature decreases toward the floor, as expectedly. Figure 7b reveals velocity vectors in the investigated office, and it is noticeable that over walls highest velocity magnitudes have occurred. This may be because of the difference in temperature and therefore density air particles over walls and ones around the heated ceiling that brings about higher amounts of air movements compared to the core of the air within the model. Figure 7c depicts radiant heat flux vectors in the chamber. It is evident that radiant heat flux vectors over the ceiling adjacent to walls are at the highest values, while ones in the middle of ceiling are less effective. It can also be understood that walls, and floor have negative flux values different than those of the ceiling, as expectedly. Moreover, Figure 7d gives total heat transfer coefficients as vectors on surfaces. One can understand through the figure that consistent with the results given via figures and tables, the most intense heat transfer coefficient values have been gained over the ceiling, and it should be noted that the coefficient vectors involve both convection and radiation heat transfer by the examined surfaces.

CONCLUSION

Suspended nanoparticles play a significant role in a wide range of applications because of their thermal and optical properties. Particle size and size distribution show significant effects in different wavelength ranges, which is very important in specific applications including solar thermal systems in addition to other applications. Radiative properties (scattering and absorption) carry the significant impact on the thermal radiative transfer, where the energy of the incident radiation is converted directly to thermal radiation. Different nanoparticles in different conditions show different radiative properties, which can be exploited in the thermal field for specific applications.

The aim of the using participating media is not only to convert concentrated radiation energy into thermal energy as efficiently as possible but also, to achieve the highest possible temperatures while retaining this high thermal efficiency.

In this study, it can observe that the nanoparticles play a significant role in the UV-Vis wavelength ranges where a large amount of the solar energy is dissipated. The extinction coefficient of the media is enhanced and affects the radiation transfer performance. Particle size and number have important effects on the thermal radiation transfer, increasing particle concentration leads to enhance the radiative properties.

NOMENCLATURE

A	Surface area of the panel (m^2)
AUST	Average unheated surface temperature (°C)
CFD	Computational fluid dynamics
c_p	Specific heat (J/kgK)
ε	Surface emissivity
E_{bi}	Blackbody emissive power of examined surface (W/m^2)
F_{i-j}	View factor between the examined and j surfaces
F_{s-j}	View factor between the radiant surface and j surfaces
F_{1-n}	View factor between the examined surface and the surroundings
h_c	Convective heat transfer coefficient (W/m^2K)
h_r	Radiative heat transfer coefficient (W/m^2K)
h_t	Total heat transfer coefficient (W/m^2K)
J_i	Thermal radiosity of examined surface (W/m^2)
J_j	Thermal radiosity of j surface (W/m^2)
Q_c	Convective heat transfer (W)
Q_r	Radiative heat transfer (W)
q_c	Convective heat flux (W/m^2)
q_r	Radiative heat flux (W/m^2)
q_t	Total heat flux (W/m^2)
R	Correlation coefficient
T_a, T_{op}, T_c, T_s, T_w	Air temperature, operative temperature, ceiling temperature, floor temperature, and temperature of walls (°C)
ΔT	Temperature difference between radiant heated wall and reference temperature at corresponding heat transfer coefficients

REFERENCES

[1] Wu, X., Zhao, J., Olesen, B. W., Fang, L., & Wang, F. (2015). A new simplified model to calculate surface temperature and heat transfer of radiant floor heating and cooling systems. Energy and Buildings, 105, 285-293.

[2] Zhao, M., Gu, Z. L., Kang, W. B., Liu, X., Zhang, L. Y., Jin, L. W., & Zhang, Q. L. (2017). Experimental investigation and feasibility analysis on a capillary radiant heating system based on solar and air source heat pump dual heat source. Applied energy, 185, 2094-2105.

[3] Evren, M. F., Özsunar, A., & Kılkış, B. (2016). Experimental investigation of energy-optimum radiant-convective heat transfer split for hybrid heating systems. Energy and Buildings, 127, 66-74.

[4] Zhao, K., Liu, X. H., & Jiang, Y. (2014). On-site measured performance of a radiant floor cooling/heating system in Xi'an Xianyang International Airport. Solar Energy, 108, 274-286.

[5] Zhang, D., Xia, X., & Cai, N. (2016). A dynamic simplified model of radiant ceiling cooling integrated with underfloor ventilation system. Applied Thermal Engineering, 106, 415-422.

[6] Bojić, M., Cvetković, D., & Bojić, L. (2015). Decreasing energy use and influence to environment by radiant panel heating using different energy sources. Applied energy, 138, 404-413.

[7] Feng, J. D., Schiavon, S., & Bauman, F. (2016). New method for the design of radiant floor cooling systems with solar radiation. Energy and buildings, 125, 9-18.

[8] Bojić, M., Cvetković, D., Marjanović, V., Blagojević, M., & Djordjević, Z. (2013). Performances of low temperature radiant heating systems. Energy and Buildings, 61, 233-238.
[9] Shin, M. S., Rhee, K. N., Ryu, S. R., Yeo, M. S., & Kim, K. W. (2015). Design of radiant floor heating panel in view of floor surface temperatures. Building and Environment, 92, 559-577.
[10] Awbi, H. B. (1998). Calculation of convective heat transfer coefficients of room surfaces for natural convection. Energy and buildings, 28(2), 219-227.
[11] Giovannelli, A., & Bashir, M. (2017). Charge and discharge analyses of a PCM storage system integrated in a high-temperature solar receiver. Energies, 10(12), 1943.
[12] Causone, F., Corgnati, S. P., Filippi, M., & Olesen, B. W. (2009). Experimental evaluation of heat transfer coefficients between radiant ceiling and room. Energy and Buildings, 41(6), 622–628.
[13] Koca, A., Acikgoz, O., Çebi, A., Çetin, G., Dalkilic, A. S., & Wongwises, S. (2017). An experimental investigation devoted to determine heat transfer characteristics in a radiant ceiling heating system. Heat and Mass Transfer/Waerme-Und Stoffuebertragung, pp. 1–13.
[14] Cholewa, T., Rosiński, M., Spik, Z., Dudzińska, M. R., & Siuta-Olcha, A. (2013). On the heat transfer coefficients between heated/cooled radiant floor and room. Energy and Buildings, 66, 599–606.
[15] Andrés-Chicote, M., Tejero-González, A., Velasco-Gómez, E., & Rey-Martínez, F. J. (2012). Experimental study on the cooling capacity of a radiant cooled ceiling system. Energy and Buildings, 54, 207–214.
[16] EN 1264-5, Water Based Surface Embedded Heating and Cooling Systems – Part 5 (2008): Heating and Cooling Surfaces Embedded in Floors, Ceilings and Walls.
[17] EN 15377-1, Heating Systems in Buildings – Design of Embedded Water Based Surface Heating and Cooling Systems – Part 1 (2008): Determination of the Design Heating and Cooling Capacity.
[18] Koca, A., & Çetin, G. (2017). Experimental investigation on the heat transfer coefficients of radiant heating systems: Wall, ceiling and wall-ceiling integration. Energy and Buildings, 148, 311–326.
[19] Awbi, H. B. (1998). Calculation of convective heat transfer coefficients of room surfaces for natural convection. Energy and Buildings, 28(2), 219–227.
[20] Acikgoz, O., & Kincay, O. (2015). Experimental and numerical investigation of the correlation between radiative and convective heat-transfer coefficients at the cooled wall of a real-sized room. Energy and Buildings, 108, 257–266.
[21] Acikgoz, O. (2015). A novel evaluation regarding the influence of surface emissivity on radiative and total heat transfer coefficients in radiant heating systems by means of theoretical and numerical methods. Energy and Buildings, 102, 105–116.
[22] Karadağ, R. (2009). New approach relevant to total heat transfer coefficient including the effect of radiation and convection at the ceiling in a cooled ceiling room. Applied Thermal Engineering, 29(8–9), 1561–1565.
[23] Demir, H., Dalkilic, A. S., Kürekci, N. A., Duangthongsuk, W., & Wongwises, S. (2011). Numerical investigation on the single phase forced convection heat transfer characteristics of TiO2 nanofluids in a double-tube counter flow heat exchanger. International Communications in Heat and Mass Transfer, 38(2), 218–228.
[24] Seyam, S., Huzayyin, A., El-Batsh, H., & Nada, S. (2014). Experimental and numerical investigation of the radiant panel heating system using scale room model. Energy and Buildings, 82, 130–141.
[25] Incropera, F. P., DeWitt, D. P., Bergman, T. L., & Lavine, A. S. (2007). Heat and Mass Transfer - Incropera 6e. Fundamentals of Heat and Mass Transfer.
[26] Rahimi, M., & Sabernaeemi, A. (2010). Experimental study of radiation and free convection in an enclosure with a radiant ceiling heating system. Energy and Buildings, 42(11), 2077–2082.

14

ENERGY AND EXERGY ANALYSIS OF A DOMESTIC REFRIGERATOR: APPROACHING A SUSTAINABLE REFRIGERATOR

N. D. Shikalgar[1,*], S.N. Sapali[1]

ABSTRACT

In the perspective of reducing the household energy consumption, current research in conventional refrigeration is concentrating on introducing innovative designs and enhances the energy efficiency of the refrigeration system. This research work compares the performance of the domestic refrigerator by employing hot-wall air cooled and box type shell and tube water-cooled condenser. The energy and exergy analysis methods help to localize exergy losses in the refrigerator. The investigation is carried out according to ISO 15502:2005. Experimental studies were conducted in the same refrigerator unit operating with a different condenser to determine the coefficient of performance, exergy efficiency, and exergy loss in all components of the domestic refrigerator. The experimental result shows that COP is increased by 18-20% and the exergy efficiency of the refrigerator with water cooled condenser unit is found to be higher by 6.89-9.13% than the one with a hot-wall air cooled condenser. The per day energy consumption of a refrigerator with a water-cooled condenser reduces by 17% in comparison with conventional refrigerator. The irreversibility of the refrigerator with the water-cooled condenser is reduced by 34 % than that of the conventional system under similar operating conditions. The total equivalent warming impact of refrigerator working with the water-cooled condenser is 16% lower than that the refrigerator with air cooled condenser. The utilization of water-cooled condenser in household refrigerators enhances energy efficiency.

Keywords: *Refrigerator, Water-cooled Condenser, Hot-wall Condenser, Energy, Exergy*

INTRODUCTION

In the view of reducing household energy consumption, current research in refrigeration is focusing on improving energy efficiency, introducing innovative designs in compact condensers and evaporators. Among the domestic appliances, household refrigerator consumes about 15% of the worldwide energy, and about 40% of the world energy consumption is from the building sector [1, 2]. The development of an energy efficient refrigerator is one of the big challenges for the Heating, Ventilation, Air-conditioning and Refrigeration (HVAC&R) sector. The majority of the research since the 1980s focused on heat transfer enhancement in condensers and evaporators. Several studies were reported on the effect of ambient temperature, optimization of mass flow rate, use of eco-friendly refrigerant etc. to make the refrigerator more energy efficient [3]. All vapour-compression refrigeration systems utilize the condenser to reject the heat absorbed by the refrigerant in the evaporator. The condenser also rejects the heat equivalent of the work of compression to the cooling medium [10]. A well-designed condenser not only enhances the energy efficiency but also reduces the space and material of the refrigerator. In a refrigeration system, the condenser holds the maximum portion of the refrigerant charge, and therefore heat exchangers with the small internal volume on the refrigerant-side should be preferred [4]. Condensers commonly used in the domestic refrigerator are of wire and tube type. The condenser temperature decrement of approximately 8°C and the corresponding decrement in power consumption of 10% are obtained by using a standard wire and-tube condenser, equipped with phase-change materials [5]. The study on spiral wire-on-tube condenser in a refrigerator shows that the decrease in condensing temperature can reduce the per day energy consumption of the refrigerator by 2.37%. This is due to the combined effect of a decrease in compressor ON time ratio and compressor power [6].

In modern refrigerators, the wire and tube condensers are replaced by the hot-wall air cooled condenser (HWAC); HWAC is often installed in the form of single serpentine tubes on the left, right and back sides. The single aluminium tube of length ranging between 20 to 25 m is fixed on the inner side of the wall of aluminium foil and gives off the heat absorbed by the refrigerant to the surrounding through walls [7]. In recent years, the bottom

This paper was recommended for publication in revised form by Regional Editor Hafiz Muhammad Ali
[1]Department of Mechanical Engineering, College of Engineering Pune, Maharashtra, India
**E-mail address: nds.mech@coep.ac.in, sns.mech@coep.ac.in*

condenser is introduced near the compressor as a part of the HWAC to evaporate the drain water continuously in a domestic refrigerator. In all refrigeration components, condenser plays an imperative task to save energy consumption. It is also observed that a reduction of 1°C condenser temperature can reduce 2% of total energy consumption [3]. HWAC condenser normally dissipates heat by convection and radiation at the rate of 250W to 400W to the surroundings. In hot climatic conditions, such an arrangement leads to 25W to 35W of heat conducted into the refrigeration compartment through the insulating material which gives rise to the energy consumption of the refrigerator [8]. Despite the fact that HWAC consumes more power as compared to a water-cooled condenser, it is preferred in small systems due to its low cost and easy maintenance [9]. Hosoz and Kilicarslan [10, 11] Depending on the type of the cooling medium, condensers can be classified as air-cooled, water-cooled and the combination of the both known as the evaporative condenser. Hosoz et al. evaluated the use of air-cooled, water-cooled and evaporative condensers for refrigeration systems and found that the condensing pressure and compressor work is decreased, consequently increasing the refrigeration capacity and compressor's life. Shiochi et al. have simulated the effect of refrigerant flow in an air conditioning system with the water cooled condenser. The result shows that one can save 20% of energy with a water-cooled air conditioning system. Henceforth, in a densely populated country, increasing use of domestic appliances leads to demanding large energy supply. Therefore, researchers and industries have their attention oriented towards the implementation of the water-cooled condenser in domestic refrigerators and air conditioning systems [12].

According to the American Society of Heating, Refrigeration and Air conditioning Engineers (ASHRAE) standards, the plate-type heat exchangers have a better heat transfer coefficient than that of the shell and coil heat exchanger [13]. Ramezanizadeh, M., et al have experimentally investigated the thermal performance of a thermosyphon-based heat exchanger is by using Ni/Glycerol–water nanofluid. Since the effective thermal conductivity of thermosyphon depends on the temperature difference between a condenser and evaporator. Effects of mass flow rates of cold and streams and inlet temperature of the hot stream on heat transfer rate are evaluated. He concluded that an increase in the mass flow rates of the streams and inlet temperature of the hot stream lead to an increase in the heat transfer rate [14]. Aprea C [15] has reported that the exergy losses in condenser and evaporator are highly influenced by a change in condensing and evaporating temperatures in a vapour compression plant. This is due to the decrease in the difference in temperature between the condenser and cooling medium. Selladurai et al. [16] has compared the performance between R134a and R290/R600a mixture on a domestic refrigerator which is originally designed to work with R134a and found the similar result as highest irreversibility in compressor followed by the condenser, expansion valve and evaporator.

S. Joseph Sekhar et al. [17] has studied the energy and exergy analysis of a domestic refrigerator with a brazed plate heat exchanger as water cooled condenser. It is observed that the per day energy consumption of a refrigerator is reduced by 21% to 27% and COP improved by 52% to 68%, as compared to the conventional refrigerator with air cooled condenser. Based on experimental studies he reported that the exergy destruction and irreversibility is reduced with a water-cooled condenser. Raveendran et al. [3, 18] experimentally evaluated the performance of refrigeration with a brazed plate heat exchanger as a condenser and found that the electric energy requirement of a refrigeration system reduces from 21% to 27% and COP increases from 52% to 68%, as compared to the conventional system. Mahmood Joybari, et al. [19] has carried out an experimental exergy analysis of the vapour compression refrigeration system applied to the domestic refrigerator with R134a to improve its performance. With exergy analysis, the optimum requirement of refrigerant charge required for alternative refrigerant R600a is determined as 50 g. It is observed that the compressor has the highest exergy destruction followed by the condenser. The highest exergy destruction observed in R 134a compressor is 0.03095 kW, whereas in 0.03612 kW in R600a compressor. Hosseinzadeh-Bandbafha, H et al. have presented a study on the assessment of energy flow and greenhouse gas emission of peanut production in Guilan province, Iran, and then the application of data envelopment analysis to determine optimum energy use pattern for saving energy and reduction of greenhouse gas emission. This reduction in greenhouse gas emission can be realized by management of diesel fuel, nitrogen and machinery consumption according to optimized input rates introduced by the data envelopment analysis. [20]. Total equivalent warming impact (TEWI) has been defined which allows a comparison of for large size refrigeration systems in relation to their impact on the environment. The TEWI index takes into account not only the direct warming effect due to refrigerant losses but also the indirect effect due to the plant efficiency and CO_2 release by utility companies supplying power to the device in question [21].

It is seen from the literature survey that there is scope for development and comparing performances of refrigeration systems employing with thermosyphon-based Box type shell and tube water-cooled condenser (BSTWC) and air cooled condenser. This research study investigates the experimental performances of the domestic refrigerator operating with two different types of condensers. Each system with a different condenser was operated under the same condensing and evaporating temperatures, and environmental conditions. Then, the performance, Characteristics such as refrigeration capacity, energy consumption of compressor, the coefficient of performance, exergy efficiency etc., of the systems were compared with each other. The experimental investigation has focused on the energy saving opportunities with help of exergy analysis tool. Investigation results show that the use of BSTWC in domestic refrigerators enhances energy efficiency.

Exergy Analysis of Vapor-Compression Refrigeration Cycle

The second law of thermodynamics (exergy analysis) deals with the quality of energy. The degradation in the quality of energy is termed as exergy loss (irreversibility), and the analysis in the loss provides ample opportunities for improvement. The second law of thermodynamics has proved to be a very powerful tool in the optimization of complex thermodynamic systems [22]. The generalized exergy balance for a control volume can be expressed as given in equation 1,

$$Ex_D = \Sigma(Ex)_{in} - \Sigma(Ex)_{out} + \left[\Sigma(Q\left(1-\frac{T_0}{T}\right)_{in} - \Sigma\left(Q\left(1-\frac{T_0}{T}\right)_{out}\right)\right] + \Sigma W_{in} - \Sigma W_{out} \quad (1)$$

where the exergy in any state is given by

$$Ex = m_r[(h - h_o) - T_o(s - s_o)] \quad (2)$$

where Ex_D is the exergy flow destruction and in general equation the first two terms are stream exergy flows, the next two terms are heat transfer exergy flows and the last two terms are work exergy flows. The domestic refrigerator is working on a vapour compression cycle as shown in Figure 1. The major components are a compressor, condenser, capillary tube, and evaporator. In this research work, the following assumptions are made:

- Steady state condition is prevailing in all the components.
- Pressure loses in the pipelines are neglected.
- Unnecessary heat gains and heat loses to and from the system are not considered.
- Kinetic energy, potential energy and energy losses are not considered.
- The mass flow rate of the refrigerant is considered to be unity.

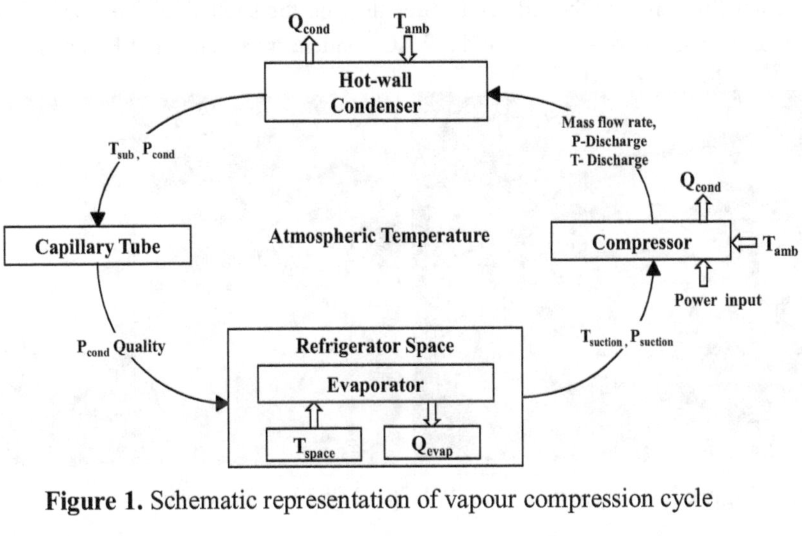

Figure 1. Schematic representation of vapour compression cycle

The exergy balance equation for each component is represented as below:

Compressor

$$(Ex_D)_{comp} = (Ex)_1 + W_{actual} - (Ex)_2 = m_r(T_o(s_2 - s_1)) \quad (3)$$

Electric power,

$$W_{actual} = \frac{m_r(h_2 - h_1)}{\eta_{mech} * \eta_{el}} \quad (4)$$

Condenser

$$(Ex_D)_{cond} = (Ex)_2 - (Ex)_3 = m_r(h_2 - T_0 s_2) - m_r(h_3 - T_0 s_3) \quad (5)$$

Expansion Device

$$(Ex_D)_{exp} = (Ex)_3 - (Ex)_4 = m_r(T_0(s_3 - s_4)) \quad (6)$$

Evaporator

$$(Ex_D)_{evap} = (Ex)_4 + Q_{evap}\left(1 - \frac{T_0}{T_r}\right) - (Ex)_4 \quad (7)$$

$$(Ex_D)_{evap} = m_r(h_4 - T_0 S_4) + Q_{evap}\left(1 - \frac{T_0}{T_r}\right) - m_r(h_1 - T_0 s_1) \quad (8)$$

The total exergy destruction in the system is the sum of exergy destruction in different components of the system and is given by

$$(Ex_D)_{Total} = (Ex_D)_{comp} + (Ex_D)_{Cond} + (Ex_D)_{exp} + (Ex_D)_{evap} \quad (9)$$

Exergy efficiency or rational efficiency computes the efficiency of a process taking the second law of thermodynamics into account.

$$\eta_{exergy} = \left(1 - \frac{(Ex_D)_{Total}}{W_{comp}}\right) \quad (10)$$

BSTWC Condenser

Nowadays the imminent need is to get the better energy efficiency of a vapour compression refrigeration system. The focus of the current research work is on the design and development of water-cooled condenser. The superiority of water-cooled condensers is his ability to absorb more heat and consume less energy. Depends on the availability of space, BSTWC condenser is designed as a box type shell and tube counter-flow heat exchanger with the consideration of refrigerant flow on the tube side and water flow on the shell side. There are distinct differences between tube configuration and geometry of BSTWC and HWAC condensers as shown in Figure 2.

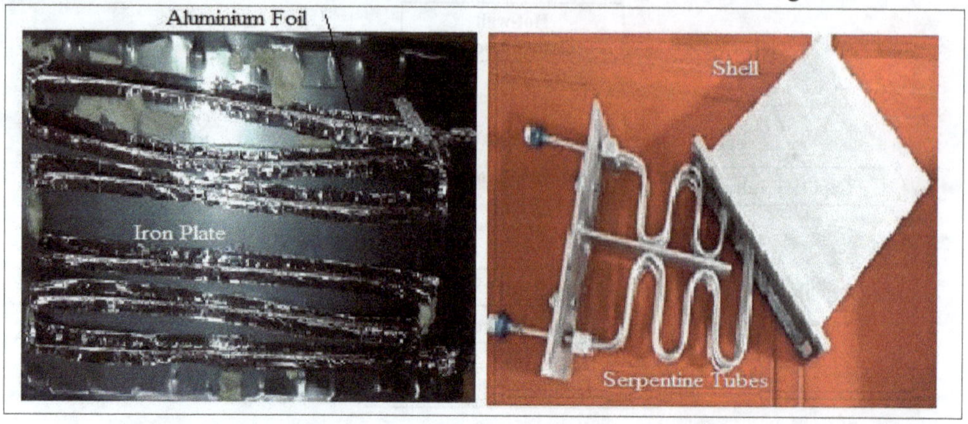

Figure 2. The geometry of HWAC and BSTWC condenser

EXPERIMENTAL SET-UP

The experimental system is a household refrigerator which has an internal volume of the refrigerator is 284 litres. The eco-friendly R600a is used as a refrigerant with smart inverter compressor, HWAC connected in parallel with BSTWC and, the capillary tube used as an expansion device. The technical specifications of a refrigerator under test are specified in Table 3. A 4W DC pump is used to circulate the water through the water-cooled condenser. The circulated water is cooled in a mini cooling tower. The actual experimental set-up is shown in Figure 3.

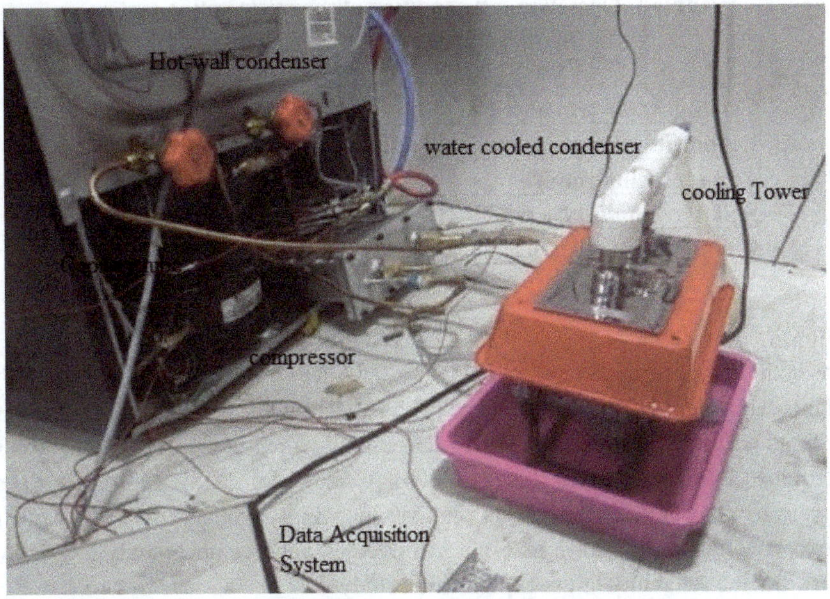

Figure 3. Experimental test set-up

The pressure gauges are used to measure compressor suction and discharge pressure with an accuracy of ±0.25%. The temperatures are measured at 16 potions with the help of K- thermocouples having an accuracy of ± 0.5°C. The thermocouples are mounted inside the freezer, food compartment, crisper compartments, compressor outlet, condenser outlet etc. as per ASHRAE refrigerator standard. The energy consumption of compressor and electrical devices is measured with a help of digital energy meter. The measuring instruments are calibrated before experimentation. The flow control valves are used to connect the refrigerator evaporator in parallel with external evaporator calorimeter. The heater is used to give a constant heating load on evaporating coils. The uncertainty analysis of the measuring instrument is presented in Table 4.

Experimental Procedure

A refrigerator test set-up is incorporated with two condensers, initially, the test set-up is flushed with nitrogen gas to check the leakage and to remove impurities from the system. The system is charged with refrigerant R-600a. According to the methodology used in the literature [3], the experimentation is carried out as per ISO 15502:2005 household refrigerating appliances test guidelines [23]. The experimentation is carried out in a test room. The test room conditions are maintained and controlled with help of an air conditioning system. In the evaporator calorimeter test setup, the calorimeter temperatures vary between −20°C to 6°C (±0.5°C). The calorimeter is an insulated chamber consists of evaporator coils, a heater connected with dimmerstat and ethylene glycol solution. The dimmerstat is needed since temperature is one of the variable parameters. The temperature of the glycol solution is varying by changing current and voltage with help of dimmerstat. The refrigeration load applied by the heater is evaluated by using $Q_{evap} = VI$, where voltage and current measurements for the heaters. This method is adapted to perform the experimentation at different evaporator temperatures. The dimmerstat is used to provide a heating load which acts as an equivalent evaporator capacity. Other parameters like voltage, current, power consumption, relative humidity, water inlet and outlet temperatures are recorded with help of a data acquisition system. The energy consumption is monitored for 8 hrs at no load condition using a digital energy meter and calculated the energy consumption for 24 hours operation. The experimental test conditions during the test are presented in Table 1. The % RH is measured and is observed in the range of 50% to 75%. The pressure and temperature readings are monitored with help of data acquisition system. During experimentation, the ambient temperature is maintained at 32°C ± 0.5°C. The refrigeration system with BSTWC condenser under test is set to keep the inlet cooling water temperature 30°C. The COP is calculated by considering the compressor work, pump work and heater load.

Table 1. Operating test condition of the refrigeration system

Factors	HWAC	BSTWC
Cooling water temperature	NA	30°C
Evaporator temperature	-20°C to 6 °C	-20°C to 6°C
Ambient temperature	32°C ± 0.5°C	32°C ± 0.5°C
Relative humidity	50% to 75%	50% to 75%
Capillary tube length	l =1800 mm	l =1800 mm

Uncertainty Analysis

If you are comparing the uncertainties in the values of two quantities, then by analyzing the absolute uncertainty you cannot tell which of the measurements is more accurate because the units of the measured quantities are different. In order to resolve this issue, the relative uncertainty which is the ratio of the absolute uncertainty and the average is determined by using weakest link rule [24]. Weakest Link Rule is used to determine the absolute uncertainty of the refrigeration system, the largest relative uncertainty from all measured relative uncertainties is considered as systems uncertainty. As shown in Table 2, the largest relative uncertainty is 1.45% for a set of sample observation. This justifies the reliability of the experimental study

Table 2. Instrument specifications

Measured Parameter	Measuring device	Make	Range	Accuracy	Relative Uncertainty %
Temperature	K- Type Thermocouple	EUREKA	-10 to 200 °C	0.5 °C	0.833
Evaporator Pressure	Pressure Gauge with Glycerine	MASS	-30 to 200 psi	1.6 %	1.16
Condenser Pressure	Pressure Gauge with Glycerine	MASS	0 to 200 psi	1.6 %	1.38
Energy Consumption	Energy Meter	Selec: EM368	1, VV I5 A	0.01kWh	0.67
Flow rate	Rotameter	EUREKA	0 to 138 kg/hr	2% of full flow	1.45

RESULTS AND DISCUSSION

The performance analysis of a refrigerator is carried out experimentally at various operating conditions with HWAC and BSTWC condenser using R600a as a refrigerant. As per the energy user guide, in order for a refrigerator to be environmentally friendly, it should contain refrigerant that won't contribute to ozone depletion or global warming. R600a also claims better energy efficiency. One of the downsides of R600a is that it's flammable, unlike R134a; however, the small quantities used in refrigeration make this a very low risk. The test results are used to analyze the variation of performance parameters such as energy consumption, the coefficient of performance (COP), exergy loss in the compressor, total exergy loss & exergy efficiency at various system components at different operating conditions. The variation of COP of a refrigerator for different evaporating temperatures is presented in Figure 4. The investigation results show that the COP increases with an increase in the evaporator temperature for both the condensers. The refrigeration capacity is measured and the COP is calculated using the following Equation (11):

$$COP = \frac{heater\ load + heat\ infiltration}{compressor\ work + pump\ work} \tag{11}$$

The BSTWC has enhanced the COP of the system in the range of 18% to 22 % as compared to a refrigeration system with HWAC due to its lower compressor work and higher evaporator capacity. In the systems with a BSTWC, the heat absorbed by the water at the condenser is rejected to the ambient air by means of a cooling tower. This lowers condenser pressure and power requirement, thus increasing the coefficient of performance and compressor life. The high heat transfer properties of the BSTWC affect on pressure ratio. The BSTWC lowers the condensing pressure; as a result, the pressure ratio reduces. It is observed that the temperature lift is small in a refrigerator with BSTWC condenser as compared with a refrigerator with HWAC.

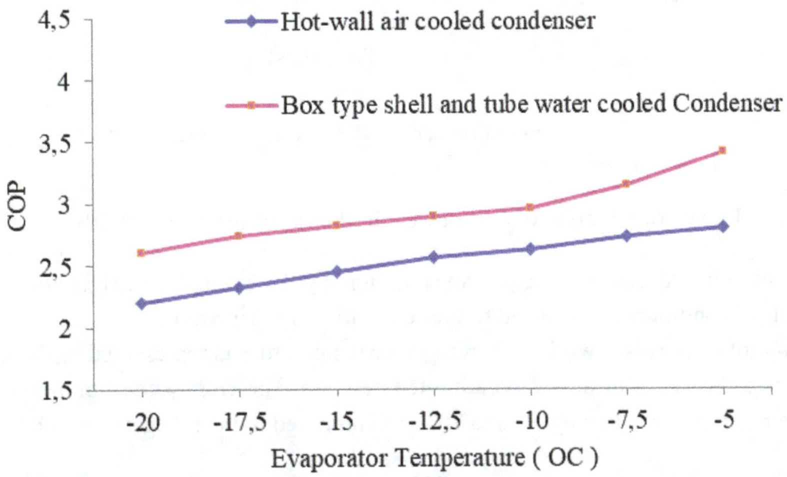

Figure 4. Variation of COP with different evaporator temperature

The energy consumption of the compressor during pull down with BSTWC with respect to time is shown in Figure 5 and Figure 6. It is observed that the pull-down time of the refrigerator is 95 minutes with BSTWC and 125 minutes with HWAC. Thus the desired set temperature is achieved faster with BSTWC than the refrigerator with the HWAC.

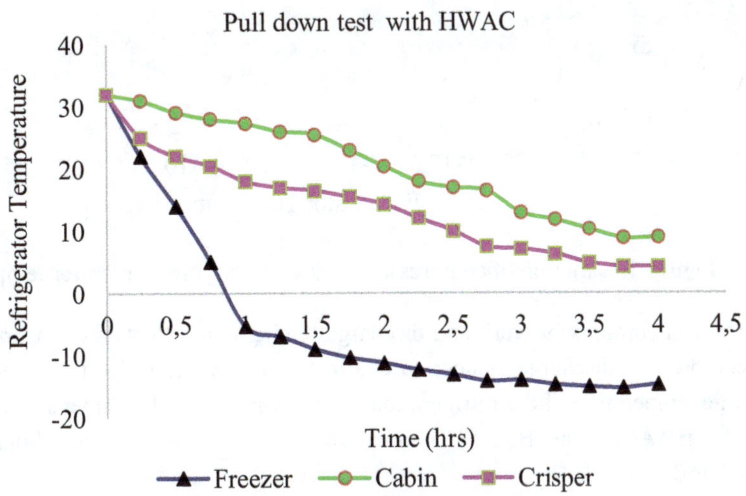

Figure 5. Variation of pull-down speed of a refrigerator with HWAC

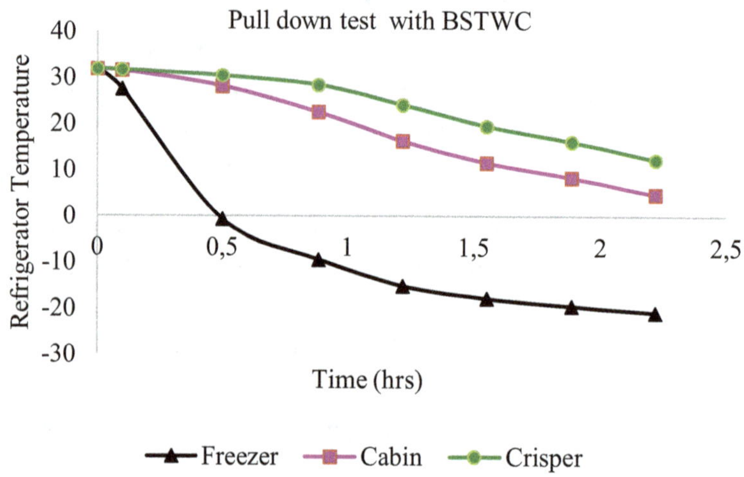

Figure 6. Variation of pull-down speed of a refrigerator with BSTWC

The effect of this reduces the energy consumption by 17% for the refrigerator working with BSTWC condenser at the no-load condition as compared to the conventional refrigerator.

The variation of compressor work with evaporator temperature is represented in Figure 7. The figure shows that compressor work increases with the evaporating temperature. The energy consumption of refrigerator working with BSTWC as a condenser is reduces by 10% to 20%, as compared to the refrigerator with the HWAC.

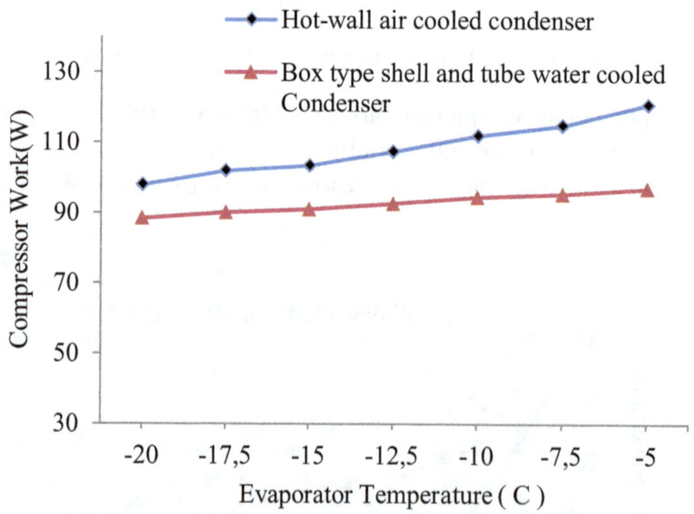

Figure 7. Variation of compressor work with varying evaporator temperature

The variation in compressor work and discharge temperature affects the performance of a refrigerator; the increase in the compressor discharge temperature affects lubricating properties. The result indicates that the compressor discharge temperature of the refrigeration system with BSTWC condenser is lower by about 29% than a refrigerator with the HWAC. The BSTWC condenser with the cooling water should be operated with water temperature below 30 °C.

Exergy analysis provided some useful information about the system such as total irreversibility distribution of the components and this analysis helps determine which component plays a key role to raise the system efficiency. Figure 8 and Figure 9 shows the variation of exergy loss in refrigeration system components with respect to an increase in evaporator temperature. The values of exergy destructions in components of the refrigeration system for both condensers are represented in table 5.

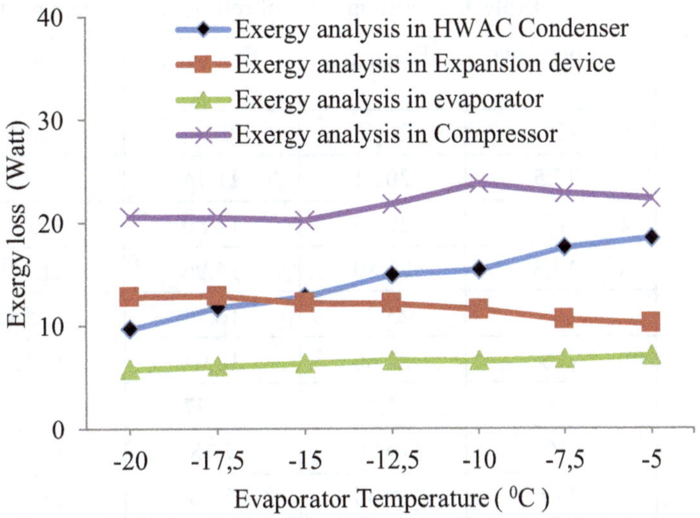

Figure 8. Variation of exergy loss in refrigerator components with HWAC condenser

In capillary tube due to a drop in pressure with an increase in evaporator temperature, tends to reduce the irreversibility with an increase in evaporator temperature. The irreversibility in the refrigeration system reduces by 8.5 % when the system operated with water cooled condenser. The exergy loss of evaporator increases with an increase in evaporator temperature when the system operated with both condensers. The exergy loss in the evaporator is 7% less when the system works with BSTWC. The irreversibility of the condenser increases with evaporator temperature. The increase in heat transfer rate due to water in BSTWC results in a reduction in exergy loss by 58% as compared with exergy loss in HWAC. The results show that the maximum exergy losses in the compressor are 26.52 W followed by the capillary at 14.65 W, condensers at 7.86 W and at the evaporator is 7.21W when the system operated with BSTWC.

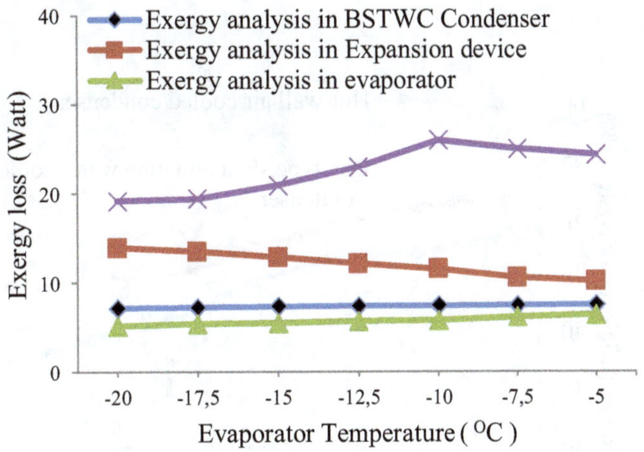

Figure 9. Variation of exergy loss in refrigerator components with BSTWC condenser

The irreversibility in compressor increases with an increase in evaporator temperature. Due to the higher temperature difference between the internal compartment and the ambient was high. The higher the temperature difference, the higher the load imposed on a refrigerator. For this reason, the ambient temperature is a significant determinant of energy consumption and so the exergy loss of the compressor was also observed. Due to an increase in work and low outlet pressure the irreversibility in the compressor when the system adopted with BSTWC is 15- 17 % more as compared to HWAC condenser.

Table 3. Exergy analysis of refrigeration components

Type of Condenser	Temperature	$(Ex_D)_{COMP}$	$(Ex_D)_{COND}$	$(Ex_D)_{EXP}$	$(Ex_D)_{EVAP}$
HWAC	-20	20.58	9.69	14.01	5.78
	-17.5	20.51	11.74	13.53	6.08
	-15	20.24	12.80	12.03	6.34
	-12.5	21.80	14.96	11.13	6.61
	-10	23.77	16.14	11.34	6.59
	-7.5	22.86	18.04	10.53	6.77
	-5	22.34	19.77	9.79	7.07
BSTWC	-20	19.20	7.18	13.97	5.15
	-17.5	19.46	7.26	13.53	5.38
	-15	20.91	7.37	12.88	5.46
	-12.5	23.01	7.41	12.13	5.69
	-10	25.98	7.45	11.55	5.76
	-7.5	25.01	7.49	10.58	6.16
	-5	24.39	7.52	10.19	6.40

The Figure 10 shows the variation of exergy efficiency with an increase in evaporator temperature. The exergy efficiency decreases with the increase in irreversibility in components at different evaporator temperature. A similar trend is reported in the literature [17, 18] when the brazed plate type of heat exchanger is used as a water-cooled condenser for the domestic refrigerator. This means that an increase in evaporator temperature affects the system positively. Comparatively, the refrigerator with BSTWC has 32% to 38% more exergy efficiency than the HWAC.

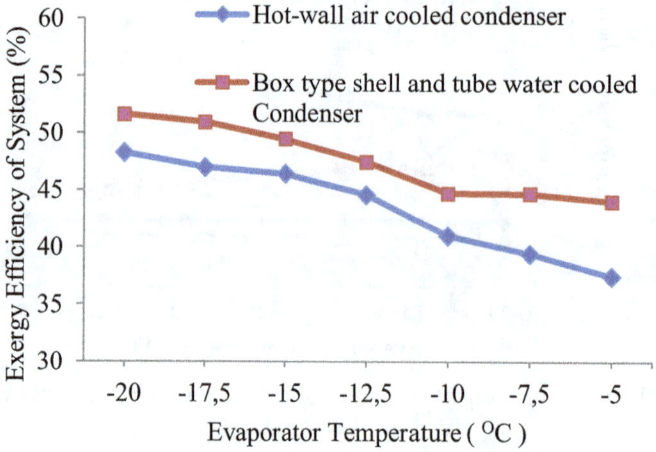

Figure 10. Variation of exergy efficiency in refrigerator components with evaporator temperature

TOTAL EQUIVALENT WARMING IMPACT

Based on the increasing contribution of refrigeration systems to climate change it is necessary to use environmentally friendly refrigerants to mitigate global warming. Three conventional environmental metrics - global warming potential (GWP), total equivalent warming impact (TEWI) and life cycle climate performance (LCCP) are used as tools to select the most environmentally friendly refrigerant. LCCP is a holistic metric to quantify the effect of the refrigerant on the total lifetime system emissions. However, in practice, the LCCP is more complex than TEWI and the contribution of additionally accounted for emissions is negligible. GWP is a useful metric to compare different refrigerants. However, it may overestimate the benefits of low GWP refrigerant to the environment, as it does not take into account many other affecting factors [21, 23]. TEWI is an important environmental index. This index is calculated using the summation of direct impact and indirect impact according to the following Equation (12) [21]:

$$\text{TEWI} = \text{GWP (direct; refrigerant leaks incl. EOL)} + \text{GWP (indirect; operation)}$$

$$\text{TEWI} = (\text{GWP}\ x\ m\ x\ L_{\text{annual}}\ x\ n) + \text{GWP}\ x\ m\ x\ (1 - \alpha_{\text{recovery}})) + (E_{\text{annual}}\ x\ \beta\ x\ n) \quad (12)$$

Global warming potentials (GWPs) express the extent to which a greenhouse gas directly contributes to radiative forcing. Table 4 shows that the TEWI of refrigerator working with BSTWC is lower than that of the refrigerator with HWAC condenser by about 16 %.

Table 4. TEWI of refrigerator

PARAMETER	HWAC	BSTWC
Refrigerant	R600a	R600a
GWP = Global Warming Potential of refrigerant, relative to CO_2	10	10
L annual = Leakage rate per year (kg)	0.066	0.066
n = System operating life (years)	15	15
Refrigerant charge (kg), m	0.058	0.058
α recovery = Recovery/recycling factor	0.8	0.8
Energy consumption per year (kWh), E	462	398
Indirect emission factor, β	0.47	0.47
TEWI Direct effect	0.5742	1.188
TEWI Indirect effect	4244.1	2806
TEWI	3257.67	2807.1

It is also observed that due to the lower refrigerant charge required in HWAC, the direct impact which is related to refrigerant leakage is less in HWAC as compared to BSTWC.

CONCLUSION

The energetic and exergetic experimental performance analysis of a domestic refrigerator with HWAC and BSTWC condenser is studied to check the possibility of energy conservation and exergy efficiency. A significant improvement in system performance and reduction in energy consumption and exergy destruction and irreversibility of a refrigerator is observed with BSTWC condenser. Following conclusions are drawn based on experimental results.

- The energy consumption of conventional domestic refrigerator is significantly reduced with BSTWC condenser. The per day energy consumption in the BSTWC condenser refrigeration system is almost 10% to 20% less than the conventional HWAC refrigeration system.
- The exergy efficiencies of the refrigerator with the BSTWC unit are found in a range of 6.89% to 9.13 %, which is more than a conventional refrigerator.

- The COP of the domestic refrigerator using BSTWC and HWAC were evaluated. COP is increased by 18-20 % with the BSTWC condenser in comparison with a refrigerator with the HWAC.
- The irreversibility of the proposed system is 34 % less than that of the conventional system under similar operating conditions.
- The societal benefit of this technology is the enhancement of energy efficiency, which leads to a reduction in energy cost. This further modification also creates new avenues for research in heat recovery in the residential building sector.

NOMENCLATURE

HWAC	Hot-wall air cooled condenser
BSTWC	Box type shell and tube water cooled condenser
COP	Coefficient of performance
HVAC&R	Heating, Ventilation, Air-conditioning and Refrigeration
ASHRAE	American Society of Heating, Refrigeration and Air conditioning Engineers
TEWI	Total equivalent warming impact
IS	Indian standard
TR	Tons of refrigeration
OD	Outer diameter
ID	Inner diameter
DC	Direct current
kWh	Kilowatt-hour
ISO	International Organization for Standardization
RH	Relative humidity
Lph	Litre per hour
Greek symbols	
h	Enthalpy (J kg^{-1})
I	Irreversibility (W)
l	length of heat exchanger
m	mass flow rate
Q	Heat transfer Rate(W)
s	Entropy (J kg^{-1} K^{-1})
T	Temperature (°C)
W	Work of compressor(W)
Ex_D	Exergy destruction
Ø	Tube diameter
η	Efficiency
Subscripts	
comp	Compressor
cond	Condenser
Exp	Expansion device
evap	Evaporator
in	inlet
out	outlet

REFERENCES

[1] Geppert, J., and Stamminger, R. (2013). Analysis of effecting factors on domestic Refrigerator's energy consumption in use, Energy Conversion and Management, 76, 794-800.

[2] Aprea, C., Greco A. and Maiorino, A. V (2012). An experimental evaluation of the greenhouse effect in the substitution of R134a with CO2, Energy, 45 (1), 753-761.

[3] Raveendran, P. Saji, and Sekhar, S. Joseph, (2016). Experimental studies on domestic refrigeration system with a brazed plate heat exchanger as condenser, Journal of Mechanical Science and Technology, 30 (6), 2865-2871.

[4] Palm, B., (2007). Refrigeration systems with a minimum charge of refrigerant, Applied Thermal Engineering, 27, 1693-1701.
[5] Sonnenrein G, Elsner A, Baumhogger E, Morbach A, Fieback K, Vrabec J(2015). Reducing the power consumption of household refrigerators through the integration of latent heat storage elements in wire-and tube condensers, International Journal of Refrigeration; 51:154–60.
[6] Zhang Z, Huang D, Zhao R, Leng Y. (2017). Effect of airflow field optimization around spiral wire-on-tube condenser on a frost-free refrigerator performance. Applied Thermal Engineering; 114:785–92.
[7] Ameen, A., Seetharamu, K.N., Mollik, S.A., Mahmud, Khizir, Quadir, G. A (2006). Numerical analysis and experimental investigation into the performance of a wire-on-tube condenser of a retrofitted refrigerator, International Journal of Refrigeration, 29,495–504.
[8] Shikalgar N. D., and Sapali, S. N., (2017). Numerical and Thermal Analysis of Condensers applied to domestic Refrigerator, Journal of International Review of Mechanical Engineering, Vol. 11 N7, PP 81- 84.
[9] Bansal, P. K., (2003). Developing new test procedures for domestic refrigerators: harmonization issues and future R&D needs—a review, International Journal of Refrigeration,26,735–748.
[10] Hosoz M., and Kilicarslan, A, (2004). Performance evaluations of refrigeration systems with air-cooled, water-cooled and evaporative condensers, International Journal of Energy Research, 28, 683-696.
[11] Hesselgreaves, J. E., The impact of compact heat exchangers on refrigeration technology and CFC replacement, International Refrigeration and Air Conditioning Conference, 135-139, 1990.
[12] Li, Y., Wu J., and Shoichi, S., (2009). Modelling and energy simulation of the variable refrigerant flow air conditioning system with a water-cooled condenser under cooling conditions, Energy and Buildings, 41, 949-957.
[13] ASHRAE Handbook, (2008). HVAC Systems and Equipment, ASHRAE
[14] Ramezanizadeh, M., et al., (2019). Experimental and Numerical Analysis of a Nanofluidic Thermosyphon Heat Exchanger, Engineering Applications of Computational Fluid Mechanics 13 (1): 40-47.
[15] Aprea C, Greco A. (2002). An exergetic analysis of R22 substitution Applied Thermal Eng.; 22:1455–69.
[16] Saravanakumar, R., and Selladurai, V.(2013). Exergy analysis of a domestic refrigerator using eco-friendly R290/R600a refrigerant mixture as an alternative to R134a, Journal of Thermal Analysis and Calorimetry, 115,933-940.
[17] S. Joseph Sekhar, Raveendran, P. Saji (2017).Exergy analysis of a domestic refrigerator with a brazed plate heat exchanger as condenser, Journal of Thermal Calorimetry, 127:2439–2446.
[18] Raveendran, P. Saji, Sekhar, S. Joseph, (2017). Performance studies on domestic refrigerators retrofitted with building-integrated water-cooled condenser, Energy and Buildings,134, 1–10.
[19] Mahmood Mastani Joybari, Mohammad Sadegh Hatamipour, Amir Rahimi, Fatemeh Ghadiri Modarres (2013). Exergy analysis and optimization of R600a as a replacement of R134a in a domestic refrigerator international journal of Refrigeration, 36,1233-1242.
[20] Hosseinzadeh-Bandbafha, H. (2018). Application of data envelopment analysis approach for optimization of energy use and reduction of greenhouse gas emission in peanut production of Iran, Journal of Cleaner Production 172: 1327-1335.
[21] Pavel Makhnatcha, Rahmatollah Khodabandeha(2014). The role of environmental metrics (GWP, TEWI, LCCP) in the selection of low GWP refrigerant, The 6th International Conference on Applied Energy – ICAE2014, Energy Procedia 61, 2460 – 2463.
[22] Ahamed, J.U., R. Saidur, H.H. Masjuki (2011). A review on exergy analysis of vapour compression refrigeration system, International Journal of Renewable and Sustainable Energy Reviews, 15, 1593–1600.
[23] Davies, Thomas W., and Caretta, Ottone,(2004). A low carbon, low TEWI refrigeration system design, Applied Thermal Engineering, 24, 1119–1128.
[24] Coleman, H.W. and Steele, W.G., (1999). Experimentation and Uncertainty Analysis for Engineers, 2nd Edition, John Wiley & Sons, Inc., New York, NY.
[25] ISO 15502: 2005, Household refrigerating appliances.

15

NUMERICAL ANALYSIS OF FLAME CHARACTERISTICS AND STABILITY FOR CONICAL NOZZLE BURNER

Julanda Al-Mawali [1], Sam M Dakka [2]*

ABSTRACT

The stability and the mean structure of methane partially premixed conical burner flames was investigated numerically using ANSYS Fluent. The study presents and discusses the stability curves of the partially premixed flame and maps the mean flame structure based on contours of mass fraction of O_2, CO and temperature. From the data obtained, it can be concluded that both premixed and non-premixed flames are less stable than the partially premixed flames. An optimum level of partially premixing was found and the flames beyond this threshold were found to be less stable. This optimum level was found, when the ratio of the mixing length to the nozzle diameter is equal to 5. At this specific degree of partially premixing, the flame exhibited triple interaction reaction zones. It was found that with an increase of the angle of the cone of the burner, the air entrainment increases which, in turns breaks the stabilization core and hence cause a reduction in the flame stability limit. The main role of the cone is to provide a protection from the surrounding environment at early phase of the reaction near the jet exit where turbulence with high intensity was observed.

Keywords: *Partially Premixed Flames, Flame Stability, Conical Burner, Aircraft Combustion Systems, Triple Reaction Zones, Mixing Length, Low Emissions*

INTRODUCTION

Aircraft engines performance cycles are characterized by wide different ranges of operating conditions which implies that flame stability is crucial in maintaining continuous operation of the engine. And therefore, the design of combustion systems has shifted towards technologies that incorporated partially premixed flames combustion systems. Partially premixed flames offer the advantage of lean premixed flames in terms of emissions reduction [1] due to relatively lower flame temperature in addition to flame stability. Emission reduction is another important factor of such technologies to meet the stringent pollutant emission regulations [2]. Therefore, special consideration of burner design in terms of burner mixing length which will impact the degree of the partial flame premixing through the flame equivalence ratio and cone angle will have strong impact on flame stability and emissions which is the subject of this investigation of optimization of the burner by selecting the degree of partially premixed flames to maintain stability and lower NOx emissions.

The study is focused on the impact of level of flame premixing and the burner cone angle on the flame behavior and flame stability and emissions. The burner design used in this study is similar to a burner developed and reported elsewhere [3]. Figure 1 shows the schematic drawing of the burner used in this investigation. The burner has a simple geometry as it contains two concentric tubes with a cone at the exit of the outer tube. The air flows through the inner tube while the fuel flows through the outer tube. The fuel and the air start to mix at the inner tube exit and they keep mixing through the mixing length "L", which can be defined as the distance between the inner tube exit to the outer tube exit. During the numerical simulation, the premixing distance was varied from 0 mm to 70 mm. When the premixing length is set at L=0, the generated flames are called non-premixed or diffusion flames. As the mixing length increases, the level of premixing between the fuel and the air increases proportionally. When the mixing length is set to be short (L/D<1.5), the generated flame is said to be partially premixed. With longer 'L' the generated flame can reach fully premixed conditions. A number of conical nozzles with cone angles of 20°, 25°, 30°, 35° and

This paper was recommended for publication in revised form by Regional Editor Jovana Radulovic
[1] Department of Engineering and Math, Sheffield Hallam University, Howard Street, Sheffield S1 1WB UK
[2] Department of Mechanical, Materials and Manufacturing Engineering, The University of Nottingham, University Park, Nottingham NG7 2RD, UK
**E-mail address: sdakka1@protonmail.com*

40°, have been used during the present investigation. The height of the conical nozzle was kept at 53 mm for all cone angles.

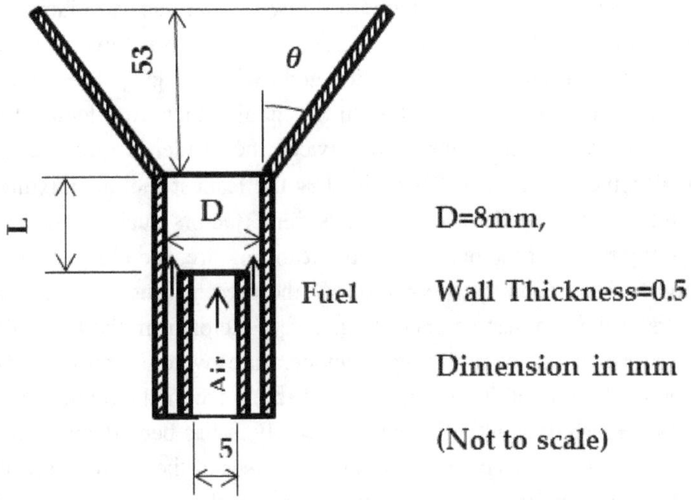

Figure 1. A schematic diagram of the burner [17]

The values of the jet velocity, equivalence ratio and the mixing length were controlled for each conical nozzle to study the flame stabilization of the burner. Based on the stability limits findings, several stable flames will be selected for different cone angle to conduct further detailed investigation.

For that, the radiative properties of the suspended particles and the media that contains those particles should be investigated deeply. As there are many particles in a particulate medium, the scattered photon from a particle may interact with those from other particles. This means, an incident beam can be scattered more than one time and they can be scattered again by the nearby particles [7, 8].

Stability is one essential feature of a good burner design. In general the burner stability is defined by maintaining a sustained flame while varying the fuel air mixture ratio, the flow speeds and the inlet temperature [4]. The current study is focused on the effect of fuel air mixing in sustaining the stability of the flames, but in general the stability of the burner will depend on a range of equivalence ratios that can sustain stable flames [5] which is bounded by two types of extinction, weak and rich extinction. The former occurs as a result of having too much fuel in the mixture whereas the latter occurs due to having too little fuel in the mixture. These two limits mainly rely on the mass flow rate of the fuel passing through the burner chamber with constant air flow rate. Numerical as well as experimental studies have demonstrated that the flame encounters sustaining problems at high velocities **[6].** Thus, having high flow velocity flow could also lead the flame to blow off. Similarly, having low velocity flow could lead to flashback phenomena. There are several reasons which could cause the burner or combustors to blow off such as the use of lean fuel-air mixture or could be as a result of very strong combustion instabilities. For the burners to operate at low level of NOx, the burners tend to utilise a very lean mixture, which means that they have to operate close to their blow-off limits [7]. The phenomena of the blow off could lead to serious problems in combustors if re-ignition system suffers a failure to re-ignite the mixture, especially under extreme conditions such as low pressure and temperature at high altitudes.

A wide range of publications is available on measuring, correlating and calculating the blow-off limits [8]. There are also numerous and different hypotheses and physical models which were established to describe the phenomena of blow off. The characteristics of the blow off for a bluff-body were determined in terms of equivalence ratio as function of the jet velocity [8]. Determining the equivalence ratio of the flame blow off starts with establishment of the flame at stoichiometric conditions and then slowly decreasing the mass flow rate of the fuel until the flame blows-off. Such phenomena happens mainly when the available time for the fuel-air reaction is less than the required time to make enough heat to cause the fresh mixture to rise to its ignition temperature [9].

Suggestion was made [10] that the blow off phenomena occurs when the flame physically abandon the burner and blow out of the burner. This problem is also known as the static stability. Such problem occurs when the flame is not able to stay attached or anchored in the burner. Furthermore, increasing the jet velocity while keeping the equivalence ratio constant, caused the flame to blow off. Similarly, reducing the equivalence ratio to a certain value while keeping the jet velocity, caused the flame to blow off. Several References were found to relate the Damkohler Number, Da, with the chemical reaction and extinction [11]. These references had calculated the time of blow off using the Damkohler Number, which in turns, enabled them to calculate the blow off limits. The Damkohlar number, Da, can be simply defined as the ratio between the residence time (t_{res}) and the chemical kinetic time (t_{chem}) [11]. Generally, the lean blow off is defined as the leanest fuel air mixture limit which allows the flame to propagate. The limit of flammability depends on several factors such as the chemical properties of the fuel-air mixture and the configuration of the burner. For the lean mixture, the blow off occurs when the velocity of unburnt mixture flow is higher than the flame speed [10]. At the present time, the phenomena of blow off are avoided by operating the burners or the combustors under large safety factor from the blow off limit. This means, operating the combustors with higher equivalence ratio (higher temperature) which in turns, leads to high level of NOx pollutions. To minimise the pollutants level of the combustors, the safety factor should be reduced to a minimum value by which the combustors can operate safely and efficiently. Recently, it has been demonstrated that the designers can monitor the blow off stability safety margins by using suitable analyses of the flame's optical signature and acoustics [12].

A study [13] was conducted on the "stabilization mechanism of highly stabilized partially premixed flames in concentric flow conical burner". The aim of the study was to investigate the stabilisation mechanism within the conical nozzle based on 2D measurements of the temperature field and flow field. Five different partially premixed flames were selected at equivalence ratio range between 2.5 and 5 and at Re number range between $8.3*10^3$ and $14.5*10^3$. A 3D PIV system was used in this study to obtain the turbulent flow field inner and outer of the conical nozzle. The obtained data shows that there is a recirculation zone at the near region inside the conical nozzle which suggests the existence of air entrainment along the wall of conical nozzle. The air stream resulting from the air entrainment is expected to be heated by the flame and therefore enhance the flame stability. Moreover, this study found the flow pattern inside the cone has a direct effect on the stabilization mechanism of the conical burner. It was also found that having conical nozzle attached to the burner improves the flame stability compared to the burner without the conical nozzle and this mainly due to the structure of the flow field inside the conical nozzle. Experimental PIV data showed the velocity vectors of flow inside the cone indicating the existence of a recirculation zone near the wall of the conical nozzle which is believed to improve the overall stability performance of the burner. As for the effect of the Equivalence ratio, this study found that there is a limited effect on the flow field structure formed inside the conical nozzle.

Another study [14] had conducted an experimental study to investigate the effect of the nozzle geometry and the mixing length on the stability and structure of the flame. The study had presented and discussed the stability curves of partially premixed flame and the maps of the mean flame structure which is based on measurements of the concentration of the O_2 and CO and the temperature. From the data obtained from the experiment, it was observed that both premixed and non-premixed flames are less stable than the partially premixed flames and an optimum level of partially premixed was found to be at L/D= 5 for the Conical Nozzle burner, and the flames beyond this level were found to be less stable. At this specific degree of partially premixing, the structure of the flame is likely to generate diffusion, lean and rich interacting reaction zones which lead to form triple flame structure. The obtained data also shows that the geometry of the conical burner has a great effect on the flame stability limits. It was found that with an increase of the angle of the cone, the air entrainment increases which, in turns breaks the stabilization core and hence cause a reduction in the flame stability limit. Regardless of the cone angle, the cone main role is to provide protection from the surrounding environment at the early phase of the reaction near the jet exit where high intensity turbulence is expected. Therefore, the study concluded that the burner with cone is considered to be more stable than that without cone.

Another study [3] showed the CO and 0_2 concentrations of different conical burner flames at different cone angles. The results illustrate and highlight the effect of the air entrainment on the flame structure as discussed earlier.

A study [15] was conducted to investigate the effect of the premixing level on the local extinction and stabilization characteristics of the methane-air flames. Simultaneous OH-PLIF/PIV techniques and Large Eddy Simulation (LES) were used in this study. The partially premixing was generated in a burner with two concentric tubes, where the degree of the partially premixed flames was generated by controlling the length of the mixing chamber. The burner was mounted with a cone at the mixing chamber exit in order to stabilize the flames. The stability limit was obtained for different Reynolds number and level of partially premixing. This study found that the flames became more stable at low Re numbers when the mixing length decreases. For high Reynold numbers, it was found that there is an optimal mixing length at which the burner generates the most stable flames. To enhance the understanding of the stability behaviour, the OH PLIF images which were taken at equivalence ratio of 1.5 and at L/D=5 were analysed [15]. The findings showed that at low Reynold Number (4000), the OH radicals were observed to be in smooth and narrow zones (Laminar Flame type). As the Re number increased beyond 4000 and below 8000 the OH radicals were found to be distributed in wider zone. Furthermore, some discontinuity within this zone was observed and believed to indicate to local flame extinction. Increasing the Re number to 12000 and beyond causes the flame to blow out.

Another study [16] was carried to investigate the effect of fuel/air premixing degree on the in interaction between the flow field and Flame Kernel. The Partial premixing level and the jet velocity were the main parameters in this study. Five different velocities ranging between 10 m/s and 20 m/s and three degrees of partial premixing were chosen to be investigated in this study at constant equivalence ratio of 2. This study used Planar Imaging Velocimetry (PIV) to take measurements of turbulent intensity and the mean flow field. Furthermore, the study had also conducted flame stability for the burner at different level of mixing, L/D=2, 4 and 6 where L is the mixing length and D is the diameter of the outer tube. The obtained result for the stability study was plotted as jet velocity (m/s) vs the jet equivalence ratio. The stability limit was obtained by reducing the fuel mass flow rate till blow off, while the air is kept constant. It was found that level of partially premixed at L/D =2 provide the most stable flame.

All previous reviewed studies have demonstrated that the mixing length (L/D) has a significant effect on the stability limit of the flame. The experimental and numerical results of these studies show that at small values of mixing length (L/D), towards diffusion conditions (L/D=0), the flames were observed to be less stable than those at higher mixing length where more air-fuel mixing is predicted. Moreover, these studies have shown that there is a specific point in the stability curve beyond which the flame is said to be less stable. Most of the reviewed studies have found that flames are most stable at mixing length (L/D=5).However, the study made by [16] found the mixing length of (L/D=2) generates the most stable flames which does not agree with other reviewed studies made by [17] and [3]. The reason behind this disagreement is believed to be as a result of a turbulence generator disk utilised in one of the studies [16]. This generator was placed to increase the level of partial premixing for the same L/D since smaller value of the mixing length is required to generate the optimum conditions for the flame stability. Thus, from the previous works, it can be concluded that the partially premixed flames are considered to be more stable compared to the non-premixed and fully premixed. This can be explained as a result of mixing air and fuel prior to combustion chamber, leads to combustion of non-homogenous mixture which is likely to generate a structure of multi reaction zones, namely, diffusion, lean and rich [18]. Such structure occurs as a result of interaction of the rich pockets with the lean pockets in the flame. The reaction zones resulting from this interaction is likely to create triple flame structure which is believed to be able to survive at high turbulence intensities. Several studies have observed the triple flame structure in the partially premixed flames [1,19,20].

The reviewed studies have also shown that the cone angle has a significant effect on flame stabiliy. Small angle generated more stable flames compared to that of large cones angles, and cone angle between 20-25 degrees were found to generate the most stable flames. This is believed to be due to air entrainment as it causes the stabilization core to squeeze and with a higher cone angle the stabilization core is expected to break, leading to flame extinction.

CAD MODELLING APPROACH

The geometry of the Conical Nozzle burner can be modelled as 3d, 2d or axisymmetric. Using the 3d model of the whole burner will require a large number of mesh nodes and elements which will lead to high processing time and this also could have negative impact on the accuracy level of the numerical results due to the limitation of the computer processor capability. The processing time of any CFD problem is proportional to number of degrees of freedoms as shown in the following expression, Processing Time α n^2 [21]. The expression illustrates the relation between the number of degrees freedom (n) and the processing time. This means that when the whole geometry is used for simulation run, it will take a processing time of n^2, while the half of the geometry will take about 4 times lower (n^2)/4 than that of the whole geometry. Since the present burner is believed to be symmetric, it was decided that the axisymmetric (2d) will be used as a modelling approach for the rest of the study. The main reason behind this selection is that the axisymmetric model will take lower processing time and will produce more accurate results. The other motivation for this selection is the mesh elements limitation by ANSYS to (500,000) and the limitations in the processor capability of the university computers. Figure 2 and 3 shows the axisymmetric model and the 3D Model of the burner and the dimensions of the axisymmetric model of the conical burner respectively.

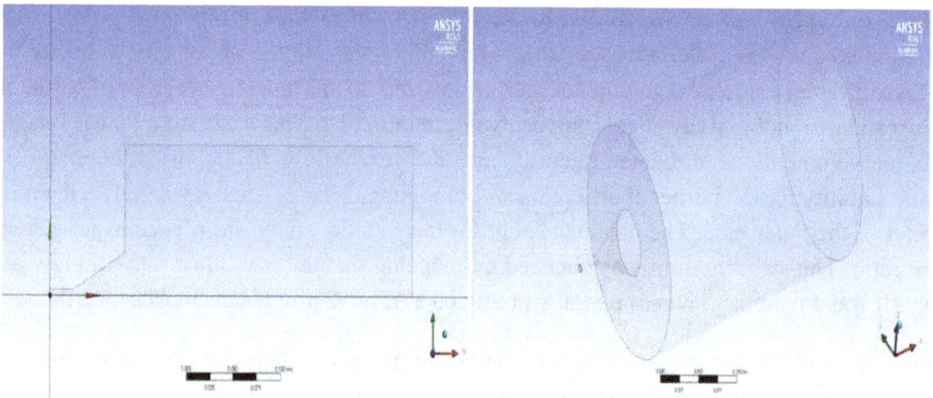

Figure 2. Axisymmetric model and 3D model of the burner respectively

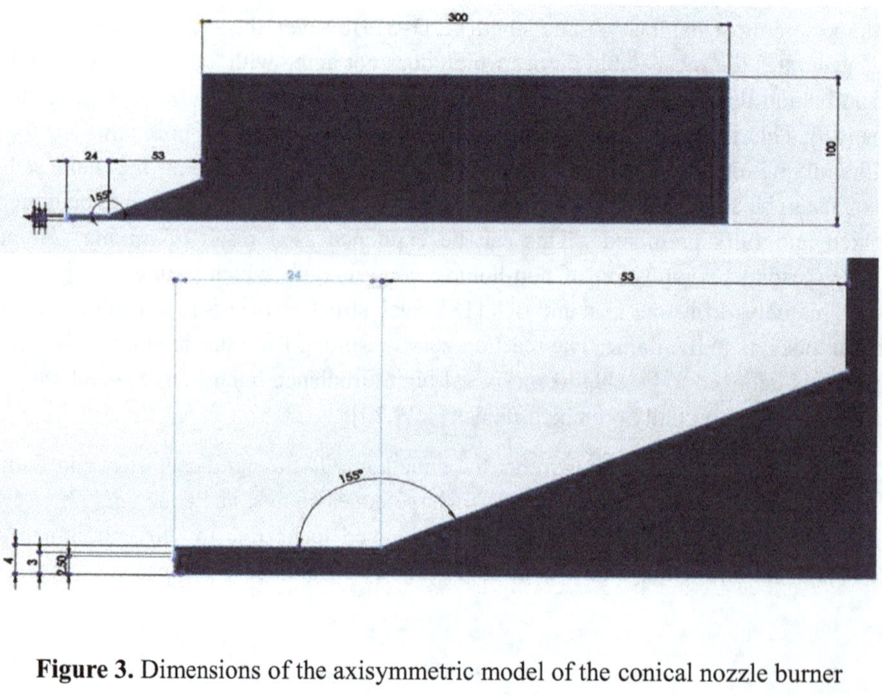

Figure 3. Dimensions of the axisymmetric model of the conical nozzle burner

Mesh Sensitivity Study

The number of mesh elements or the mesh grid size is considered to be very essential in numerical simulations as the accuracy level of flow description depends mainly on the grid size. Therefore, developing or creating the ideal model which is able to fulfil the engineering solutions without using a very fine mesh size is one of the top challenges designers and engineers are encountering in the present time. As the mesh grid gets smaller (finer), it leads to a high computational cost as it uses a model that requires a very fine mesh such as the Direct Numerical Simulation (DNS). In comparison, as the mesh size gets larger (Coarser), it leads to low computational cost as it utilises models such as Reynolds Average Navier-Stokes (RANS) which result in very poor description of the simulated flow especially when simulating turbulent flow [22]. Thus, utilising the right mesh size is very significant.

In the present study, different mesh sizes were used in order to investigate the effect of the grid size on the level of accuracy of the numerical results and then selected the right grid size to be used throughout the study. The simulation was run for turbulent partially premixed flame using standard k-epsilon model for methane-air combustion at stoichiometric condition (Φ=1) at 300 k, ambient pressure. Reviewing the literature, it was found that the adiabatic flame temperature for the methane air combustion at stoichiometric condition is between 2240 to 2245 K [23,24].

This study conducted mesh sensitivity analysis by running the same boundary conditions at different numbers of elements and monitoring the Adiabatic Flame temperature of partially premixed flame at each run. It can be seen from the Figure 4 that the flame temperature changed by 27 % from 2791 to 8426 elements number, while it only changed by 2.5 % from 20582 to 72356 elements number. Thus, it can be concluded that the mesh convergence occurs between 20582 to 72356 elements number. Based on the mesh sensitivity study, the mesh with 20582 elements number was selected for rest of the study. The main reason behind this selection is that the mesh with 20582 would produce results with high level of accuracy and with minimum computing time. It was also found that the predicted flame temperature results at the selected mesh is close to what it was found in the literature, as shown in Figure 4.

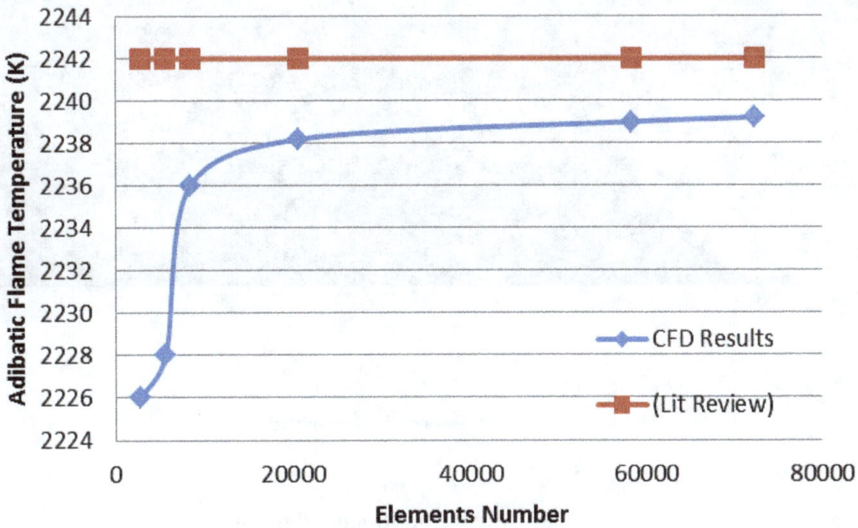

Figure 4. Mesh sensitivity analysis

Mesh Skewness

The skewness of the mesh is a good indicator of the mesh quality level. Table 1 shows the mesh quality measurements based on the mesh skewness recommended by ANSYS. For our selected mesh, the max skewness was found to be 0.438, as shown in Figure 5.This means that the selected mesh is of very good quality based on the ANSYS measurement table.

Table 1. Skewness mesh metrics spectrum

Excellent	Very Good	Good	Acceptable	Bad	Unacceptable
0 - 0.25	0.25 – 0.5	0.5 - 0.8	0.80 – 0.94	0.95 -0.97	0.98 – 1.00

Mesh Metric	Skewness
Min	3.2535e-004
Max	0.43877
Average	1.5638e-002
Standard Deviation	1.8312e-002

Figure 5. Mesh skewness of the selected model

Figure 6 shows the final mesh that was used through the study. The edge sizing with bias and face sizing were used to generate a structured mesh. More refine mesh was created near the cone angle as string turbulence was expected at early stage of the flame.

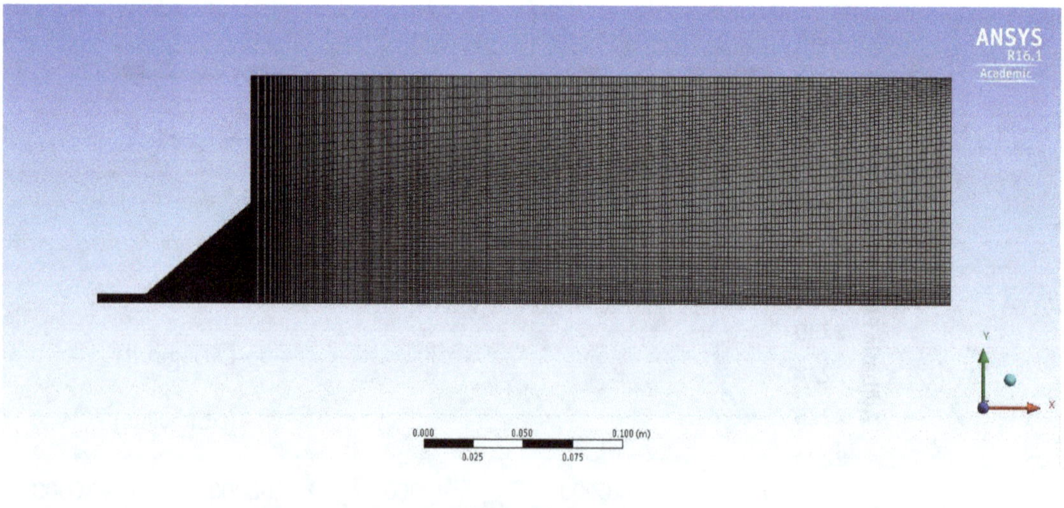

Figure 6. Structured mesh

When it comes to the mesh, it is also very essential to select the right size of the mesh near the wall in order to ensure a flow simulation with level of accuracy [25]. The Y^+ value can be calculated to determine the right distance between the wall and the first node near the wall. The Y^+ for the present burner was found to be 0.00029. This value was then entered in the first layer thickness of the inflation. Table 2 illustrates final mesh details utilised in the simulation.

Table 2. Final mesh details

	Parameter	Value
1	Nodes number	29039
2	Elements number	28303
3	Max Skewness	0.442
4	Orthogonal Quality	0.76
5	Mesh Method (1)	Edge sizing with bias
6	Mesh Method (2)	Face Meshing
7	Y^+	0.00029

CFD Solver Set-up

CFD Solver:

The methods used by the Computational Fluid Dynamic are usually categorised to two schemes:

- Pressure-based
- Density-based.

The pressure based method is considered to be the basis for most of the present commercial CFD codes which is also a very suitable to be used for the low Mach number and incompressible flows, while the density-based method is well-suited for the transonic and supersonic flow fields [22]. The pressure-based method uses the pressure correction and the momentum as its primary variables and it has two algorithms, namely, segregated solver, which is used to solve momentum and pressure in sequential manner, and Coupled solver which is used to solve for momentum and pressure in simultaneous manner. The density based method solves energy, momentum and continuity equations in form of vector and it also uses the equation of state to solve the pressure. This solver is able to utilise either explicit or implicit solution approach [26].

The pressure based solver can be used for wide range of flows from high Mach number compressible flows to low Mach number incompressible flow. This method requires less storage and also allows high flexibility during the solving procedure. On the other hand, the density-based method is applicable for the flow regimes where a strong coupling exists between momentum, density and energy [22]. Such solver is well-suited to be used for high Mach compressible flow with shock interactions and hypersonic flows. Since, the present study is directed to investigate low speed incompressible flows and have storage limitations; the pressure-based solver was selected for the current study. This study has also used the steady time over the transient as the former is considered to run faster as it ignores number of cross terms and higher order terms that deals with time. The transient simulation is also known to have higher requirements in terms of disk storage and CPU time [27] .Two dimensional space is selected to be axisymmetric as the model approach of the burner was selected earlier to be axisymmetric as explained in the modelling approach section. Table 3 illustrates the selected parameters for the simulation set-up.

Table 3. Simulation set-up parameter

No	Parameter	Selected Value
1	CFD Solver	Pressure-Based
2	Viscous Model	Realizable, k-ε Model
3	Species Model	Partially Premixed Combustion
4	Boundary Conditions	UDF
5	Solutions Methods	Second Order Upwind
6	Solutions Controls	Retained as default
7	Residual	$1\text{x}10^{-5}$

RESULTS AND DISCUSSION

This section presents and discusses the results which were obtained from the numerical approach. At the beginning of the this section, Cold simulations of the air and fuel were run at different conditions in order to investigate the effect of mixing length and the fuel velocity on the premixing level between the fuel and air. This will help to develop a better understating of the generated flames as the flame characteristics depends mainly on condition of the jet flow at the exit of the mixing chamber. For sake of verification, the calculated equivalence ratios will be then compared to the Computational equivalence ratios at different mass flow rates of fuel.

Stability study will then be made to investigate the effect of the mixing length and conical nozzle geometry on the stability of the flame. This was made to determine the optimal configuration of the burner beyond which the flame is said to be less stable. Once the optimum configuration is determined, two sets of flames were then selected at the optimum configuration to conduct further analysis of the flame mean structure.

Cold Simulation (Non-Reactive)

Effect of the Mixing Length on the Degree of Partially Premixing:

The mixing length (L/D) has a direct influence on the gradient of concentration in the mixture at the jet exit and because of this; the mixing length has a control over the local level of partially premixing as well as scalar dissipation in the combustion chamber [28]. The process of the air/fuel mixing inside the concentric tubes and the level of partially premixing at various mixing length (L/D) could be more understood from numerical results presented by [28]. At the region near the fuel inlet, it can be observed the development of an axisymmetric vortex ring. Such phenomenon occurs due to the difference in velocity across the interface between two flows, which is also known as the Kelvin-Helmholtz instability [29].

The generated Kelvin-Helmholtz vortices results in a big scale roll-up of air pockets and fuel which leads to more efficient mixing of the air and fuel. The numerical results presented in Figure 7 shows a good agreement with LES results [28].

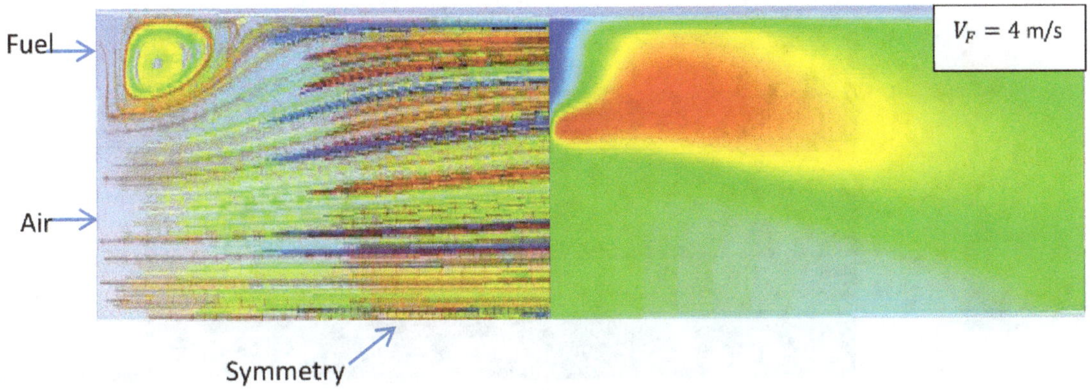

Figure 7. Path lines and turbulence intensity

For further analysis of the mixing length effect on the concentration gradient in the fuel air mixture, a 3D Model of the mixing chamber was used to run a cold simulation. Once the simulation was converged, several planes of O2 and CH4 were taken for different mixing lengths. It was found that as the mixing length (L/D) increases, the degree of the partially premixing increases proportionally. At smaller values of L/D, there is almost no premixing which in turns lead to generating diffusion flame. As the mixing length increases, more fuel air mixture is predicted till it gets to fully premixed. This can be seen in Figure 8.

Effect of the Fuel Velocity on the Degree of Partially Premixing:

Beside the mixing length effect, the fuel velocity has also a significant effect on the gradient concentration in the mixture. Thus, three different velocities where simulated while keeping the other parameters constant. From the numerical result, shown in Figure 9, it can be seen that increasing the fuel velocity leads to an increase in the level of the partially premixing

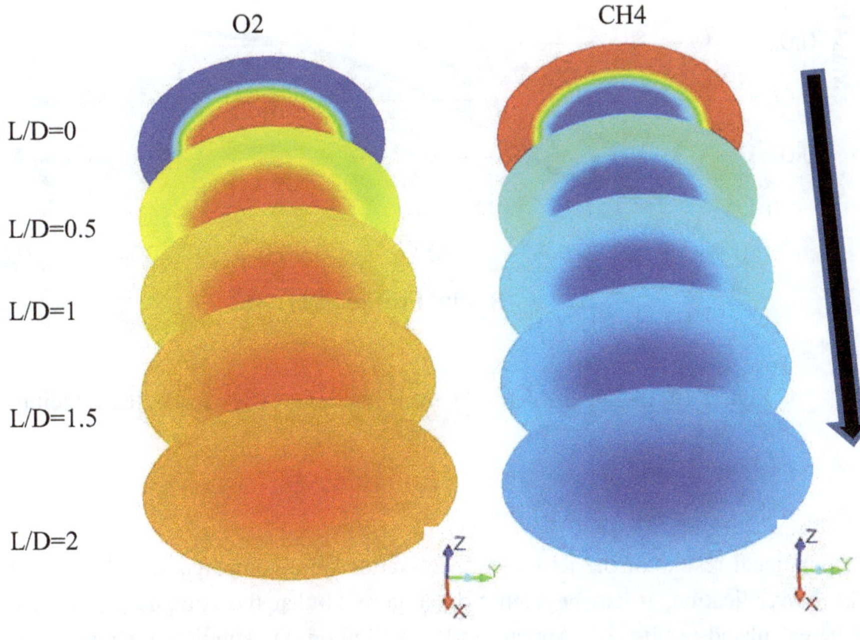

Figure 8. O2 and Ch4 concentrations for different mixing lengths

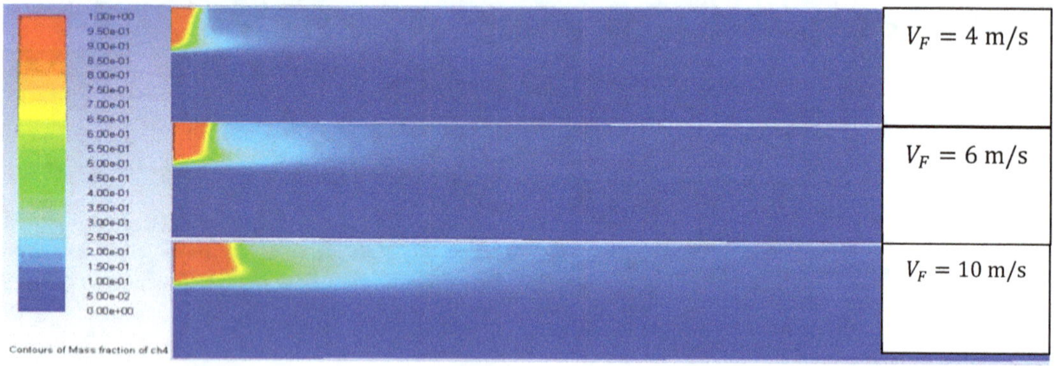

Figure 9. Contours of the mass Fraction of CH4 for varied velocities.

To further analyse of the effect of fuel velocity, radial profile of the mass fraction of CH4 was plotted at L/r=6 for three different fuel velocities. It can be clearly seen that the level of the mixing is higher at higher fuel velocity, as shown in Figure 10.

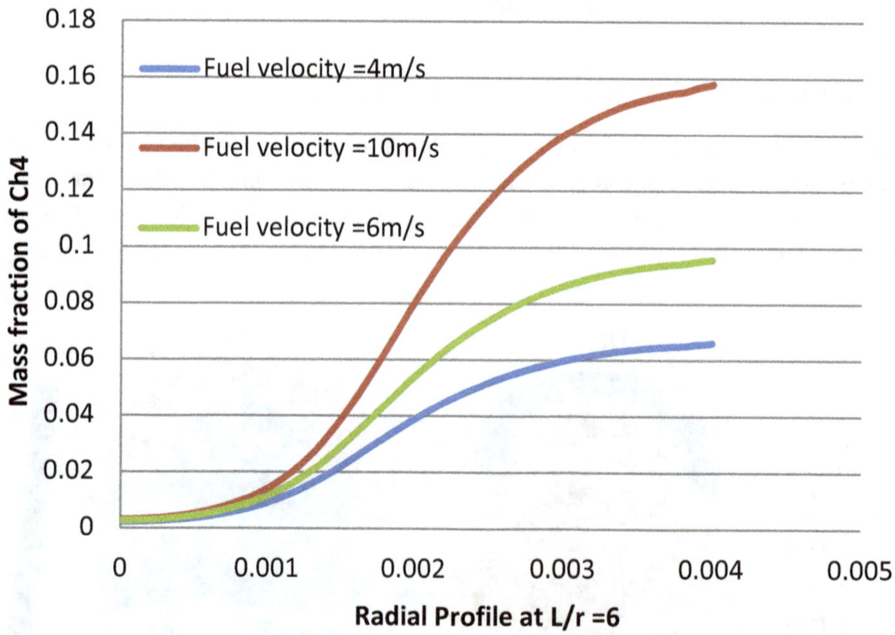

Figure 10. Radial profile of the mass fraction of CH4 for varies velocities

Verification of the Computational Results

The computational results of the jet equivalence ratios were compared with the calculated values in cold flow for the sake of verification. It can be seen from Figure 11 that the computational results are with a good agreement with the calculated results (See Appendix B). In addition, the simulation shows a good prediction of the equivalence ratio at low fuel flow rate, however as the fuel flow rate increases the gap between the computational and the calculated values increases slightly. This could be due to a slight lack in the quality of the mesh or numerical error during the calculations.

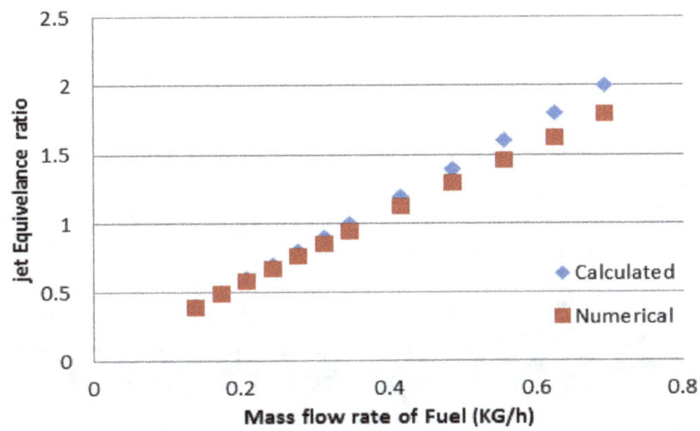

Figure 11. Comparison between computational and calculated results for different fuel flow rates

Reactive Simulations

Stability Limit of the Conical Nozzle Burner:

The stability characteristics of the present burner were studied based on the two following parameters:

- The level of partial premixing.
- Cone angle of the nozzle.

The level of premixing was conducted by modifying the mixing length, L. In this study the mixing length was altered from 0 cm to 9 cm. The stability limit of the burner is defined in this study as the point above which the flame is considered to be sustained and stable. The stability point was obtained by reducing the rate of the fuel flow while keeping the air flow rate constant. This point was recorded at the lowest flow rate (lowest equivalence ratio Φ) which corresponds to unsustainable or extinguished flame. Both attached flames and lifted flame were observed and presented in this study. This run was carried out at 6 kg/hr of air flow rate and the data shown in Figure 13 was obtained using pure methane (100% CH_4). Figure 12 represents the data as jet equivalence ratio vs the L/D ratio, where L is the mixing the length (variable) and D is the nozzle diameter (Fixed value and equal to 8 mm), for different cone angles.

Generally, it was observed from Figure 12 that there is a specific point at all curves which offers better stability limit with lowest Equivalence ratio. This point was noticed to occur at the L/D of 5 for all cone angles and beyond which the flames were observed to be less stable. It was also found that the stability of the flames at lower value of the (L/D<5) towards diffusion conditions are lower than that for the (L/D > 5) where more mixing of the fuel and air is predicted due the higher mixing length.

Mixing the air and the fuel prior to the combustion chamber, leads to combustion of non-homogenous mixture which is likely to generate a structure of multi reaction zones, namely, diffusion, lean and rich [18]. Such structure occurs as a result of interaction of the rich pockets with the lean pockets in the flame. The reaction zones resulting from this interaction is likely to create triple flame structure which is able to survive at high turbulence intensities. Several studies had observed the triple flame structure in the partially premixed flames [1,19,20].

Thus in the present study, there is a high possibility of forming triple flame structure at L/D=5 and this, could be the main reason behind the high stability level at this specific level of premixing. As it can be seen from Figure 12 that the stability curve at the non-premixed conditions is steeper which proves that partially premixed flames can survive better than the non-premixed and premixed flames. Furthermore, the non-premixed has the worst stability characteristics, as shown in Figure 12.

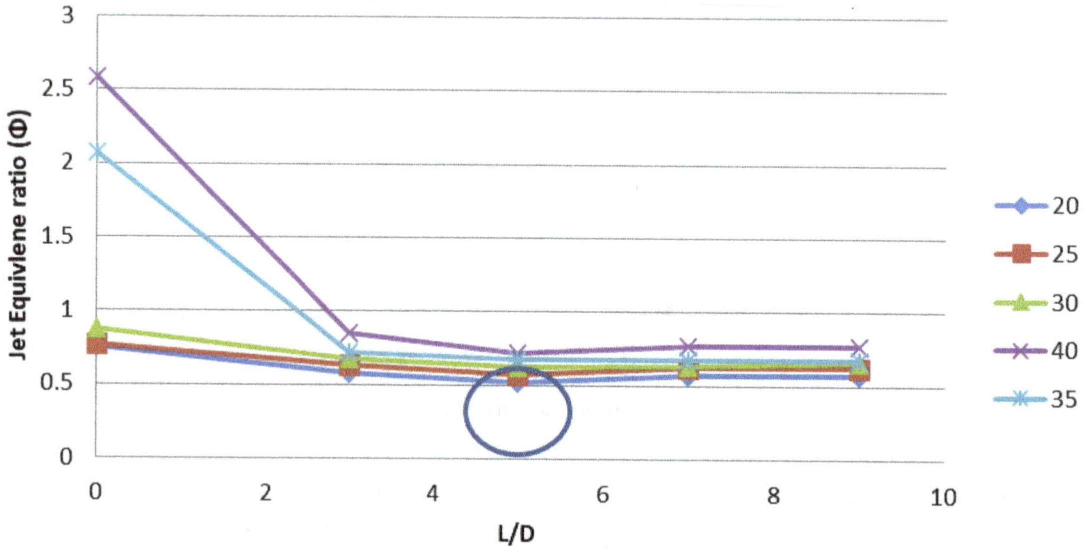

Figure 12. "Stability Curve " Equivalence ration vs the L/D

Mean Flame Structure

Two sets of flames were selected at different conditions to conduct further analysis and develop a better understanding of the flame behaviours at different cone angles , while keeping L/D=5.

First Flames Set

Table 4. First set of flames

Flame	U_J (m/sec)	Φj	RD
PP20I	32	0.519	0.5
PP25I	30.5	0.57	0.5
PP30I	27.8	0.622	0.5

Three different partially premixed flames, as shown above in table 4, were selected to study the flame characteristics at the same level of stability (L/D=5). These flames were selected at different cone angles with Relative Deviation kept constant. The relative deviation (RD) can be calculated using the following equation.

$$RD = \frac{\Phi_j - \Phi_E}{\Phi_E} \tag{1}$$

Where the Φ_j represent the jet equivalence ratio and Φ_E stands for the equivalence ratio at extinction limit at L/D= 5. 50% was selected as RD for the selected flames.

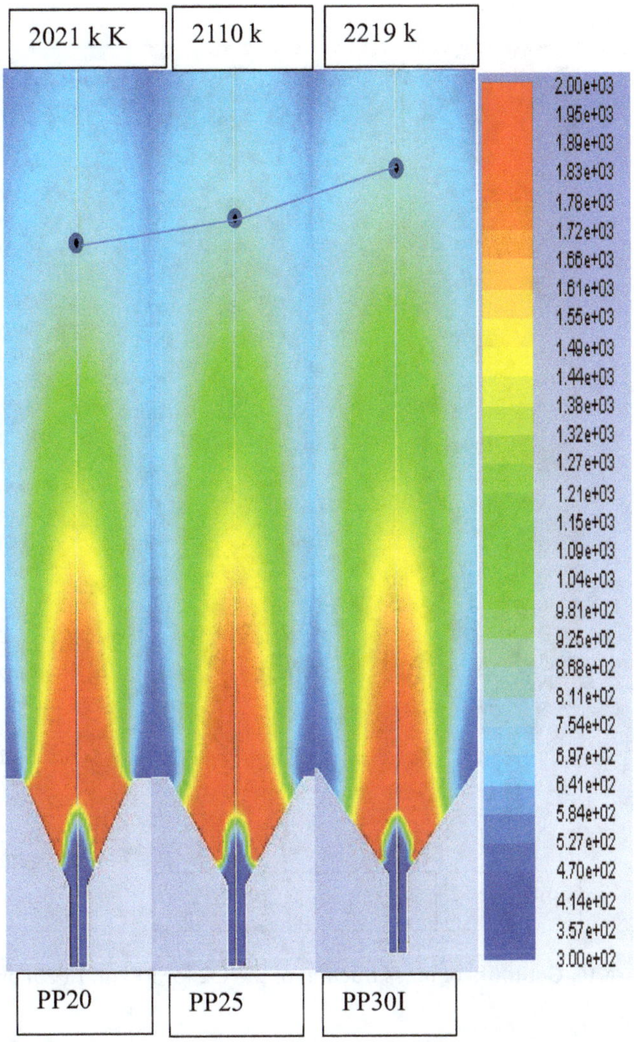

Figure 13. Contours of temperature for PP20I, PP25I and PP30I

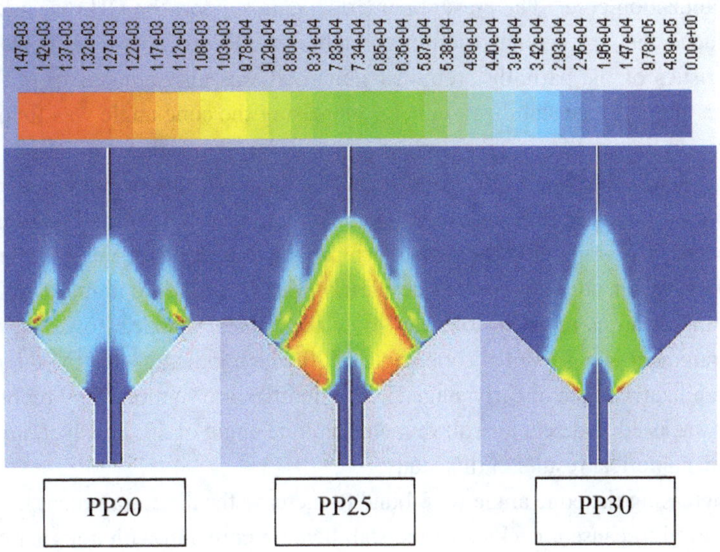

Figure 14. Contours of the mass fraction of CO

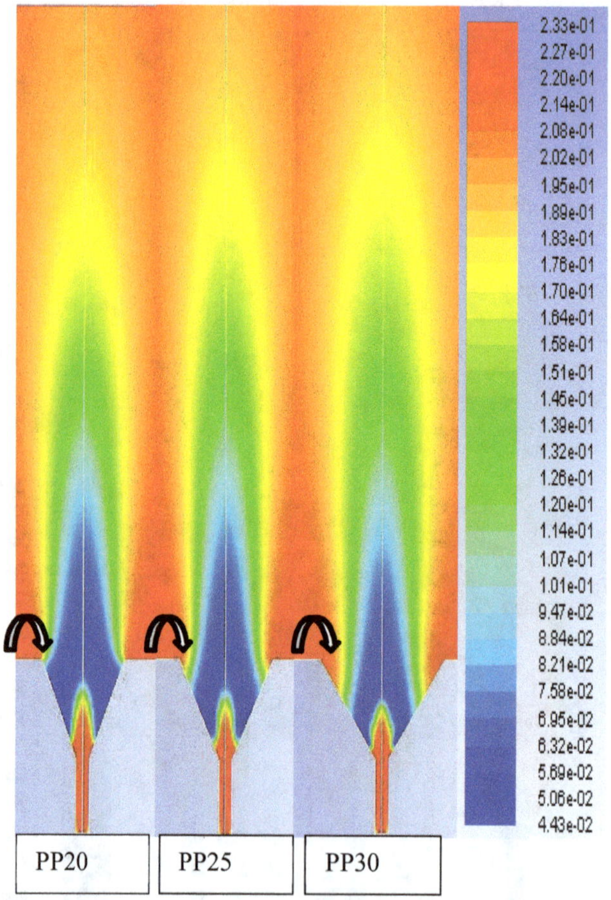

Figure 15. Contours of mass fraction of the O2 for PP20I, PP25I and PP30I

OH and Rayleigh measurements taken by [1] for the same burner at 26 cone angle found that the partially premixed flames are mainly stable by the presence of the intense reaction core within the conical nozzle which is also known as the stabilization core. The present numerical data support the OH measurements reported by [1] and thus it can be concluded that the intene reaction or stabilisation core is the main responsible factor for highly stabilization characteristics of the partially premixed flames. It was also noticed from the numerical data that the cone angle has a direct effect on the stabilization core. Increasing the cone angle was found to make the stabilization core narrower as shown in Figure 14.

Moreover, increasing the cone angle was found to increase the air entrainment inside the conical nozzle. The air entrainment is believed to cause a necking of the stabilization core and with a higher cone angle the stabilization core is expected to break. This can cause the flame to blow off as a result of the expected extinction of the intense reaction core. The effect of air entrainment can be clearly seen from the contours of O2 mass fraction, where the increase of the cone angle causes more air to enter inside the conical nozzle. However, the numerical result shows that almost no air entrainment is observed at cone angle of 20 which indicates that the there is no effect caused by the surrounding on the stabilization core at early stage. These findings are supported by reported experimental data [17] where a smoke tracer was used to detect air entrainment for cone angle of 20, however, no smoke tracer was detected which agrees with the numerical results of this study.

Moreover, increasing the cone angle was found to increase the air entrainment inside the conical nozzle. The air entrainment is believed to cause a necking of the stabilization core and with a higher cone angle the stabilization core is expected to break. This can cause the flame to blow off as a result of the expected extinction of the intense reaction core. The effect of air entrainment can be clearly seen from the contours of O2 mass fraction, where the increase of the cone angle causes more air to enter inside the conical nozzle (Figure 15). However, the numerical result shows that almost no air entrainment is observed at cone angle of 20 which indicates that the there is no effect caused by the surrounding on the stabilization core at early stage. These findings are supported by reported experimental data [17] where a smoke tracer was used to detect air entrainment for cone angle of 20, however, no smoke tracer was detected which agrees with the numerical results of this study (Figure 16).

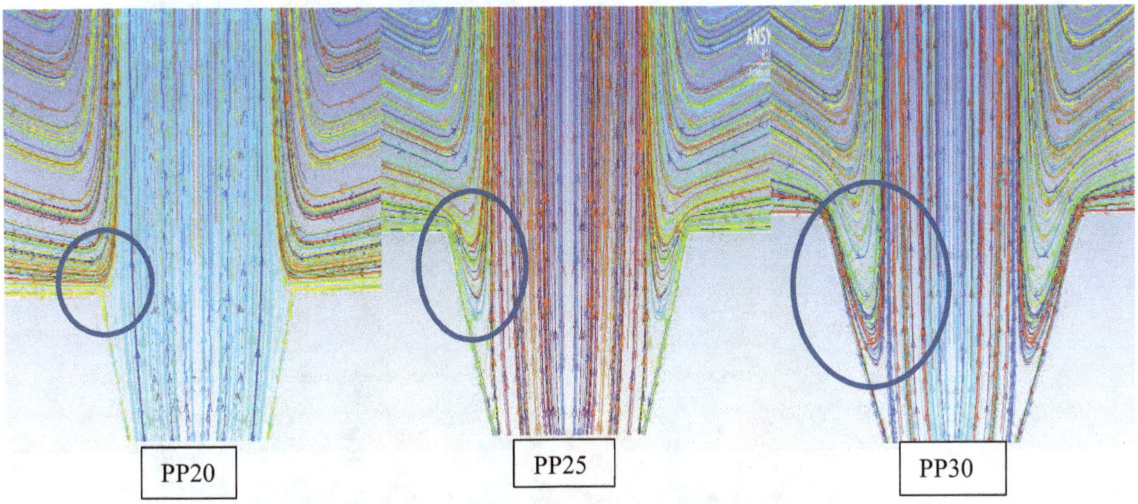

Figure 16. Path lines for different cone angles

Figure 17, shows the numerical results, for three different flames with same stability, mixing length equal to 5, for 20, 25 and 30 degrees cone angles. Experimental observations for similar flames, reported [17] that as the cone angle increased, the flame attachment to the wall decreased and the same findings can be seen from the numerical results illustrated in Figure 17 of this study.

Second Flames Set

The flames shown in the table 5 are selected to study the effect of changing the cone angle on the flame structure while the air and fuel flow rates are kept constant.

Table 5. Second Set of the Flames.

Flame	U_J (m/sec)	Φj	RD
PP20II	32	0.933	0.797
PP25II	32	0.933	0.636
PP30II	32	0.933	0.5

This set of flames will cover the axial profile (along the central line) and the radial profile across the rim of the nozzle for the selected flames. These profiles will be taken for the temperature and species concentration (O2 and CO). As stated earlier, these flames are mainly selected to investigate the influence of the cone angle at fixed jet equivalence ratio of 0.933 and jet velocity of 32 m/s with varied Relative Deviation.

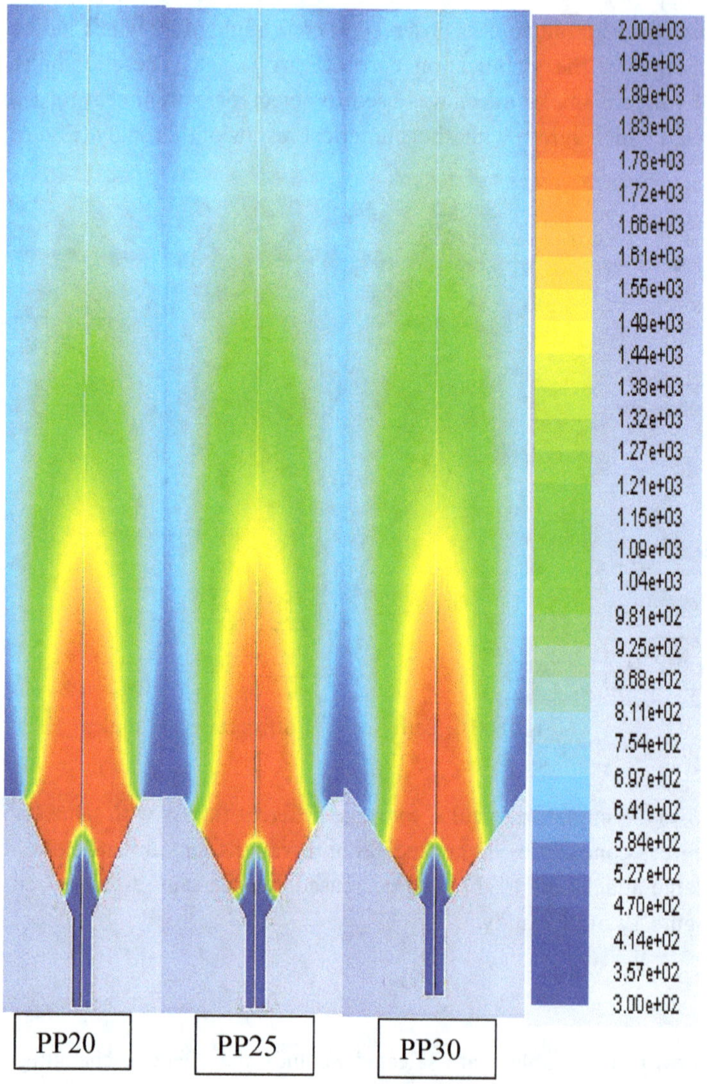

Figure 17. Numerical data demonstrated, increase in the cone angle from 20° (left) to 30° (right) decreased the flame wall attachment

Figure 18 illustrates the axial and radial profile for the selected flames covering the temperature and the concentration of the O2 and CO. The data plotted in Figure 19 shows that increasing the value of the cone angle causes the stabilization core to be longer and narrower, where the temperature and the concentration of the CO are high and the concentration of O2 is low. This effect can be explained, as discussed earlier, as a result of the air entrainment into the nozzle that causes the flame to squeeze and therefore elongate it. This result shows a good agreement with the data available in the public domain [17].

Flow Field

Figure 19a shows the velocity vectors of flow inside the cone indicating the existence of a recirculation zone near the wall of the conical nozzle which is believed to improve the overall stability performance of the burner. These recirculation zones were found to form for the cone angle lower than 30 degree and beyond which the recirculation zone is broken due to the excessive air entrainment, as shown in Figure 19b.

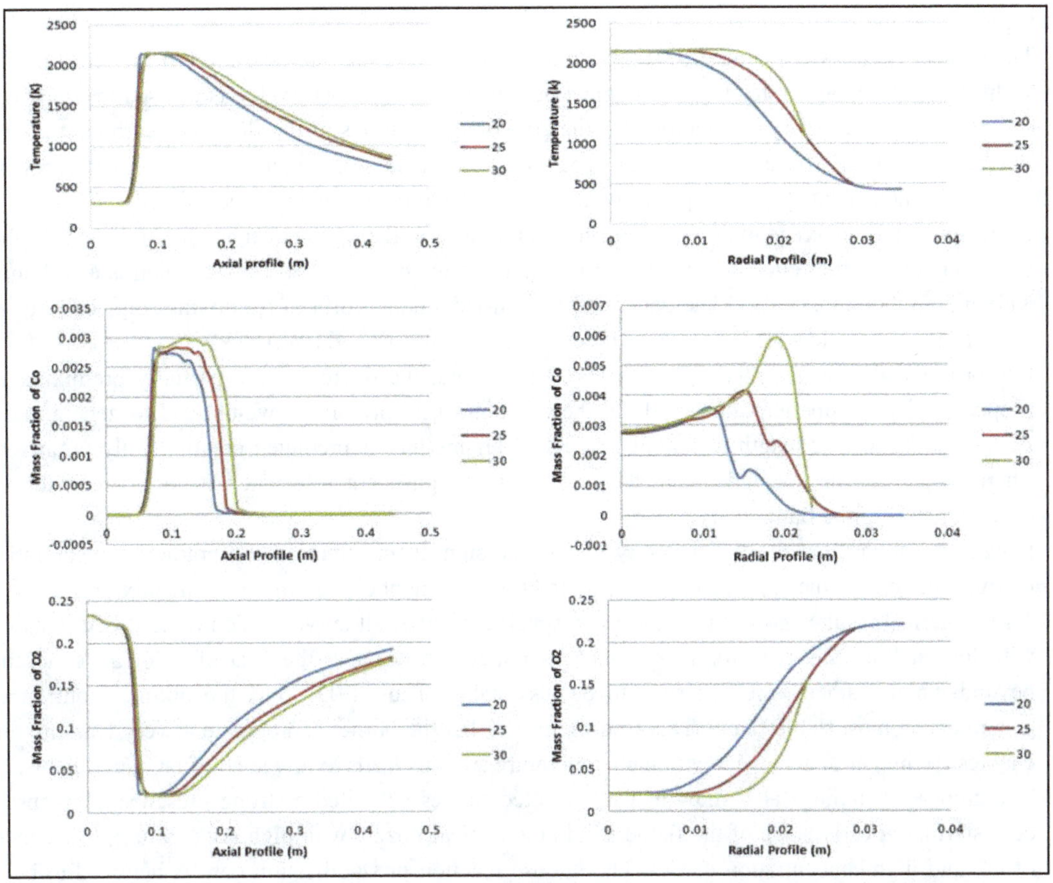

Figure 18. Temperature, CO and O2 concentration for the axial and radial profiles

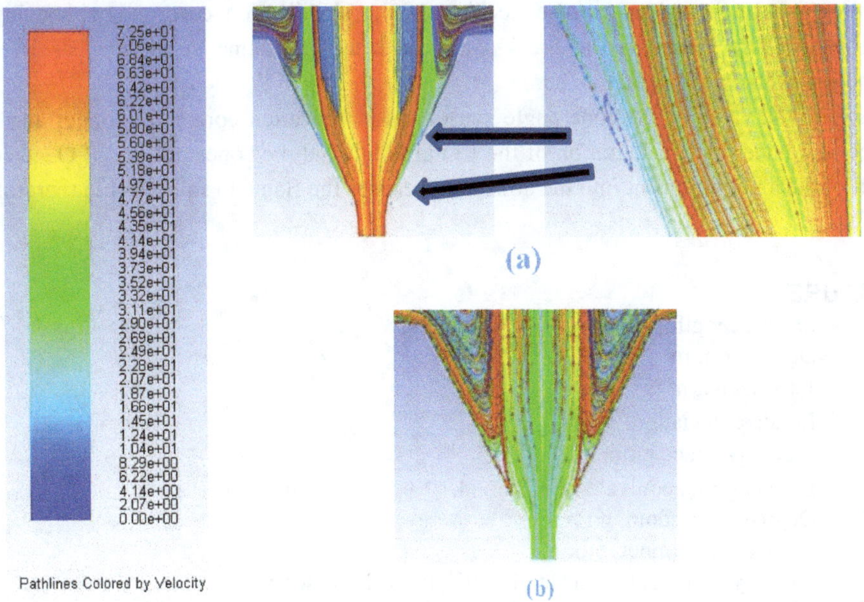

Figure 19. (a) Path lines of the velocity for the recirculation zone of 25 degrees, (b) Path line inside the burner for 35 degrees

The numerical findings of this study [30] compared favourably with the experimental study [13] which showed a development of recirculation zone at early stage of the flame inside the conical nozzle.

CONCLUSION

The present study used a numerical approach to conduct a thorough investigation of the effect of the mixing length and the cone angle on the stability of a conical burner flame. FLUENT 16.1 was utilised in this investigation in order to carry out the necessary calculations for the governing equations of the flow with turbulent model of K-Ɛ and the Partially Premixed Combustion model. The numerical data was found to be in good agreement with measurement experimental data [3, 17, 19]. This present study led to the following conclusions:

i. Development of an axisymmetric vortex ring was noticed at the region near the fuel inlet. Such phenomena occur due to the difference in velocity across the interface between two flows, which is also known as the Kelvin-Helmholtz instability. The generated Kelvin-Helmholtz vortices results in a big scale roll-up of air pockets and fuel which leads to more efficient mixing of the air and fuel.
ii. It was found that as the mixing length (L/D) increased, the degree of the partially premixing increased proportionally. At smaller values of L/D, there is almost no premixing which lead to generating diffusion flame. As the mixing length increased, more fuel air mixture is predicted creating fully premixed flames. Therefore, it can be concluded that the L/D plays a significant role when it comes to determining the characteristics of the flame.
iii. It was also found that the fuel velocity has also a significant effect on the gradient concentration in the mixture. Increasing the fuel velocity resulted in higher gradient concentration in the mixture.
iv. The numerical results showed that there is a specific point at all curves which offered better stability limit with lowest Equivalence ratio. This point was noticed to occur at the L/D of 5 for all cone angles and beyond which flames were observed to be less stable. Thus, L/D= 5 is the optimal configuration that generates the most stable flame. This finding means that the burner can operate at very lean mixture which enables the burner to achieve low flame temperature and this leads to low level of NOx pollutants.
v. The numerical findings also show that all selected flames exhibited a strong presence of intense reaction core starting at early stage of the flame inside the conical nozzle with high temperature, low concentration of O2 and high concentration of CO. The intense reaction or stabilisation core is believed to be the main responsible factor for highly stabilization characteristics of the partially premixed flames.
vi. Increasing the cone angle was found to increase the air entrainment inside the conical nozzle. The air entrainment is believed to cause a necking of the stabilization core and with a higher cone angle the stabilization core is expected to break. This can cause the flame to blow off as a result of the expected extinction of the intense reaction core.
vii. Increasing the value of the cone angle causes the stabilization core to be longer and narrower, where the temperature and the concentration of the CO are high and the concentration of O2 is low. This occurs as a result of the air entrainment into the nozzle that causes the flame to squeeze and therefore elongate it.

NOMENCLATURE

L	Mixing length, mm
D	Diameter, mm
U_j	Jet velocity, m/s^2
RD	Relative deviation, nd
Φ_j	Jet equivalence ratio, nd.
Φ_E	Extinction jet equivalence ratio, nd.
N	Degree of freedom, nd
DA	Damkolhar number, nd.
PP20I	Partially premixed, Cone angle 20° , First flame set, nd
PP25I	Partially premixed, Cone angle 25° , First flame set, nd
PP30I	Partially premixed, Cone angle 30° , First flame set, nd
PP20II	Partially premixed, Cone angle 20° , Second flame set, nd
PP25II	Partially premixed, Cone angle 25° , Second flame set, nd
PP30II	Partially premixed, Cone angle 30° , Second flame set, nd

Greek symbols

Φ	Equivalence ratio
Subscripts	
J	Refers to jet
E	Refers to Extinction

APPINDEX A

User-Defined Function (UDF)

User-Defined Function (UDF) is defined as a function which can be programmed by the users. This programmed function is then loaded into the FLUENT in order to improve the standard features of the code. This tool can be used to define users' own boundary conditions. This function can be written in C programming language by utilising any TYPE of text editor.

By default, Fluent sets the inlet velocity to enter with uniform velocity across the inlet boundary. This mean that the velocity at the mid of the tube will be the same as the velocity near the edge of the tube. However, this is not what actually happens in the tube. In real world, the velocity is at highest in the mid of the tube and it decreases gradually towards the wall t(velocity=0).This occurs due to no slip condition. Thus, the Fluent Default setting for the inlet boundary conditions will lead to a poor simulation of the flow through the tube. Thus, in order to enhance the standard features of the code for the boundary conditions, this study has used a C code for the parabolic Velocity inlet for the air and fuel inlet. The C code was then loaded into the Fluent as UDF. This UDF was then selected in the velocity boundary window in FLUENT.

The difference between the Defaults setting and the UDF for the velocity profile across the inlet can be seen in Figure A1.

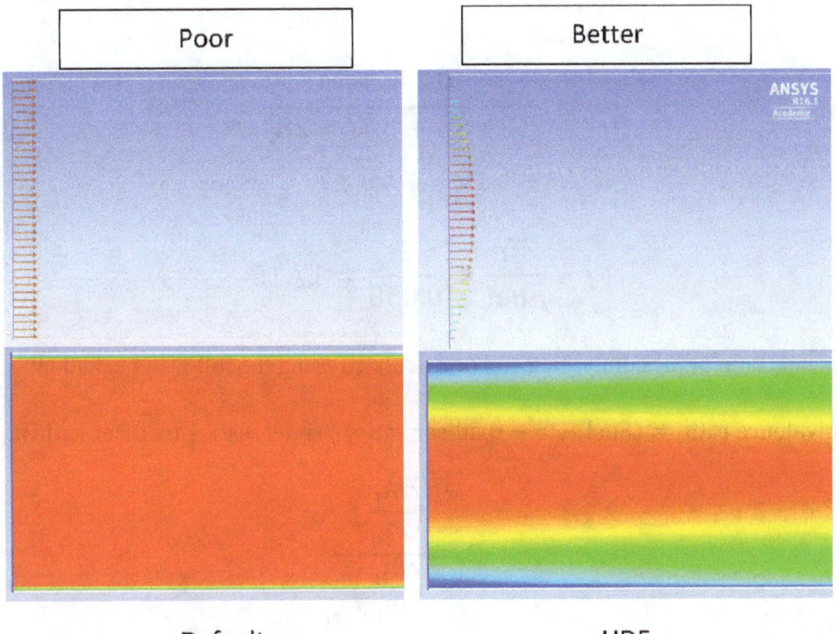

Figure A1. Difference between Default setting and UDF for velocity profile across the inlet.

APPINDEX B

This section covers several calculations that will be used for the boundary conditions at the set-up stage and will contribute in the selection process for the right viscous model for the tested flows. These calculations include:

1. Calculating the Mass flow rate for air and fuel inlet for different Equivalence ratios (Φ).

2. Calculating the air and fuel inlet velocities using the mass flow rate.

Mass Flow rate Calculation

This section covers several calculations that will be used for the boundary conditions at the set-up stage and will contribute in the selection process for the right viscous model for the tested flows. These calculations include:

1. Calculating the Mass flow rate for air and fuel inlet for different Equivalence ratios (Φ).

2. Calculating the air and fuel inlet velocities using the mass flow rate.

Mass Flow rate Calculation:

Methane-air reaction occurs when the methane reacts with the oxygen and results in carbon dioxide and water plus the released heat [23]. The following expression represents the chemical reaction for the Methane-air:

$$CH_4 + 2\ (O_2 + 3.76N_2) = CO_2 + 2H_2O + 7.52N_2$$

This reaction is said to be stoichiometric when the fuel is completely burned. The stoichiometric fuel to air ratio of the methane air reaction can be calculated as following:

$$\frac{Fuel}{Air} = \frac{CH_4}{2(O_2 + 3.76N_2)}$$

$$\frac{Fuel}{Air} = \frac{(12+4)}{2(2*16+3.76*28)} = 0.058$$

$$\frac{Air}{Fuel} = \frac{1}{0.058} = 17.16$$

The equivalence ratio of the combustion can be calculated using the following equation:

$$\text{Equivalence ratio} = (\text{Fuel} - \text{to} - \text{oxidiser ratio})\ /(\text{Fuel} - \text{to} - \text{oxidiser ratio})_{\text{stoc}}$$

$$\emptyset = \frac{\left(\frac{m_f}{m_a}\right)_{act}}{\left(\frac{m_f}{m_a}\right)_{sto}}$$

$$\left(\frac{m_f}{m_a}\right)_{sto} = 0.058$$

$$\left(\frac{m_f}{m_a}\right)_{act} = 0.058 \times \emptyset$$

$$\mathrm{m_f} = 0.058 \times \emptyset \times \mathrm{m_a} \quad \text{(B.1)}$$

Equation (B.1) will be used to calculate the required mass flow rate of fuel to achieve different equivalence ratios at constant air mass flow rate of 5.98 Kg/h. For example, the required mass fuel rate for equivalence ratio ($\emptyset =$ **2**) can be calculated as following:

$$m_f = 0.058 \times 1 \times 5.98 \frac{Kg}{h} = 0.3468 \frac{kg}{h} = 9.633 * 10^{-5} kg/s$$

Calculation the fuel and air inlet velocity

The air and fuel inlet velocities can be calculated using the mass flow rate equations, as shown below:

$$m_a = p * A * V \tag{B.2}$$

where,

p= density (kg/m^3) where Air density=1.18 kg/m^3 and methane density =0.66kg/m^3 [23].

A= Area (m^2).

$\boldsymbol{m_a}$= flow rate of air = 9.63*10^{-3} kg/s for equivalence ratio of one.

$\boldsymbol{m_f}$= flow rate of fuel = 0.001661 kg/s for equivalence ratio of one.

Area of air inlet = 1.96*10^{-5} m^2

Area of fuel inlet=2.2*10^{-5}

This equation was used to calculate the fuel and air inlet velocity for a range of equivalence ratio, as shown in the table B.1 below. These were then used during the simulation process.

Table B.1. Air and fuel inlet velocity for wide range of equivalence ratios (Φ)

	Φ	M_a (kg/s)	V_a (m/s)	M_F (kg/s)	V_F (kg/s)
1	2	0.001661	71.81771	0.000193	13.2697
2	1.8	0.001661	71.81771	0.000173	11.94273
3	1.6	0.001661	71.81771	0.000154	10.61576
4	1.4	0.001661	71.81771	0.000135	9.288788
5	1.2	0.001661	71.81771	0.000116	7.961818
6	1.0	0.001661	71.81771	9.63E-05	6.634848
7	0.9	0.001661	71.81771	8.67E-05	5.971364
8	0.8	0.001661	71.81771	7.71E-05	5.307879
9	0.7	0.001661	71.81771	6.74E-05	4.644394
10	0.6	0.001661	71.81771	5.78E-05	3.980909
11	0.5	0.001661	71.81771	4.82E-05	3.317424
12	0.4	0.001661	71.81771	3.85E-05	2.653939

REFERENCES

[1] Mansour, M. (2003). Stability characteristics of lifted turbulent partially premixed jet Flames, Combustion and Flame, 133, 236-274.

[2] INFORSE-EUROPE (2010), Regulation of air pollution from small biomass combustion, 17-19, [online] http://www.inforse.org/europe/pdfs/biomass-pollution-regulation-proposal.pdf

[3] Mansour, M. (2005). Mixing and nozzle geometry effects on flame structure and stability, Portugal, 6-10 October 2005. Lisbon, 23-34. http://scholar.cu.edu.eg/sites/default/files/ahmedabdelhafez/files/mixing_and_nozzle_geometry_effects_on_flame_structure_and_stability.pdf

[4] Miao, J. Leung, W. and Cheung, K. (2013). Flame Stability and Structure of Liquefied Petroleum Gas-Fired Inverse Diffusion Flame with Hydrogen Enrichment, International Journal of Mechanical, Aerospace, Industrial, Mechatronic and Manufacturing Engineering, 1, 121-129.

[5] Schefer, W. Wicksal, D.M. and Agrawal A.K. (2002) Combustion of hydrogen-enriched methane in a lean premixed swirl burner, Proceedings of the combustion institute, 29,1, 843-851

[6] Shoshin, Y. (2011). Stabilization of laminar inverted ultra -lean hydrogen-methane-air flames, Combustion Technology, 602-613, [online]. http://www.combustion-institute.it/proceedings/MCS-7/papers/PEC/PEC-11.pdf

[7] Alhashimi, M. (2012). Flashback and Blowoff Characteristics of Gas Turbine Swirl Combustor, 36-37. http://orca.cf.ac.uk/24193/1/2012%20Abdulsada%20M%20PhD.pdf

[8] Chaparro, A. and Cetegen, B. (2006). Blowoff characteristics of bluff-body stabilized conical premixed flames under upstream velocity modulation, Combustion and Flame, 144, 1-2, 318–335.

[9] Nair, S. and Muruganandam M. (2004). Lean Blowout Detection in a Single Nozzle Swirl Cup Combustor, 42nd AIAA Aerospace Sciences Meeting, 5, 138.

[10] Lieuwen, T. (2006), Chapter 3.1.1: Static and Dynamic Combustion Stablity, In: Gas Turbine Handbook. USA, Department of energy, office of fossil energy, 124-145.

[11] Noble, D. and Shareef, A. (2006) Syngas Mixture Composition Effects Upon Flashback and Blowout, ASME Conference Proceedings, 13, 357-368

[12] Sarli, V. and Bendetto, A. (2007). Laminar burning velocity of hydrogen–methane/air premixed flames," International Journal of Hydrogen Energy, 32, 5, 637–646.

[13] Mansour, M. (2012). The stabilization mechanism of highly stabilised partially premixed flames in a concentric flow conical nozzle burner, Experimental thermal and fluid science, 43, 55-62.

[14] El-Mahallawy, F. (2005). Mixing and nozzle geometry effects on flame structure and stability, Combustion Science and Technology, 2, 60-12.

[15] Baudoin, E. (2011). Effect of partial premixing on stabilization and local extinction of turbulent methane/air flames. Chia Laguna, 4, 20-35.

[16] Elbaz, M.A. (2013). An Experimental Study of the Effect of Partial Premixing Level on the Interaction between the Flame Kernel and Flow Field, Computer Science Journals, 1, 121-145.

[17] El-Mahallawy, F. (2007). Mixing and nozzle geometry effects on flame structure and stability, Combustion Science and Technology, 2, 249-263.

[18] Peters, N. (2000). Turbulent Combustion, vol.II. 1st ed., Cambridge, Cambridge University Press. United Kingdom

[19]Mansour, M. (2004) The flow field structure at the base of lifted turbulent partially premixed jet flames, Experimental Thermal and Fluid Science, 7, 771-779.

[20] Joedicke, A. (2005) The stabilization mechanism and structure of turbulent, Proceedings of the Combustion Institute, 1, 901-909.

[21] Dutt, A. (2015). Effect of Mesh Size on Finite Element Analysis of Beam, International Journal of Mechanical Engineering (SSRG-IJME), 12, 8-9.

[22] Charles, M. (2006). Comparison of Pressure-Based and Density-Based Methods for low Mach number CFD computations, 3, 12-15.

[23] Pupo, R. (2011). Adiabatic Flame Temperature for Combustion of Methane, Journal of Mathematical Modelling, 2, 3-7.

[24] Khalil, R. (2010) Effect of Pressure and Inlet Velocity on the Adiabatic Flame Temperature of a Methane-Air Flame, Jordan Journal of Mechanical and Industrial Engineering, 4, 21-28.

[25] Jonathon, Z. (2009). Modelling Fluid Flow Using Fluent," 35-38.

[26] Mathur, M.J. (2007). A pressure-based method for unstructured meshes, Numerical Heat Transfer, Part B: Fundamentals, 20-21.
[online] http://www.tandfonline.com/doi/pdf/10.1080/10407799708915105?needAccess=true

[27] Salim, M. (2011). Comparison of RANS, URANS and LES in the Comparison of RANS, URANS and LES in the Prediction of Airflow and Pollutant Dispersion," the World Congress on Engineering and Computer Science, 40-43.
[28] Baudoin, E. (2012). Effect of partial premixing on stabilization and local extinction of turbulent methane/air flames, Flow Turbulence Combust, 9, 269-284.
[29] Manera, J. (2008). Kelvin-Helmholtz instabilities occurring at a nacelle exhaust," Aeroacoustics Conference, 14, 235-264.
[30] Almawali J. (2016, November). Numerical Analysis of Flame Characteristics and Stability for the Conical Nozzle Burner, Master thesis, Sheffield Hallam University.

16

EMISSIONS AND PERFORMANCE CHARACTERISTICS OF AN SI ENGINE WITH BIOGAS FUEL AT DIFFERENT CO_2 RATIOS

Yasin Karagöz*

ABSTRACT

Nowadays increasing emission regulations and environmental concerns have led to the use of alternative fuels in internal combustion engines. Biogas is an alternative fuel that can be produced with clean, green and natural resources. In this study, a single cylinder, four stroke SI engine was operated at stoichiometric conditions, 2800 rpm constant engine speed with biogas fuel at different ratios. As biogas fuel, methane (CH_4) and CO_2 were used during engine tests. Test results which were taken by using only gasoline were compared with the results obtained by using biogas fuel at different CO_2 ratios (10%, 20%, 30% and 40% by volume). Although it was determined specific CO emissions decrease and THC emissions slightly increases with the increasing CO_2 ratio, it is foreseen that the increase in THC emissions can be prevented by the use of three way catalytic converter. Although the NOx emissions increase with the use of biogas compared to gasoline, it was significantly reduced by the increased CO_2 rate in the biogas fuel, the CO_2's heat carrying capacity and reducing the O_2 availability of the fuel. On the other hand, because of the high cost of purification process of biogas and after-treatment equipment, it is foreseen that use of biogas at high ratio of CO_2 in SI engines will be advantageous in terms of NOx emissions.

Keywords: *: Biogas, CO_2, Methane, SI Engine, Emissions, Combustion*

INTRODUCTION

Nowadays, as the number of vehicles increases, the percentage of diesel vehicles in the vehicle market increases [1]. In 2016, the market of diesel vehicles have 49% share among all vehicles in European countries [2]. In this case, the emission of NOx and PM emissions increases, affecting the environment and human health negatively [3]. Due to this reason, several steps have been taken in recent years to reduce the number of diesel vehicles. As an alternative to diesel vehicles, electric and Fuel Cell vehicles are targeted to be presented to the market [4]. However, because of the high cost and short life spans of PEM fuel cells and high cost of purification of hydrogen and hydrogen storage problems avoids the use of fuel cells in vehicles [5]. On the other hand, because of the need of improvement at current battery technology and the inadequacy of charge stations prevents the widespread use of these systems in the near future [6]. Due to this reason, it is foreseen that fuel cells will be used to supply very small units' batteries and increase their range which can be used as range extender in small powers and operated as both serial hybrids and parallel hybrids when necessary. In the near future, the use of environmentally friendly, low-cost and alternative gas fuels in SI engines will gain great importance. Because of this reasons, biogas emerges as a clean and environmentally friendly alternative fuel containing methane, CO_2 and other gases that can be produced from natural sources [7]. Due to the methane in its content, biogas is a environmentally friendly fuel, which has low C/H ratio [8]. Because of the methane in its content, the use of biogas in in SI engines provides great benefit in terms of emissions [9].

Related studies in the literature on the use of biogas in SI engines are given below:

Karagoz et al. [10] used an ANN model for a spark ignition engine with use of different methane (CH_4) ratios to estimate engine performance data. The study shows that with the increase in methane concentration, positive effect is seen in volumetric efficiency but a dramatic decrease occurs in thermal efficiency. Despite that there isn't any observable change in brake specific fuel consumption.

Wang et al. [11] studied the effect of methane content to NO_x emissions and indicated thermal efficiency in a biogas fuelled spark ignition engine. It is included that increase in methane content in biogas have positive effect in both indicated thermal efficiency and NO_x emissions.

Shin et al. [12] studied the effect of using biogas in a small spark ignition engine at different compression ratio and various CO_2 dilution (0%-50%) on engine performance. The results show that, engine performance increases by using biogas with decreasing CO_2 dilution.

This paper was recommended for publication in revised form by Regional Editor Jaap Hoffman
1 Department of Mechanical Engineering, University of Lagos, Akoka- Yaba, Lagos, Nigeria
**E-mail address: ta.akinshilo@gmail.com*

Nadaleti et al. [13] made experiments tos see the effect of biogas-syngas dual fuel in a SI engine on engine emission characteristics. The results show that the use of biogas fuel decreases NO_x emissions with a increase HC emissions.

Porpatham et al. [14] investigated the effect of increased oxygen concentration of intake air on engine performance, emissions and combustion in a biogas fuelled spark ignition engine. At higher oxygen levels, an observable improvement was seen in brake thermal efficiency and brake power. With higher oxygen concentration, NO_x emissions increases, but HC and CO_2 emissions decreases. There is also an improvement in thermal efficiency as a result of enchancement in combustion.

Yadav et al. [15] studied exhaust emissions characteristics of a biogas operated automotive SI engine experimentally. Experiments were performed with %100 biogas fuel via petrol. With the comparison of petrol, the results shows that the use of biogas makes reduction in NO_x and CO emissions and causes negligible CO emissions.

Considering the work done so far in the literature, it has been found that there is not enough work on the performance and emissions effects of using biogas with different CO_2 contents directly on SI engines. In addition, the emission and performance impact of using biogas at different CO_2 ratios in SI engines has not been investigated because the naturally purifiable biogas fuel refinement is costly and CO_2 in the content of biogas has a high heat transfer capacity and low O_2 availability. It is known that biogas usage in SI engines may have some negative impact on thermal efficiency and mean effective pressure. However, decreasing the toxic emissions, such as NO_x, to the limits determined by the stringent emission regulations requires high cost after-treatment equipments. Because of being a good alternative to reduce the emissions, the research of biogas usage which has different CO_2 ratios on SI engines is performed in this study.

TEST METHODOLOGY

Schematic diagram of test system is shown in Figure 1. A single cylinder SI engine is used in the experiments. Engine is loaded with an 8 kW AC dynamometer. Fuel system consists of a methane gas tank which is connected to intake and exhaust system of the test engine. Details of the test system is given in the next subsection.

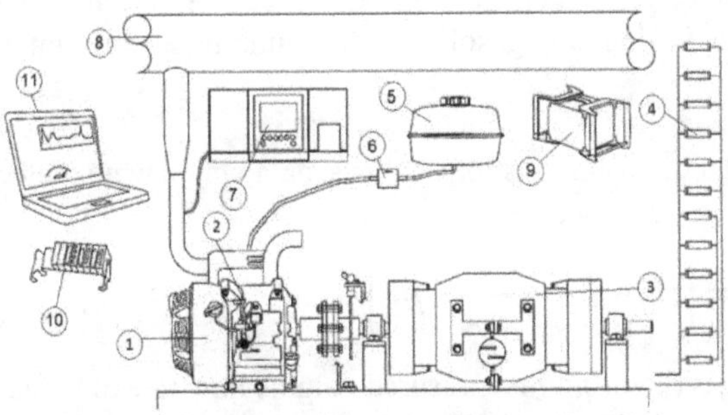

1. Test engine 2. Pressure sensor 3. Dynamometer 4. External resistive load banks 5. Gasoline tank 6. Gasoline flowmeter 7. Exhaust gas analyser 8. Exhaust vantilation system 9. Kibox combustion analyser 10. NI data acquisition device 11. Personal computer for data recording and post processing

Figure 1. Schematic diagram of engine test system

Test Setup

Experiments were performed on a single cylinder, 270 cc, four stroke, naturally aspirated, SI engine. Engine was operated first with only gasoline then with methane at different CO_2 contents (10%, 20%, 30%, 40%). All the tests were carried out at constant speed (2800 rpm) and repeated three times.

A miniature oval gear type fuel meter was used to measure gasoline consumption. New-flow, TLF series hotwire type mass flow meter was preferred to measure methane and CO_2 gas consumption. Mass flow rate of intake air was measured by a mass flow meter. AVL Digas 4000 was used to carry out the emission measurements. Calculation method which is announced by VDMA exhaust emission legislation for diesel and gas engines was used for the calculation of specific emissions by using gasoline, $CO_2 + CH_4$ and intake air mass flow rates [16].

Crank angle position and engine speed was obtained with use of a proximity sensor and an incremental type encoder. In order to measure in-cylinder pressure, Kistler 6118B in-cylinder pressure transducer, which can also be used as spark plug was fixed to the engine head instead of original spark plug of the SI engine. Then obtained in-cylinder pressure data were processed by Kistler Kibox combustion analyzer device.

Engine and Dynamometer

A 270 cc, single cylinder, four stroke, naturally aspirated, SI engine was used during the tests. Methane and

CO_2 gas tanks were installed to the intake port of the test engine. Test engine was loaded by a 8 kW AC dynamometer. Detailed information of test system can be seen in Table 1. Tests were performed in Yıldız Technical University, Internal Combustion Engines Laboratory. All of the measurement devices and test equipments in Internal Combustion Engines Laboratory of Yıldız Technical University are calibrated and certificated by the authorized companies in Istanbul.

Table 1. Engine specs and dynamometer features

Engine Specs	
Engine manufacturer	Honda
Number of cylinders	1
Bore × stroke [mm]	77 × 58
Cylinder volume [cm³]	270
Aspiration	Natural
Compression ratio	8.5 : 1
Net power [kW]	6.3 @3600 rpm
Recommended speed range [rpm]	2000-3500
Cooling	Air
Dynamometer Features	
Power [kVA] [kW]	8

Test Procedure

Two independent tests were performed in this study sequentially,

- Firstly, CO_2 and CH_4 gas tanks were equipped to the intake system of the test engine. Engine tests were performed with methane. CO_2 was fed into intake manifold as supplementary fuel at 4 different ratios (10%, 20%, 30%, 40%). All tests were conducted at constant speed (2800 rpm). Then,
- The SI engine were operated using only gasoline as fuel running at constant engine speed (2800 rpm).

DATA PROCESSING

In order to obtain engine performance and combustion analysis, data which were collected form sensors were used. For the calculation, following equations were used

$$\dot{m}_B = \left[\dot{v}_B \frac{2}{3} x \frac{1}{1000x60} x \rho_B\right] \quad (1)$$

where, $\dot{m}_B$ is mass flow rate of biogas (kg/m³), $\dot{v}_B$ is measured volume flow rate of biogas, ρ_B is density of biogas (kg/m³).

The equations [17] used to calculate the indicated engine power and indicated engine work are given as:

$$W_i = P_{mi} \, x \, V_H = \int_0^{720} P \, dV \quad (2)$$

$$N_i = \frac{W_i \, x \, n}{60 \, x \, 2} \quad (3)$$

where W_i is indicated work done per cycle (J), P_{mi} is mean indicated pressure ($_p$a), V_H is cylinder displacement (m³), P is pressure which was measured from the cylinder (Pa), V is calculated instant cylinder volume (m³), N_i is indicated engine power (kW).

Indıcated engine torque can be calculated as given in equation (8) by the use of indicated power value [17].

$$M_i = \frac{60}{2\pi n} N_i \quad (4)$$

where n is the engine speed (rpm) and M_i is indicated engine torque (Nm).

In order to obtain engine performance characteristics, the calculated engine power value, measured pressure and flow rates of gasoline, methane and CO_2 can be used.

Equation (5) is used for the calculation of indicated thermal efficiency [17].

$$\eta_{Ti} = \frac{N_i}{\dot{m}_g \times LHV_g + \dot{m}_H \times LHV_B} \quad (5)$$

where, η_{Ti} is indicated thermal efficiency, $\dot{m}_g$ and $\dot{m}_{CO2}$ are mass flow rates of gasoline and CO_2 respectively (kg/s) and $LHVg$ and LHV_{CO2} are lower heating values of gasoline and CO_2 respectively (kJ/kg).

For the calculation of emission values, which were also measured by volume and mass flow rate of exhaust gases during tests, measured flow rates of intake air, methane, CO_2 and gasoline can be used as seen equation (6) [15]. Thereby, by using these values and indicated engine power, specific emissions can be calculated as given in equation (7) [15].

$$\dot{m}_{ex} = \dot{m}_{ia} + \dot{m}_H + \dot{m}_g \tag{6}$$

$$EP_i = EV_i \ x \ \frac{M_i}{M_{ex}} \ x \ \frac{\dot{m}_{ex}}{N_i} \tag{7}$$

where $\dot{m}_{ex}$ and $\dot{m}_{ia}$ is mass flow rates of exhaust gases and intake air, respectively (g/h), EP_i is specific emission value of relevant gases (g/kWh), EV_i is relevant gases and total exhaust gases ratio by volume for CO and by ppm for THC and NO_x emissions, M_i and M_{ex} are molar mass of relevant gases and exhaust gases respectively (kg/kmol) and N_i is indicated engine power (kW).

The total propagated uncertainty of the Precision and Systematic (Bias) errors measurement determined by Kline and McClintock analysis [18] is:

$$(W_R)_{P,B} = \left[\left(\frac{\partial R}{\partial x_1} w_1\right)^2 + \left(\frac{\partial R}{\partial x_2} w_2\right)^2 + \cdots + \left(\frac{\partial R}{\partial x_n} w_n\right)^2\right]^{1/2} \tag{8}$$

And the computation of the total uncertainty W_R is done with:

$$W_R = \sqrt{(W_R)_P{}^2 + (W_R)_B{}^2} \tag{9}$$

where $(W_R)P,B$ denotes the propagated uncertainty for either precision error $(W_R)P$, or systematic (Bias) error $(W_R)B$ functions. $x_1, x_2, \ldots, x_n$ are the measured variables and $w_1, w_2, \ldots w_n$ are the corresponding uncertainties of the variables The measurement uncertainty values of test setup were given in Table 2.

Table 2. Accuracy values of test devices and obtained total uncertainty values

Parameter	Device	Accuracy
Engine speed	Incremental encoder	±5 rpm
Gasoline flow rate	Biotech VZS-005	±1 % (of reading)
CO	AVL Digas 4000	0.01 % Vol.
THC	AVL Digas 4000	1 ppm
NO_x	AVL Digas 4000	1 ppm
Calculated results		Uncertainty value
Indicated thermal efficiency		±1.18 ÷ 1.32 %

RESULTS AND DISCUSSION

ITE

The efficiency of useful energy conversion from the fuel consumed by the engine is defined by indicated thermal efficiency. In other words, the engine specific fuel consumption is directly affected by ITE [17]. The variation in indicated thermal efficiency due to biogas usage at different CO_2 ratios is given in Figure 2.

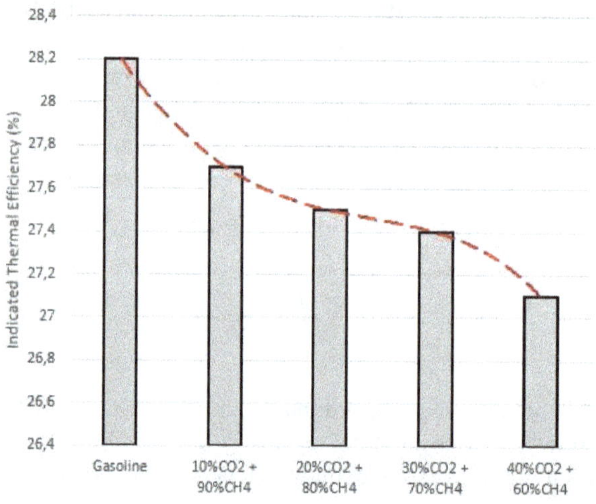

Figure 2. Effect of different level of CO2 and CH4 addition on indicated thermal efficiency

As comparison with the results when the engine is operated with only gasoline, ITE value is reduced at all CO_2 ratios with biogas usage as fuel. ITE value decreased 1.8% at 10% CO_2 ratio, 2.4% at 20 CO_2 ratio, 2.8% at 30% CO_2 ratio and decreased 3.9% at 40% CO_2 ratio.

The heat carrying capacity of CO_2 in the biogas causes reduction of in-cylinder combuion pressure values [20]. This situation explains the reduction of ITE values with biogas usage which contains CO_2.

IMEP

Through the cycle, in-cylinder pressure of an engine changes continuously. In order to estimate an average value of in-cylinder pressure variation, mean effective pressure parameter is defined [19]. Because of independency of engine speed and size, using mean effective pressure is a good way to compare engines in terms of design and output [19]. Indicated mean effective pressure (IMEP) is obtained by using indicated work to define mean effective pressure value of the engine [17]. In this study, indicated mean effective pressure (IMEP) parameter is used to compare engine performance of gasoline and biogas blends.

Figure 3 gives the test results about the variation of IMEP with the usage of only gasoline as fuel and with only biogas at four different CO_2 ratios (%10, %20, %30 and %40) at constant engine speed (2800 rpm).
In comparison with the results when the engine is operated with only gasoline, IMEP value is reduced at all CO_2 ratios with biogas usage as fuel. IMEP value decreased %8.9 at % 10 CO_2 ratio, %9.4 at 20 CO_2 ratio, %10.5 at %30 CO_2 ratio and decreased %12.2 at %40 CO_2 ratio.

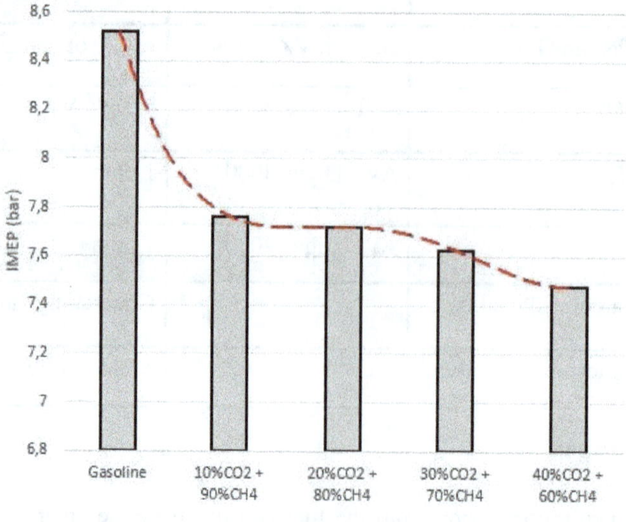

Figure 3. Effect of using CH_4 at different CO_2 levels on indicated mean effective pressure

As seen in Figure 3, with use of biogas and increasing CO_2 content, IMEP values decreases. The main reason of this result is the decrease in maximum in-cylinder pressure because of operating the engine at near stoichiometric conditons. Also, because of using gaseous fuel, low density of methane, methane's rendering high cylinder volume and dilution of combustion chamber by CO_2 (EGR effect) are the other reasons of the decrease in IMEP values.

CO

Carbon monoxide is a colorless, odorless, poisonous gas, which is mostly generated when engine is operated with rich fuel ratio [19]. When there is not enough O_2 to convert all carbon to CO_2, some fuel does not get burned and some carbon ends up as CO [19].

Figure 4 gives the test results about the variation of CO emission with the usage of only gasoline as fuel and with only biogas at four different CO_2 ratios (10%, 20%, 30% and 40%) at constant engine speed (2800 rpm).

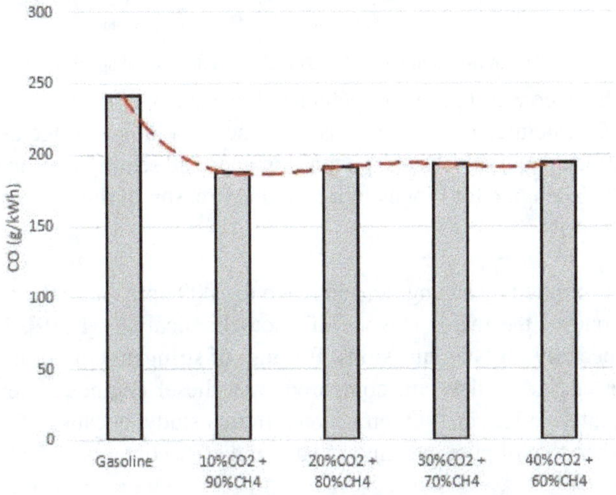

Figure 4. Effect of different level of CO2 and CH4 addition on specific CO emissions

As the comparison with only gasoline usage as fuel, CO emissions are low with only biogas usage at all CO_2 ratios. CO emission reduces about 22,5% at 10% CO_2 ratio, reduces about 20,8% at 20% CO_2 ratio, reduces about 20% at 30% CO_2 ratio and reduces about 19,5% at 40% CO_2 ratio from CO emission released by engine operated with only gasoline.

Methane composes the main structure of biogas. Methane has low C/H ratio and low carbon concentration in its composition. Because of its low C structure, methane increases the oxygenation of the fuel [21]. So, this situation explains the reduction in CO emission with use of biogas as fuel.

Methane has low C content, so it provides low CO emissions. As CO_2 content of mixture increases, CO emissions slightly increases. However, methane is a reactive gaseous fuel, and with the use of methane, a nearly homogenous mixture occurs in combustion chamber. So, the changes in CO emissions at all CO contents are nearly negligible.

THC

As one of the reasons of lung and blood diseases, hydrocarbons are another harmful emission [22]. So, for human health, it is essential decrease in THC emission. THC is the definition of the hydrocarbon emissions which are emitted from tailpipe. It contains both partially burned and unburned hyrocarbons. So, CH_4 emissions are counted in total hyrocarbons and more accurate results are obtained. The measured THC emission values with only gasoline fuel and with biogas at four different ratios (10%, 20%, 30% and 40%) at constant engine speed (2800 rpm) are given in Figure 5.

As compared with only gasoline usage as fuel, hydrocarbon emissions were increased considerably with the use of biogas at all CO_2 ratios. THC emissions increased at about 2.4% at 10% CO_2 ratio, increased about 6.6% at 20% CO_2 ratio, increased about 8.7% at 30% CO_2 ratio and increased about 11.2% at 40% CO_2 ratio in comparison with gasoline.

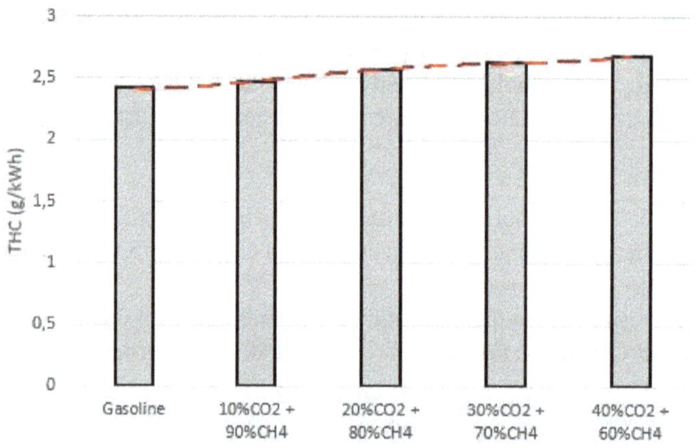

Figure 5. Effect of different level of CO_2 and CH_4 addition on specific THC emissions

Biogas contains high concentration of methane. Because of the high quenching distances and zones of methane and its volatile HC structure, THC emissions increase with biogas usage as fuel [23]. For optimal usage of biogas, crevice zones, electrode gaps, valve seats, piston ring gaps and sealing gaps in the engine should be minimized. The unsuitable design of the SI engine for biogas fuel is another reason of the increase in THC emissions.

NO_x

Oxides of nitrogen, which is emitted by an engine, can reach to 2000 ppm value. Nitrogen in the air is the main source of NO_x. Released NO_x is one of the major reasons of photochemical smog [19]. Nowadays, emission regulations become more stringent, especially for NO_x emissions. Because of stringent emission regulations, SI engines are being named as positive ignition engine as they are compared with diesel engines in terms of released amount of NO_x emissions. Methane has additive effect on NO_x emissions. In this study, because of high methane percentage of fuel, NO_x emissions are higher than that of pure gasoline at 10% and 20% contents.

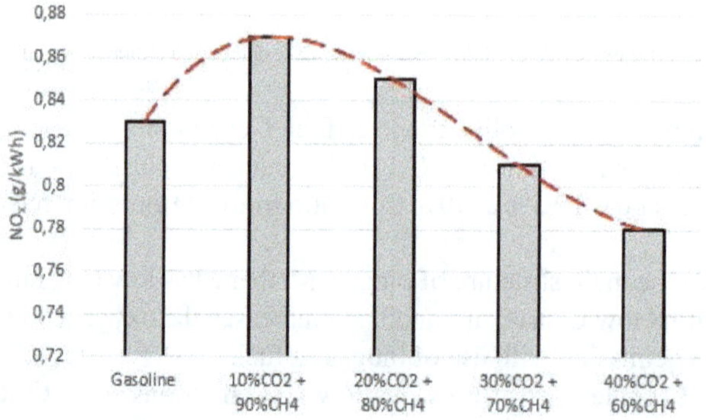

Figure 6. Effect of different level of CO_2 and CH_4 addition on specific NO_x emissions

NO_x emissions are shown in Figure 6 with biogas at different CO_2 ratios (10%, 20%, 30% and 40%) and with only gasoline at constant engine speed (2800 rpm).

When compared to only gasoline usage, NOx emissions raise about %4.8 with biogas at %10 CO_2 ratio, and raises about 2.4% at 20% CO_2 ratio. Then, NO_x emissions decreases at about 2.4% with biogas at 30% CO_2 ratio and decreases at about 6% at 40% CO_2 ratio.

As seen in the results, with the usage of biogas, NOx emissions increase because of the increased local peak in-cylinder pressure, but when CO_2 ratio in the biogas reaches to 30% and upper values, there can be seen a decrease in NO_x emissions, The main reason of this decrease is the heating capacity of CO_2 and its decreasing O_2 availability in combustion process [24].

CONCLUSIONS

In this study, in a single cylinder, 270 cc, naturally aspirated, four stroke SI engine, at stoichiometric condition and constant engine speed (2800 rpm), effect of biogas fuel at different CO_2 ratios (10%, 20%, 30% and 40%) on engine performance and emissions was investigated with comparison of only gasoline usage.

The main conclusions were summarized below:

- With the increasing CO_2 ratio, there has been a decrease in indicated mean effective pressure. The main reason of this situation is stoichiometric working conditions and the low heating value (by volume) of biogas. ITE value decrease with use of biogas and increasing ratio of CO_2.
- With use of biogas and increasing CO_2 content, IMEP values decreases. The main reason of this result is the decrease in maximum in-cylinder pressure because of operating the engine at near stoichiometric condition. Also, because of using gaseous fuel, low density of methane, methane's rendering high cylinder volume and dilution of combustion chamber by CO_2 (EGR effect) are the other reason's of the decrease in IMEP values.
- CO emissions decrease with biogas usage. The low C/H ratio of methane, which is main component of biogas, is the main cause of the decrease in CO. THC emissions increases as the CO content in biogas increases. The unsuitable design of the engine for methane fuel, the crevice zones in combustion chamber and the structure of methane which is already an unburned HC are the other reasons of increase in THC emission.
- Because of the increase in local peak in-cylinder pressure values with biogas usage, NOx emissions increase also. However, when CO_2 ratio in biogas reach 30% and upper values, because of CO_2's heating capacity and decreasing O_2 availability, NOx emission values are under the values which are taken with only gasoline usage. Although it has some negative impact on engine performance, biogas usage in SI engines can simulate EGR effect, so it can be possible to control the NO_x emissions.

NOMENCLATURE

CA Crank angle
CH_4 Methane
CO Carbon monoxide
CO_2 Carbon dioxide
EGR Exhaust gas recirculation
HC Hydrocarbons
IMEP Indicated Mean Effective Pressure
ITE Indicated Thermal Efficiency
NO_x Oxides of nitrogen
PM Particulate matter
SI Spark ignition
THC Total unburned hydrocarbons

REFERENCES

[1] European Environment Agency. (2017). Monitoring CO_2 emissions from new passenger cars and vans in 2016. Report No 19/2017. Copenhagen: EEA.

[2] European Automobile Association. (2017). Economic and Market Report: key takeaways about the EU industry for 2016. ACEA
Journal of Thermal Engineering, Research Article, Vol. 5, No. 6, Special Issue 10, pp. 131-140, December, 2019

[3] European Environment Agency. (2016). Air Quality in Europe – 2016 Report. Report No 28/2016. Copenhagen: EEA.

[4] Koten, H. (2018). Performance analysis of a diesel engine within a multi-dimensional framework. Journal of Thermal Engineering (JTEN), Vol. 4, 2075-2082.

[5] Environment Pollution (Prevention and Control) Authority for the National Capital Region (EPCA). (2017). Comprehensive Action Plan for air pollution control. Report No. 70. Delhi: EPCA.

[6] Demirci, A., Koten, H., Gumus, M. (2018). The effects of small amount of hydrogen addition on performance and emissions of a direct injection compression ignition engine. Thermal Science. 22. 4-4. 10.2298/TSCI170802004D

[7] Porpatham, E., Ramesh, A., Nagalingam, B. (2013). Effect of swirl of a biogas fuelled spark ignition engine. Energy Conversion and Management, 76, 463-471.

[8] Nunes de Faria, M., Bueno, J., Ayad, S., Belchior, C. (2017). Thermodynamic simulation model for predicting the performance of spark ignition engines using biogas as fuel. Energy Conversion and Management, 149, 1096-1108.

[9] Nadaleti, C.W., Przybyla, G. (2018). Emissions and performance of a spark-ignition gas engine generator operating with hydrogen-rich syngas, methane and biogas blends for application in southern Brazillian rice industries. Energy, 154, 38-51.

[10] Karagoz, M., Kurtgoz, Y., Deniz, E. (2017) Biogas engine performance using ANN. Engineering Science and

Technology, an International Journal, 20, 1563-1570.
[11] Wang, C., Kan, X., Zhou, D., Yang, W., Zhai, X. (2018) An investigation on utilization of biogas and syngas produced from biomass waste in premixed spark ignition engine. Applied Energy. 212, 210-222.
[12] Kwon, E., Song, K., Kim, M., Shin, Y., Choi, S. (2017). Performance of small spark ignition engine fueled with biogas at different compression ratio and various carbon dioxide dilution. Fuel, 196, 217-224.
[13] Nadaleti, C.W, Przybyla, G. (2018). SI engine assessment using biogas, natural gas and syngas with different content of hydrogen for application in Brazillian rice industries: Efficiency and pollutant emissions. Int J. Hyydrogen Energy, 1-14.
[14] Porpatham, E., Ramesh, A., Nagalingam, B. (2018) Experimental studies on the effects of enhancing the concentration of oxygen in the inducted charge of a biogas fuelled spark ignition engine. Energy, 142, 303-312.
[15] Yadav, S.D., Kumar, B., Thipse, S. (2013). Characteristics of biogas operated automotive SI engine to reduce exhaust emissions for green development. (No. 2013-26-0012). SAE Technical Paper.
[16] VDMA Engines and Systems. (2017). Exhaust emission legislation diesel- and gas engines. Frankfurt: VDMA.
[17] Heywood, J.B. (1988). Internal combustion engine fundamental. McGraw-Hill Education.
[18] Kline, S.J., McClintock, F.A. (1953). Describing uncertainties in single-sample experiments. Mech Eng, 75, 3-8.
[19] Pulkrabek, W. (2004). Engineering fundamentals of the internal combustion engines. Prentice Hall, New Jersey.
[20] Chuayboon, S., Prasertsan, S., Theppaya, T., Maliwan, K., Prasertsan, P. (2014). Effects of CH_4, H_2 and CO_2 mixtures on SI Gas engine. Energy Procedia, 52, 659-665.
[21] Koten, H., Gul, M. Z. (2014). Multidimensional Modelling of Compressed Bio Gas (CBG) Engine For Ultra Low Emission. International Journal of Pure and Applied Sciences, 26(2), 41-49.
[22] Degobert, P. (1995). Automobiles and Pollution. Warrendale: SAE Society of Automotive Engineers Inc., Paris: Editions Technip.
[23] Koten, H., Yilmaz, M., Gul, M.Z. (2014). Compressed biogas-diesel dual-fuel engine optimization study for ultralow emission. Advances in Mechanical Engineering. Article ID 571063. http://dx.doi.org/10.1155/2014/571063
[24] Koten, H. (2017). Usage of biogas in a diesel engine: Compressed bio gas (CBG). LAP Lambert Academic Publishing. ISBN-13:978-3-330-07965-6.

Permissions

All chapters in this book were first published in JTE, by Yildiz Technical University; hereby published with permission under the Creative Commons Attribution License or equivalent. Every chapter published in this book has been scrutinized by our experts. Their significance has been extensively debated. The topics covered herein carry significant findings which will fuel the growth of the discipline. They may even be implemented as practical applications or may be referred to as a beginning point for another development.

The contributors of this book come from diverse backgrounds, making this book a truly international effort. This book will bring forth new frontiers with its revolutionizing research information and detailed analysis of the nascent developments around the world.

We would like to thank all the contributing authors for lending their expertise to make the book truly unique. They have played a crucial role in the development of this book. Without their invaluable contributions this book wouldn't have been possible. They have made vital efforts to compile up to date information on the varied aspects of this subject to make this book a valuable addition to the collection of many professionals and students.

This book was conceptualized with the vision of imparting up-to-date information and advanced data in this field. To ensure the same, a matchless editorial board was set up. Every individual on the board went through rigorous rounds of assessment to prove their worth. After which they invested a large part of their time researching and compiling the most relevant data for our readers.

The editorial board has been involved in producing this book since its inception. They have spent rigorous hours researching and exploring the diverse topics which have resulted in the successful publishing of this book. They have passed on their knowledge of decades through this book. To expedite this challenging task, the publisher supported the team at every step. A small team of assistant editors was also appointed to further simplify the editing procedure and attain best results for the readers.

Apart from the editorial board, the designing team has also invested a significant amount of their time in understanding the subject and creating the most relevant covers. They scrutinized every image to scout for the most suitable representation of the subject and create an appropriate cover for the book.

The publishing team has been an ardent support to the editorial, designing and production team. Their endless efforts to recruit the best for this project, has resulted in the accomplishment of this book. They are a veteran in the field of academics and their pool of knowledge is as vast as their experience in printing. Their expertise and guidance has proved useful at every step. Their uncompromising quality standards have made this book an exceptional effort. Their encouragement from time to time has been an inspiration for everyone.

The publisher and the editorial board hope that this book will prove to be a valuable piece of knowledge for researchers, students, practitioners and scholars across the globe.

List of Contributors

Akinbowale T. Akinshilo
Department of Mechanical Engineering, University of Lagos, Akoka- Yaba, Lagos, Nigeria

N. Kizilova
Warsaw University of Technology, Warsaw, Poland
V.N. Karazin Kharkov National University, Faculty of Mathematics and Informatics, Kharkov, Ukraine

A. Korobov
Kharkov National University, Faculty of Radiophysics, Biomedical Electronics and Computer Systems, Ukraine

Balpetek and E. KavakAkpinar
Department of Mechanical Engineering, Firat University, Elazig, Turkey

H. Abbasi, F. Pourfayaz and A. Kasaeian
Faculty of New Sciences and Technologies, University of Tehran, Tehran, Iran

F. Rahimpour
School of Chemical Engineering, College of Engineering, University of Tehran, Tehran, Iran

E. B. Öğüt
Department of Vocational School, Kocaeli University, Kocaeli, TURKEY

T. Mohapatra and S. K. Rout
Department of Mechanical Engineering, C. V. Raman College of Engineering, Bhubaneswar, Odisha, India

M. Kassas, W. M. Hamanah, O. Al-Tamimi and C. B. Ahmed
Department of Electrical Engineering, KFUPM, Dhahran, Saudi Arabia

A. Sahin and B. S. Yilbas
Department of Mechanical Engineering, KFUPM, Dhahran, Saudi Arabia

K. Pavan Kumar and K. Sudheer Kumar
Department of Mechanical Engineering (MCT), M.G.I.T., Hyderabad, INDIA

R. Siddhardha
Department of Mechanical Engineering, G.V.P.S.O.E., Visakhapatnam, INDIA

Ramabhadri Raju
Department of Mechanical Engineering, S.R.K.R., Bhimavaram, INDIA

A. S. Dalkilic, O. Acikgoz and E. Ekici
Department of Mechanical Engineering, Yıldız Technical University, İstanbul, Turkey

S. Wongwises
Department of Mechanical Engineering, King Mongkut's University of Technology and Thonburi, Bangkok, Thailand

F. Pourfayaz, A. Kasaeian and M. Mehdizadeh Fard
Department of Renewable Energies and Environment, Faculty of New Sciences and Technologies, University of Tehran, Iran

O. Al-Tamimi, M. Kassas, W. M. Hamanah and B. S. Yilbas
Department of Electrical Engineering, KFUPM, Dhahran, Saudi Arabia

A. T. Akinshilo
Mechanical Engineering Department; University of Lagos, Akoka-Yaba, Nigeria

A. O. Ilegbusi
Mechanical Engineering Department; Yaba College of Technology, Yaba, Nigeria

Ozgen Acikgoz
Department of Mechanical Engineering, Yildiz Technical University, Istanbul, Turkey

N. D. Shikalgar and S. N. Sapali
Department of Mechanical Engineering, College of Engineering Pune, Maharashtra, India

Julanda Al-Mawali
Department of Engineering and Math, Sheffield Hallam University, Howard Street, Sheffield S1 1WB UK

Sam M Dakka
Department of Mechanical, Materials and Manufacturing Engineering, The University of Nottingham, University Park, Nottingham NG7 2RD, UK

Yasin Karagöz
Department of Mechanical Engineering, University of Lagos, Akoka- Yaba, Lagos, Nigeria

Index

T

V

W

www.ingramcontent.com/pod-product-compliance
Lightning Source LLC
LaVergne TN
LVHW060513210926
848485LV00005B/1076